DENSITY FUNCTIONAL THEORY
BASICS AND APPLICATIONS

DENSITY FUNCTIONAL THEORY

BASICS AND APPLICATIONS

BY

ANELE WRIGHT

KNOWLEDGE
BAKERS

KNOWLEDGE BAKERS

ISBN: 978-93-9001-398-2 (HB)

Published by : **Knowledge Bakers**
Pune, India
Email: info@knowledgebakers.com

PREFACE

In the realm of theoretical condensed matter physics and quantum chemistry, Density Functional Theory (DFT) stands as a cornerstone methodology, offering powerful insights into the electronic structure and properties of materials with remarkable efficiency and accuracy. From fundamental principles rooted in quantum mechanics to practical applications spanning materials science, chemistry, and beyond, DFT has revolutionized our understanding of atomic and molecular systems, paving the way for groundbreaking discoveries and technological innovations. As we embark on this journey through the fundamentals and applications of Density Functional Theory, it becomes evident that its impact extends far beyond the confines of theoretical research, shaping our understanding of matter at the atomic and subatomic levels and driving progress in diverse fields of science and technology.

This book, titled "Density Functional Theory: Basics and Applications," emerges as a comprehensive guide to understanding the theoretical foundations, computational methodologies, and practical applications of DFT. Through collaboration with leading experts and researchers in the field, we aim to provide readers with a thorough exploration of the principles, techniques, and advancements that underpin the success of DFT as a versatile and predictive tool for studying complex materials and chemical systems.

As editors of this volume, we have curated a collection of chapters that cover a wide range of topics within Density Functional Theory, including its mathematical formulation, computational algorithms, practical implementation, and applications in materials science, catalysis, nanotechnology, and beyond. Each chapter offers insights from both theoretical and computational perspectives, drawing on cutting-edge research, case studies, and interdisciplinary connections to illuminate the diverse facets of DFT and its relevance to contemporary scientific inquiry and technological innovation.

Furthermore, this book seeks to showcase the transformative impact of DFT in addressing grand challenges such as the design of novel materials with tailored properties, the understanding of complex chemical reactions, and the development of sustainable energy technologies. By highlighting recent advancements, emerging trends, and future directions in DFT research, we aim to inspire readers to explore new frontiers in theoretical and computational science, pushing the boundaries of knowledge and fostering collaboration across disciplines and research communities.

As we embark on this enlightening journey through the world of Density Functional Theory, we extend our gratitude to the contributors whose expertise and dedication have enriched this volume. We hope that "Density Functional Theory: Basics and Applications" serves as a valuable resource for students, researchers, and practitioners alike, empowering them to unravel the mysteries of matter and harness the predictive power of DFT to address pressing scientific and technological challenges in the quest for a deeper understanding of the natural world.

Anele Wright

CONTENTS

SECTION-2
NEW PERSPECTIVES

SECTION-3

APPLICATIONS

Section 1

Recent Advances

Chapter 1

Fundamentals of Density Functional Theory: Recent Developments, Challenges and Future Horizons

Abstract

Density Functional Theory (DFT) is a powerful and commonly employed quantum mechanical tool for investigating various aspects of matter. The research in this field ranges from the development of novel analytical approaches focused on the design of precise exchange-correlation functionals to the use of this technique to predict the molecular and electronic configuration of atoms, molecules, complexes, and solids in both gas and solution phases. The history to DFT's success is the quest for the exchange-correlation functional, which utilizes density to represent advanced many-body phenomena inside one element formalism. If a precise exchange-correlation functional is applied, it may correctly describe the quantum nature of matter. The estimated character of the exchange-correlation functional is the basis for DFT implementation success or failure. Hohenberg-Kohn established that every characteristic of a system in ground state is a unique functional of its density, laying the foundation for DFT, which is being utilized to explore the novelty of materials. This chapter is aimed to present an overview of DFT by explaining the theoretical background, commonly used approximations as well as their recent developments and challenges faced along-with new horizons.

Keywords: DFT, Kohn-Sham equations, exchange-correlation functionals, challenges

1. Introduction

Density functional theory (DFT) is a low-cost, time-saving quantum mechanical (QM) theory, used to compute many physical characteristics of solids with high precision. The research in this field ranges from the development of novel analytical approaches focused on the design of precise exchange-correlation functionals to the use of this technique to predict the molecular and electronic configuration of atoms, molecules, complexes, and solids in both gas and solution phases. The history to DFT's success is the quest for the exchange-correlation functional, which utilizes density to represent advanced many-body phenomena inside one element formalism. If a precise exchange-correlation functional is applied, it may correctly describe the quantum nature of matter. The estimated character of the exchange-correlation functional is the basis for DFT implementation success or failure. DFT's early breakthroughs concentrated on the most fundamental issues in chemistry, such as the opportunity to generate functionals that could describe both molecular geometries as well as dissociation energy. The fact that every feature of a system in ground state is a unique ground state density functional was demonstrated by Hohenberg-Kohn, laying the foundation for DFT, which is now used to explore novelty of materials. This chapter is aimed to present an overview of DFT by describing the theoretical foundations, widely used approximations, current advances, and issues addressed, as well as future horizons.

2. Fundamentals of DFT

The Schrodinger Equation [1] for a many body system may be simplified to Kohn-Sham equation, which is a single particle independent Schrodinger equation, and can be numerically solved with density functional theory. This computational process produces physical characteristics of solids; however, this hypothesis is based on electron density rather than wave functions, for which scientist Walter Kohn was given the Nobel Prize in 1998 [2]. Despite the fact that no exchange-correlation effects had been documented at the time, Thomas and Fermi claimed in 1927 that total density is the essential parameter in many body problems [3, 4]. The theorems of Hohenberg, Kohn, and Sham laid the groundwork for DFT in 1964, stating that the functional of a many-body problem's (non-degenerated) ground state electron charge density may completely characterize all properties in absence of magnetic field [5].

2.1 The Hohenberg - Kohn (HK) theorems

Hohenberg and Kohn [6] stated seemingly two simple theorems in 1964 that enabled the implementation of DFT.

Theorem I: The external potential, $V_{ext}(\vec{r})$ is a unique functional of electron density $\rho(\vec{r})$, having a unique association among potential and electron density for a many body system; $V_{ext}(\vec{r}) \Longrightarrow \rho(\vec{r})$, whereas this electron density can be used to describe the entire information of the system.

In order to establish a mathematical relation, let us assume external potentials as v(r) and v(r′), whereas the change between these potentials is always identical since the ground state electron density is comparable at entire parts of the crystal, that is, v(r′) - v(r) = constant. According to theory, electrons move in a field produced by external potential V_{ext} and interact with one-another in addition to their external potential, and the corresponding Hamiltonian of energy can be written as;

$$H = T + V_{ext} + U \tag{1}$$

Where T, U, and V_{ext} represents the K.E of electrons, coulomb interaction, and external potential respectively. Quantum mechanically the factors T, U, and V_{ext} can be expressed as;

$$T = \frac{1}{2}\int[\nabla\psi^*(r)\nabla\psi(r)]dr \tag{2}$$

$$V = \int[v(r)\psi^*(r)\psi(r)]dr \tag{3}$$

$$U = \frac{1}{2}\int\left[\psi^*(r')\psi^*(r)\psi(r')\psi(r)\frac{1}{|r\text{-}r'|}\right]drdr' \tag{4}$$

The solution of Hamiltonian for Eq. (1) can be expressed as;

$$H\psi(r_1, r_2, \ldots \ldots \ldots r_N) = E\psi(r_1, r_2, \ldots \ldots \ldots r_N) \tag{5}$$

The $\psi(r_1, r_2, \ldots \ldots \ldots r_N)$ is a ground state N interacting particle's wave-function. Suppose an additional potential $v'(r)$ with changed Hamiltonian H' and wave-function $\psi'(r)$ where the ground state density $\rho(r)$ must remain the same for both cases. The Hamiltonian for this many-body system can be written as; $H'\psi' = E'\psi'$.

Following a thorough exploration of the situation, established on $v(r)\text{-}v(r') =$ constant, it can be concluded, that $\psi(r)$ and $\psi'(r)$ are different; as a result, they both fulfill distinct Schrodinger wave equations. According to variational principle, it is an irrefutable reality that no wave function may produce energy lower than that of the ground stateψ (r) and this fact can be characterized as $\langle\psi|H|\psi\rangle < \langle\psi'|H|\psi'\rangle$ where $E = \langle\psi|H|\psi\rangle$.

Employing essential property of ground state:

$$\langle\psi'|H|\psi'\rangle = \langle\psi'|H'|\psi'\rangle + \int\left[v\left(\vec{r}\right)\text{-}v'\left(\vec{r}\right)\right]\rho\left(\vec{r}\right)dr \tag{6}$$

Alternatively, by swapping;

$$\langle\psi|H'|\psi\rangle = \langle\psi|H|\psi\rangle + \int\left[v'\left(\vec{r}\right)\text{-}v\left(\vec{r}\right)\right]\rho\left(\vec{r}\right)dr \tag{7}$$

By adding above equations we get;

$$E + E' < E' + E \tag{8}$$

The Eq. (8) confirms clear disagreement, and two unlike potentials, $v(r)$ as well as $v'(r)$ will certainly provide different density $\rho(r)$ and $\rho'(r)$ respectively. As a result, details relating density and external potential are needed to determine the Hamiltonian information. Also, T and U are known for N-partials systems so $\rho\left(\vec{r}\right)$ may be employed to find ground state H and E. The functional association of minimum energy state and corresponding resulting density is;

$$E[\rho(r)] = T[\rho(r) + V[\rho(r)] + U[\rho(r)] \tag{9}$$

Theorem II: The true ground state density of an electron corresponds to electron density that minimizes the overall energy of the functional.

Consider, $\rho(r)$ is the density which corresponds to ground state while $\rho'(r)$ to any other state of a many-body system. The functional for total energy in this context is given as; $E[\rho'] > E[\rho]$. Also, assume that $F[\rho(r)]$ is a general functional that is valid for fixed electrons at all external potentials. Mathematically this can be written as;

$$F\rho(r)] = T[\rho(r)] + U[\rho(r)] \tag{10}$$

Also,

$$E[\rho(r)] = \int[v(r)\rho(r)]dr + F[\rho(r) \tag{11}$$

In order to have minimum energy functional, the corresponding density ρ(r) must be essentially a ground state density.

$$E[\psi'] = (\psi', V\psi') + (\psi', T + U)\psi' \quad (12)$$

Assumingψ is ground state function associated to ρ(r) for external potential v(r), the ρ′(r) will correspond to higher energy in accordance with the variational principle.

$$E[\psi'] = \int [v(r)\rho'(r)]dr + F[\rho'(r)] > E[\psi] = [v(r)\rho(r)]dr + F[\rho(r)] \quad (13)$$

As a result, provided the density functional is accurately described, one may easily compute the ground state density as well as energy in an identified external potential. Furthermore, it also demonstrates that ρ(r) minimize the energy functional E[ρ(r)].

2.2 The Kohn-Sham (KH) equations

The theorems given by Hohenberg-Kohn are exact; however not very useful in real calculations [6]. The equation given by Kohn-Sham [7] turned DFT into an applied tool. They converted the difficult problem of electrons interacting together in external effective potential (V_{ext}) into the electrons that are non-interacting in V_{ext}, and the total energy for a ground state of interacting electrons in fixed potential, v(r) is;

$$E[\rho(r)] = V[\rho] + U[\rho] + G[\rho] \quad (14)$$

Where universal density functional G[ρ] holds exchange-correlation, and is expressed as;

$$G[\rho(r)] = T_s[\rho] + E_{xc}[\rho] \quad (15)$$

$$E[\rho(r)] = T_s[\rho] + V[\rho] + U[\rho] + E_{xc}[\rho] \quad (16)$$

The kinetic energy for a many body system having non-interacting electrons is denoted by $T_s[\rho]$, while V[ρ] is the external potential produced by core having positive charge, U[ρ] is coulomb potential as a result of electron–electron interactions, and $E_{xc}[\rho]$ is the energy due to exchange-correlation effects.

$$T_s[\rho(r)] = -\frac{\hbar^2}{2m}\sum_i^N \int \varphi_i^*(r)\nabla^2\varphi_i(r)d^3r = T\left[\varphi_i \sum(\rho)\right] \quad (17)$$

and

$$U[\rho] = \frac{q^2}{2}\iint\left[\frac{\rho(r)\rho(r')}{|r-r'|}\right]drdr' \quad (18)$$

$$V[\rho] = \int v(r)\rho(r)dr \quad (19)$$

The exchange correlation energy $E_{xc}[\rho]$ for a many-body system produced by ρ(r) is given by;

$$E_{xc}[\rho] = \int [\rho(r)\varepsilon_{xc}\rho(r)]dr \tag{20}$$

and

$$E_{xc}[\rho] = E_x[\rho]_{exchange} + E_c[\rho]_{correlation} \tag{21}$$

The E_x term denotes the reduction in energy as an outcome of anti-symmetrization, and it may be represented through a single particle orbital as:

$$E_x = \int [\rho(r)\varepsilon_x\rho(r)]dr \tag{22}$$

and

$$E_c = \int [\rho(r)\varepsilon_c\rho(r)]dr \tag{23}$$

and

$$\varepsilon_x[\varphi_i\rho(r)] = \frac{-q^2}{r}\sum_{j,k}\int d^3r'(r)\varphi_k^*(r)\frac{\varphi_j^*(r')\varphi_k^*(r')\varphi_k(r)}{|r-r'|} \tag{24}$$

Where the single term in the summation refers to the energy of a molecule 'j' at site 'r' in relation to a molecule 'k' at 'r''. The system's energy is further reduced owing to mutual avoidance of the interacting particles, such as electrons that are anti-parallel and lower their energy by evenly arranging their moments. Kohn-Sham mapping of interacting and non-interacting system is shown in **Figure 1**.

$$\varepsilon_c = \sum_{j<k}\frac{q^2}{|r-r'|} = \frac{q^2}{2}\int d^3r\int\left[\frac{\rho(r)\rho(r')-\rho(r)\delta(r-r')}{|r-r'|}\right]d^3r' \tag{25}$$

The energy of ground state may be obtained by differentiating Eq. (14) with respect to $\rho(r)$

$$0 = \frac{\delta E[\rho]}{\delta\rho(r)} = \frac{\delta T_s[\rho]}{\delta\rho(r)} + \frac{\delta U[\rho]}{\delta\rho(r)} + \frac{\delta V[\rho]}{\delta\rho(r)} + \frac{\delta E_{xc}[\rho]}{\delta\rho(r)} = \frac{\delta T_s[\rho]}{\delta\rho(r)} + v(r) + V_c(r) + E_{xc}(r) \tag{26}$$

By employing density $\rho_s(r)$, the minimum state for a non-interacting many-body system is;

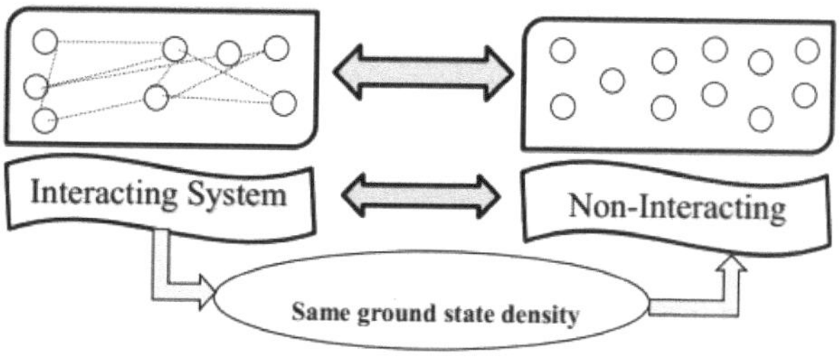

Figure 1.
Kohn-Sham mapping of interacting and non-interacting system.

$$0 = \frac{\delta E_s[\rho]}{\delta \rho_s(r)} = \frac{\delta T_s[\rho]}{\delta \rho_s(r)} + \frac{\delta V_s[\rho]}{\delta \rho_s(r)} + \frac{\delta V_s[\rho]}{\delta \rho_s(r)} = \frac{\delta T_s[\rho]}{\delta \rho_s(r)} + v_s(r) \tag{27}$$

Equating Eqs. (26) and (27), the potential V_s can be obtained as;

$$v_s = V(r) + V_c(r) + V_{xc}(r) \tag{28}$$

The equation for a one-particle system that is non-interacting in potential $v_s(r)$ can be derived from the equation of interacting electrons of the system in the presence of v(r).

$$\left[-\frac{\hbar^2}{2m}\nabla^2 + v_s(r)\right]\varphi_k(r) = E_k\varphi_k(r) \tag{29}$$

The ρ(r) of an original system is replicated by orbitals, where f_k is the kth orbital occupation, and can be expressed as;

$$\rho(r) = \rho_s(r) = \sum_k^N f|\varphi_k(r)|^2 \tag{30}$$

2.3 Exchange-correlation potential

The consequences of KS scheme revealed that the minimum energy state can be established by limiting energy of the energy functional, and it can be done using an agreeable solution of a set of single-particle equations. In the KS scheme, just one critical difficulty is that E_{xc} (exchange-correlation energy) cannot be found exactly. If E_{xc} is determined accurately, it is a precise solution for a many-body problem. There is currently no such exact solution exists, hence approximations are employed to estimate E_{xc} with LDA and GGA being the most commonly used approximations.

3. Commonly used exchange-correlation approximations

In this part, we will go through some of the major advances that lead to contemporary DFT in order to lay a foundation that will help us to comprehend both the theory's foundations and limits. Bloch (1929) was the first to write about the exchange contribution, and it has become well-known as a result of quantum Monte-Carlo simulations of uniform gases [8], which are parameterized in simple formulations [9, 10]. The Local Density Approximation (LDA) [11], proposed by Kohn and Sham, asserts that the exchange-correlation functional at any point in space is simply dependent on that location's spin density. LDA is quite correct for geometries, but it often over-binds atoms/molecules roughly by 1 eV per bond, rendering it ineffective for thermo-chemistry [12]. The Generalized Gradient Approximation (GGA) [13, 14] is an extension to the LDA component that includes terms that are dependent on density derivatives. Perdew was the first to apply real-space cutoffs to make GGAs, which led to the development of the PW86 functional model [13]. The PW91 functional [15] was the pinnacle of this comprehensive development, and it produces useful precision for binding energies, as proven in 1993 of around 6–10 kcal/mol [16]. PBE [17] is the most widely used GGA to investigate materials today, whereas BLYP [18] and Lee-Yang-Parr correlation [19] is the most generally employed GGA in chemistry. A hybrid GGA [20] is one that combines a normal GGA plus a Hartree-Fock component, in which the kinetic energy density is also employed to define the

GGA component. The GGA, Hartree-Fock, and kinetic energy density components are all present in a meta-hybrid, while hybrid or meta-hybrid component of a double-hybrid includes an involvement from second-order Moller-Plesset perturbation theory [21]. The Density Functional (DF) consists of a part of GGA, LDA, Hartree-Fock exchange or hybrids, and/or a meta-GGA, commonly known as the exchange-and-correlation (XC) functional (meta-GGA or meta-hybrid). Furthermore, the addition of an orbital-dependent correlation, it may also be reliant on virtual Kohn-Sham orbitals (double-hybrids) [22]. A comparison of simplicity versus accuracy of existing approximations in DFT is shown in **Figure 2**.

The functionals currently utilized in DFT simulations constitute a natural hierarchy, and no systematic approach to the precise functional can be claimed. The available functional form is clearly improving, resulting in a considerably more accurate representation of ground state properties. The most important recent advancements are those that include the non-local aspect of the exchange potential in some way. **Table 1**, summarizes the present hierarchy.

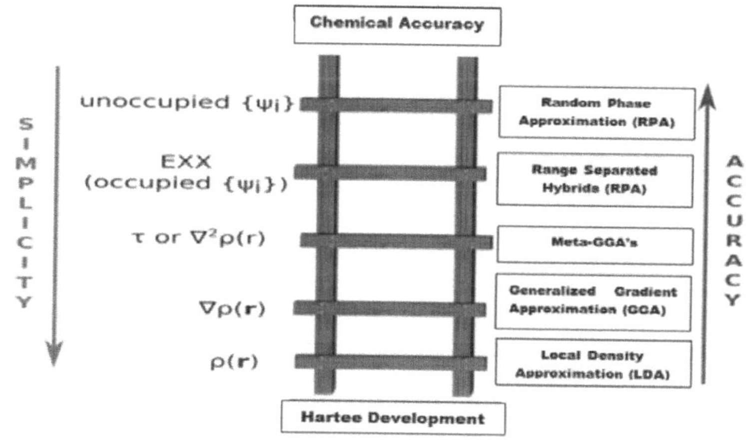

Figure 2.
A comparison of simplicity versus accuracy of existing approximations in DFT [23].

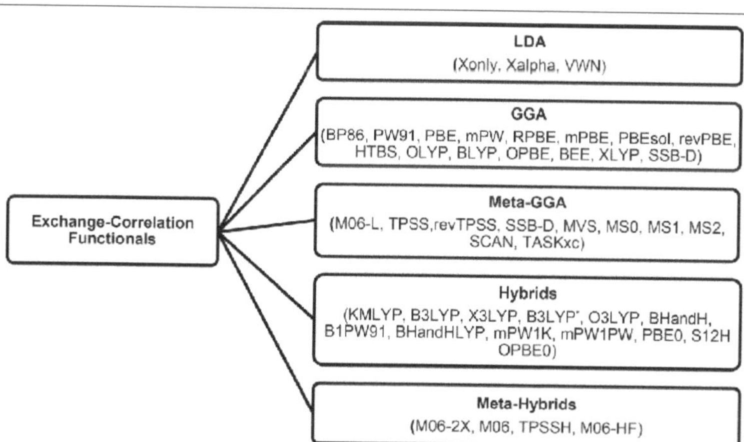

Table 1.
Commonly used E_{xc} functionals.

4. Recent developments

This section focuses on the evolution of new functionals in DFT during the last decades.

4.1 Random phase approximation (RPA-type Functionals)

The exchange correlation energy (E_{XC}) can be calculated using DFT fluctuation dissipation in the form of coupling constant and frequency [24–26]. The direct random-phase approximation (RPA) [27, 28] or time dependent TD-Hartree, are the results of ignoring the exchange kernel of TDFT. A fifth-rung approximation is generated as a result of this methodology, and this can be expensive to examine, although the relative burden is always reducing [29, 30]. It only examines bubble diagrams in the many-body expansion of the energy, so direct RPA over-correlates systems by ignoring extra contributions at higher levels that diminish correlation. It also has issues with self-interaction since, even when just one electron is involved, it yields low correlation energies, and the dissociation energies of molecules are erroneous [23].

4.2 Meta-GGA's

The meta-GGA [31] is a novel component that extends beyond density and gradient, and is commonly used to indicate the KS orbitals' kinetic energy density. The objective of a successful meta-GGA is to achieve hybrid accuracy without incurring the computational expense of the exact exchange contribution. The incorporation of atom-centered basis functions, the cost of accurate exchange is reasonable, however, it can be costly while using periodic boundary conditions in addition of basis sets. Perdew and colleagues, and plenty of others, have worked on meta-GGAs for decades, with multiple failed attempts [32]. SCAN (strongly constrained and suitably normed semi-local density functional) [33], the most current effort has undergone a number of conventional tests and looks to have a good chance of becoming part of the pantheon of widely employed functionals. The G3 data-set [34] is a common collection of chemical compounds that LDA over-binds around 3 eV, while PBE reduces it to approximately 1 eV, and SCAN around 1/4 eV. On the S22 data-set [35] of weakly bonded systems, SCAN has 2–3 times less errors than PBE does, while SCAN decreases miscalculations of lattice constant and other parameters on the LC20 data [36] set around 0.05 Å, and to around 0.01 Å in PBE. The PBE [37], on contrary to SCAN, only improves underestimation of chemical barrier height by 30 percent, while hybrids on the other hand are frequently 2–3 times superior with conventional varieties. Therefore, one can conclude that SCAN achieves accuracies comparable to hybrid functionals for several characteristics at a fraction of the computing cost [38].

4.3 Range separated hybrids (RSH)

Andreas Savin was the first to create the range separation hypothesis, which is quite precise [39, 40], through which coulomb repulsion may be easily expressed by combining a short-ranged input with a long-ranged involvement that do not have coulomb singularity at zero separation, and decays quicker than the inverse of the separating distance. In KS equation generalizations, one contribution is treated as an interaction, while the other is compensated by a redefined XC contribution. The HSE06 functional [41] is a hybrid with a range separation that manages long-ranged exchanges with an approximation, short-ranged exchanges with accuracy in an extended insulator [42], and this combination frequently yields exact gaps for moderate-gap semiconductors and insulators [37].

4.4 Weak interactions

Over the last two decades, tremendous progress has been made in addressing the challenges associated with weak van der Waals (vdW) interactions. Traditional functionals do a good job at manipulation of covalent, ionic, and metallic interactions due to their semi-local nature, but they fall short when it comes to longer, weaker bonds, and cannot offer weak binding that drops off as a function of R^6 (R is the distance among two atoms) [38]. To account these impacts, modifications must be applied to the conventional functionals and this can be accomplished in one of three ways. There is a succession of approximations produced by Langreth and Lundqvist and collaborators [43] for the evolution of explicit non-local functionals of electron density, while these approximations are generated non-empirically, notably beginning with contributions of correlation energy. Additionally, these functionals may be useful for any materials, ranging from solids to molecules, and have been designed by supposing systems that contain a gap [44]. RPA, which incorporates approximations to the vdW forces by default, as well as the Becke and Johnson technique [45], leverages the exchange hole's dipole moment to approximate C_6, as well as higher coefficients.

4.5 Gaps of solids

The inadequacy of traditional approximations to anticipate band gaps of semiconductors and insulators is a critical flaw. The LDA undervalues gap between bulk Si and Ge by a factor of two, making Germanium a metal, whereas GGAs performs a bit good but underestimate as well. The ability to give precise and dependable gaps has always been a strong suit of the GW approach [46]. In the last two decades, precise gap computation utilizing hybrid functionals such as HSE06 [41] has been a huge success, and is accomplished through the use of a generalized KS scheme [47]. In this case, rather than using pure Kohn-Sham theory, the orbital reliant element of the functional is considered as in Hartree-Fork approach to overcome flaws of other E_{xc} functionals [48].

5. Challenges for DFT

In principle, DFT is exact; however its effectiveness depends on the development as well as advancement in exchange-correlation (E_{xc}) functionals which may be achieved by optimizing against larger data-sets and using improved functional arrangements that are more flexible and contain more elements. Smoothness has also been prioritized in recent enhancements, which helps to alleviate problems like grid-size convergence and self-consistent field iterations. In this section, we will go through some of DFT's challenges that may differ from those that appears to be "solved" to those that are still being explored. There are numerous more that are less well-known, and yet crucial to DFT's future growth as well as use.

5.1 Strong correlation

DFT's inadequacy for strongly correlated systems utilizing typical approximations has been acknowledged since its inception, and this can be investigated as well as linked to standard approximation localization or delocalization inaccuracies when integer or half-integer electron quantities are found in distinct locations [49]. In quantum computational physics and chemistry, the Kohn-Sham gap among two states becomes too narrow, and the wave function of a many-body system is very nearly equal to mixing of two slater determinants, which is referred to as static correlation. The failure of approximations under these situations cause the challenges, not the KS scheme itself, as demonstrated by the two-site Hubbard model, in which the precise KS system is simple to design, even when one deal with strongly correlated systems [50]. This problem can be addressed by breaking the symmetry of evenly spaced atomic chains into multiple solutions, and one of which will have the least amount of energy [51]. This is such a significant issue; hence, a great deal of research has been done on it, particularly by Weitao Yang's group [52], but also by Scuseria [53] and Becke [54].

5.2 Development of uniformly better and simple functionals

One of the biggest problems for DFT is to preserve some aspect of simplicity as its foundations. When DFT functionals get as complicated as full configuration interaction, one of the theory's most significant properties, namely simplicity, is lost, which is particularly true in terms of computational environment. This simplicity, however, must not be at the expense of accuracy, nor should it become an exclusively empirical approach. The precise representation of binding energies and geometries of simple molecules was one of DFT's first major hurdles in chemistry. Becke, Perdew, Langreth, and Parr presented the density's first derivative in the form of generalized gradient approximation in the 1980s, which was the first step towards chemists being able to correctly use DFT. In the early 1990s, Becke described the proportion of Hartree-Fock exact exchange (HF) which is included in the functionals, and as a result of this effort, B3LYP [55], the utmost extensively utilized of all the functionals, was developed, and has demonstrated outstanding performance in variety of systems. Despite the introduction of new concepts into more current functionals of varying complication, it remains the prevalent, and DFT will likely benefit from developing functionals that improves on B3LYP [56].

5.3 Dispersion and reaction barriers

To provide a comprehensive chemistry explanation, it is indispensable to go beyond explaining a molecule in equilibrium geometry to similarly explain weakly interacting atoms or molecules, and chemical reaction transition states. It's challenging to describe reaction barriers with LDA or GGA functional since they consistently underestimate the difficulty of transitioning from one condition to another. Formerly the functionals may be utilized to represent potential energy surfaces, and this systematic imperfection must be corrected. Transition states, covalent bonding, and van der Waals attraction are all challenging to represent precisely and effectively, though efforts are to be made to address these problems. This is especially true when DFT becomes more widely applied to biologically important regions, where all of these interactions might occur at the same time [57].

5.4 Static correlation and delocalization errors

The enactment of DFT, as evidenced by significant errors for one-electron systems, is another important issue. In DFT, a single electron system has no exceptional role; in fact, one electron can interrelate with itself, as the self-interaction error has long proved. Of course, there is no self-interaction in the accurate

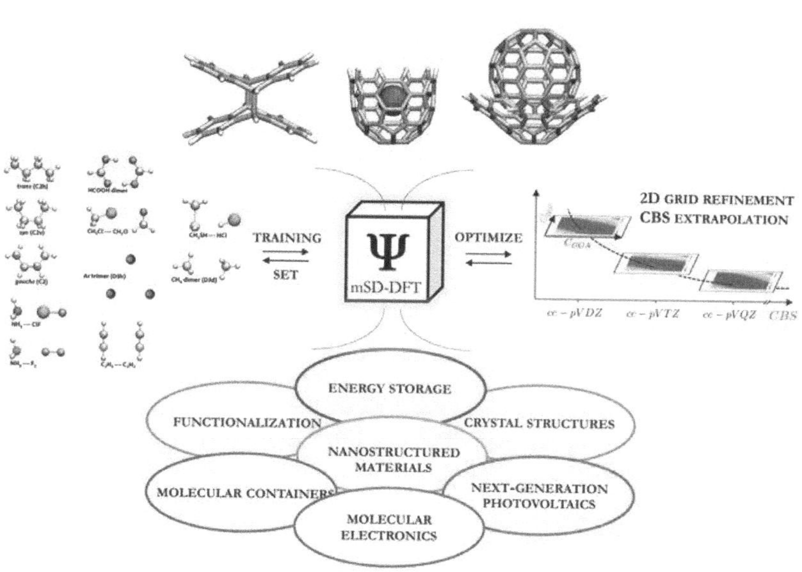

Figure 3.
Potential application areas of DFT [61].

functional; the exchange energy precisely cancels the coulomb energy of single electron. In increasingly complicated systems, they can be linked to systematic flaws like static correlation and delocalization error, and despite most recent advancements, even the simplest systems can contain mistakes in most recent functionals [58]. Hence, these basic systems should not be overlooked since they hold the vital knowledge of functionals that can lead to advancements [57].

6. New horizons

The applications of warm dense matter vary from modeling planetary interiors to inertial confinement fusion [59], which is a completely new field for DFT, and has been exploded in the last decade, with considerable temperatures on the electronic scale of roughly 10^5 K but not to the point that the Thomas- Fermi hypothesis or classical performance takes precedence. This domain is so "new" that temperature-dependent exchange-correlation energy of a uniform gas, which is the input to thermal LDA, is just now being computed with remarkable precision [60]. **Figure 3** summarizes some of the potential application areas of DFT.

7. Concluding remarks

Density Functional Theory is a powerful and commonly employed quantum mechanical tool for investigating various aspects of matter. This field's research ranges from the development of novel analytical approaches focused on the design of precise exchange-correlation functionals to the use of this technique to predict the molecular and electronic configuration of atoms, molecules, and solids in both gas and solution phases. Designing and evolution of more efficient density functionals is a continuous endeavor since there are still challenges to be resolved, and getting all of the attributes correct at a reasonable computing cost is a quantum fantasy. The future research will focus on developing even more consistently precise density functionals for specific applications, allowing researchers to take use of DFT's comparatively high accuracy at cheap processing cost, and the possibility of even more improvements awaits.

References

[1] Tsutsumi Y. Schrodinger equation. Funkcialaj Ekvacioj. 1987;30:115-125.

[2] Kohn W, and Sham LJ. Self-consistent equations including exchange and correlation effects. Physical review. 1965;140(4A):A1133

[3] Thomas LH. The calculation of atomic fields. Mathematical proceedings of the Cambridge philosophical society 1927;542-548

[4] Fermi E. Statistical method to determine some properties of atoms. Rend. Accad. Naz. Lincei. 1927;6:5.

[5] Lundqvist S, and March NH, editors. Theory of the inhomogeneous electron gas. 2013.

[6] Gilbert TL. Hohenberg-Kohn theorem for nonlocal external potentials. Physical Review B. 1975;12 (6):2111.

[7] Kohn W, and Sham LJ. Self-consistent equations including exchange and correlation effects. Physical review. 1965;140(4A):A1133.

[8] Ceperley DM, and Alder BJ. Ground state of the electron gas by a stochastic method. Physical review letters. 1980;45 (7):566.

[9] Vosko SH, Wilk L, and Nusair M. Accurate spin-dependent electron liquid correlation energies for local spin density calculations: a critical analysis. Canadian Journal of physics. 1980;58(8): 1200-1211.

[10] Perdew JP, and Wang Y. Accurate and simple analytic representation of the electron-gas correlation energy. Physical review B. 1992;45(23):13244.

[11] Gross EK, and Kohn W. Local density-functional theory of frequency-dependent linear response. Physical review letters. 1985;55(26): 2850.

[12] Jones RO, and Gunnarsson O. The density functional formalism, its applications and prospects. Reviews of Modern Physics. 1989;61(3):689.

[13] Perdew JP, and Yue W. Accurate and simple density functional for the electronic exchange energy: Generalized gradient approximation. Physical review B. 1986;33(12):8800.

[14] Perdew JP, and Wang Y. Accurate and simple analytic representation of the electron-gas correlation energy. Physical review B. 1992;45(23):13244.

[15] Burke K, Perdew JP, and Wang Y. Derivation of a generalized gradient approximation: The PW91 density functional. Electronic density functional theory. 1998;81-111.

[16] Gill PM, Johnson BG, Pople JA, Frisch MJ. et.al. An investigation of the performance of a hybrid of Hartree-Fock and density functional theory. International Journal of Quantum Chemistry. 1992;44:319-331.

[17] Perdew JP, Burke K, and Ernzerhof M. Generalized gradient approximation made simple. Physical review letters. 1996;77(18):3865.

[18] Becke AD. Density-functional exchange-energy approximation with correct asymptotic behavior. Physical review A. 1988;38(6):3098.

[19] Lee C, Yang W, and Parr RG. Development of the Colle-Salvetti correlation-energy formula into a functional of the electron density. Physical review B. 1988;37(2):785.

[20] Sousa SF, Fernandes PA, and Ramos MJ. General performance of

density functionals. The Journal of Physical Chemistry A. 2007;111(42): 10439-10452.

[21] Pulay P, and Saebø S. Orbital-invariant formulation and second-order gradient evaluation in Møller-Plesset perturbation theory. Theoretica chimica acta. 1986;69(5):357-368.

[22] Goerigk L, and Grimme S. Double-hybrid density functionals. Wiley Interdisciplinary Reviews: Computational Molecular Science. 2014; 4(6):576-600.

[23] Datta S, and Jana D. Semiconductor Physics: A Density Functional Journey. arXiv preprint arXiv:2010.13050.

[24] Langreth DC, and Perdew JP. The exchange-correlation energy of a metallic surface. Solid State Communications. 1975;17(11): 1425-1429.

[25] Gunnarsson O, and Lundqvist BI. Exchange and correlation in atoms, molecules, and solids by the spin-density-functional formalism. Physical Review B. 1976;13(10):4274.

[26] Harris J, and Jones RO. The surface energy of a bounded electron gas. Journal of Physics F: Metal Physics. 1974;4(8):1170.

[27] Furche F. Molecular tests of the random phase approximation to the exchange-correlation energy functional. Physical Review B. 2001;64(19):195120.

[28] Heßelmann A, and Görling A. Random-phase approximation correlation methods for molecules and solids. Molecular Physics. 2011;109(21): 2473-2500.

[29] Furche F. Developing the random phase approximation into a practical post-Kohn–Sham correlation model. The Journal of chemical physics. 2008; 129(11):114105.

[30] Eshuis H, Yarkony J, and Furche F. Fast computation of molecular random phase approximation correlation energies using resolution of the identity and imaginary frequency integration. The Journal of chemical physics. 2010; 132(23):234114.

[31] Adamo C, Ernzerhof M, and Scuseria GE. The meta-GGA functional: Thermochemistry with a kinetic energy density dependent exchange-correlation functional. The Journal of Chemical Physics. 2000;112(6):2643-2649.

[32] Sun J, Ruzsinszky A, and Perdew JP. Strongly constrained and appropriately normed semilocal density functional. Physical review letters. 2015;115(3): 036402.

[33] Butler KM, Saxena J, Jain A, Fryars T, Lewis J, Hetherington G. et.al. Minimizing power consumption in scan testing: Pattern generation and DFT techniques. In2004 International Conference on Test 2004;355-364.

[34] Curtiss LA, Redfern PC, and Raghavachari K. Assessment of Gaussian-3 and density-functional theories on the G3/05 test set of experimental energies. The Journal of chemical physics. 2005;123(12):124107.

[35] Jurecka P, Sponer J, Cerny J, Hobza P. et.al. Benchmark database of accurate (MP2 and CCSD (T) complete basis set limit) interaction energies of small model complexes, DNA base pairs, and amino acid pairs. Physical Chemistry Chemical Physics. 2006;8(17)

[36] Sun J, Marsman M, Csonka GI, Ruzsinszky A, Hao P, Kim YS, Kresse G, Perdew JP. et.al. Self-consistent meta-generalized gradient approximation within the projector-augmented-wave method. Physical Review B. 2011;84(3): 035117.

[37] Perdew JP, Yang W, Burke K, Yang Z, Gross EK, Scheffler M,

Scuseria GE, Henderson TM, Zhang IY, Ruzsinszky A, Peng H. et.al. Understanding band gaps of solids in generalized Kohn–Sham theory. Proceedings of the national academy of sciences. 2017;114(11):2801-2806.

[38] Yip S. Handbook of Materials Modeling: Methods: Theory and Modeling. Andreoni W, editor. Springer; 2018.

[39] Savin A. On degeneracy, near-degeneracy and density functional theory. Second international congress on theoretical chemical physics-ICTCP II 1996.

[40] Leininger T, Stoll H, Werner HJ, Savin A. et.al. Combining long-range configuration interaction with short-range density functionals. Chemical physics letters. 1997;275(3-4):151-160.

[41] Heyd J, Scuseria GE, and Ernzerhof M. Hybrid functionals based on a screened Coulomb potential. The Journal of chemical physics. 2003;118 (18):8207-8215.

[42] Janesko BG, Henderson TM, Scuseria GE. Screened hybrid density functionals for solid-state chemistry and physics. Physical Chemistry Chemical Physics. 2009;11(3):443-454.

[43] Dion M, Rydberg H, Schröder E, Langreth DC, Lundqvist BI. et.al. Van der Waals density functional for general geometries. Physical review letters. 2004;92(24):246401.

[44] Berland K, Cooper VR, Lee K, Schröder E, Thonhauser T, Hyldgaard P, Lundqvist BI. et.al. van der Waals forces in density functional theory: a review of the vdW-DF method. Reports on Progress in Physics. 2015;78(6):066501.

[45] Johnson ER, and Becke AD. Van der Waals interactions from the exchange hole dipole moment: application to bio-organic benchmark systems. Chemical physics letters. 2006;432 (4-6):600-603.

[46] Aryasetiawan F, and Gunnarsson O. The GW method. Reports on Progress in Physics. 1998;61(3):237.

[47] Seidl A, Görling A, Vogl P, Majewski JA, Levy M. et.al. Generalized Kohn-Sham schemes and the band-gap problem. Physical Review B. 1996;53(7): 3764.

[48] Kummel S, and Kronik L. Orbital-dependent density functionals: Theory and applications. Reviews of Modern Physics. 2008;80(1):3.

[49] Cohen AJ, Mori-Sánchez P, and Yang W. Insights into current limitations of density functional theory. Science. 2008;321(5890):792-794.

[50] Carrascal DJ, and Ferrer J. Exact Kohn-Sham eigenstates versus quasiparticles in simple models of strongly correlated electrons. Physical Review B. 2012;85(4):045110.

[51] Wagner LO, Baker TE, Stoudenmire EM, Burke K, White SR. et.al. Kohn-Sham calculations with the exact functional. Physical Review B. 2014;90(4):045109.

[52] Zheng X, Cohen AJ, Mori-Sánchez P, Hu X, Yang W. et.al. Improving band gap prediction in density functional theory from molecules to solids. Physical review letters. 2011;107(2):026403.

[53] Motta M, Ceperley DM, Chan GK, Gomez JA, Gull E, Guo S, Jiménez-Hoyos CA, Lan TN, Li J, Ma F, Millis AJ. et.al. Towards the solution of the many-electron problem in real materials: Equation of state of the hydrogen chain with state of the art many-body methods. Physical Review X. 2017;7(3):031059.

[54] Johnson ER, and Becke AD. Communication: DFT treatment of

strong correlation in 3d transition-metal diatomics. The Journal of chemical physics. 2017;146(21):211105.

[55] Stephens PJ, Devlin FJ, Chabalowski CF, Frisch MJ. et.al. Ab initio calculation of vibrational absorption and circular dichroism spectra using density functional force fields. The Journal of physical chemistry. 1994;98(45):11623-11627.

[56] Paier J, Marsman M, and Kresse G. Why does the B3LYP hybrid functional fail for metals?. The Journal of chemical physics. 2007;127(2):024103.

[57] Cohen AJ, Mori-Sanchez P, and Yang W. Challenges for density functional theory. Chemical reviews. 2012;112(1):289-320.

[58] Perdew JP, McMullen ER, and Zunger A. Density-functional theory of the correlation energy in atoms and ions: a simple analytic model and a challenge. Physical Review A. 1981;23(6):2785.

[59] Smith JC, Sagredo F, and Burke K. Warming up density functional theory. Frontiers of Quantum Chemistry 2018; 249-271.

[60] Groth S, Dornheim T, Sjostrom T, Malone FD, Foulkes WM, Bonitz M. et. al. Ab initio exchange-correlation free energy of the uniform electron gas at warm dense matter conditions. Physical review letters. 2017;119(13):135001.

[61] Roch LM, and Baldridge KK. General optimization procedure towards the design of a new family of minimal parameter spin-component-scaled double-hybrid density functional theory. Physical Chemistry Chemical Physics. 2017;19(38):26191-26200.

Chapter 2

Modern State of the Conventional DFT Method Studies and the Limits Following from the Quantum State of the System and Its Total Spin

Abstract

At present, the density functional theory (DFT) approach became the most widely used method for study molecules and solids. In the atmosphere of such great popularity, it is particularly important to know the limits of the applicability of DFT methods. In this chapter, I will discuss the modern state of DFT studies basing on the last publications and will consider in detail two cases when the conventional DFT approaches, in which used only electron density and its modifications by gradients, cannot be applied. First, the case related to the total spin S of the state. As I rigorously proved for an arbitrary N-electron state by group theoretical methods, the electron density does not depend on the total spin S of the state. From this follows that the Kohn-Sham equations have the same form for states with different S. The critical survey of elaborated DFT procedures, in which the spin is taken into account, shows that they modified only exchange functionals, and the correlation functionals do not correspond to the spin of the state. The point is that the conception of spin in principle cannot be defined in the framework of the electron density formalism, and this is the main reason of the problems arising in the study by DFT approaches the magnetic properties of the transition metals. The possible way of resolving spin problems can be found in the two-particle reduced density matrix formulation of DFT. In the end, it will be considered the case of the degenerated states, in which, as follows from the adiabatic approximation, the electron density may not be defined, since electronic and nuclear motions cannot be separated, since, the vibronic interaction mixed them.

Keywords: DFT, problem of spin in DFT, methods taking into account spin-multiplet structure, problem of degenerate states in DFT, vibronic interaction

1. Introduction

The DFT method developed by Walter Kohn with his collaborators more than 55 years ago is widely used for study molecular systems and solids. First, Kohn published with Hohenberg [1] their famous theorem on which the DFT theory is based. Then Kohn with Sham [2] obtained their well-known Kohn-Sham equation. The number of citations on these first Kohn and collaborators publications is increased every year [3]. In 2010 papers [1, 2] were cited 11,000 times and in 2012 14,000 times. So, we can expect that at present the number of citations will be larger.

Last years, the number of Congresses on Nanosciences and Nanotechnologies is considerably increased. In one of the reports made in Las Vegas on October 2016 by Miyazaki, it was claimed that using the modern computational facilities they were able to apply the DFT molecular dynamic simulation to systems with million atoms, see Ref. [4]. Certainly, it can be done only using the traditional Kohn-Sham (KS) formalism based on the electron density $\rho(r)$ and its modifications by gradients.

The electron density is the diagonal element of the spinless one-particle reduced density matrix,

$$\rho(\boldsymbol{r}_1) = N \sum_{\sigma_1, \ldots, \sigma_N} \int |\Psi(\boldsymbol{r}_1\sigma_1, \ldots, \boldsymbol{r}_N\sigma_N)|^2 dV^{(1)} \tag{1}$$

In Eq. (1), the spin projections σ_i span over the whole spin space and the spatial coordinates are integrated over the N − 1 electrons excepting the first. If Ψ is defined in 4N-dimensional Gilbert space, $\rho(r)$ is defined in a three-dimensional space. Evidently, the calculations, in which only $\rho(r)$-formalism is used, will be considerably faster than Ψ-formalism is used, and the $\rho(r)$-formalism can be applied to larger systems.

Let us consider another point: what we lose, if we use ρ-formalism with the modulus of $|\Psi|^2$ squared, as in Eq. (1), instead of the wave function formalism. It is evident that in the transition from the wave function formalism to the probability density $|\Psi|^2$, we lose the phase of the wave function. Due to the insensitivity of the probability density to the symmetry of the state (we will discuss it in the following sections), we also lose the symmetry characteristic of the wave function and cannot determine the Pauli permitted states, on which molecular spectroscopy is based. The diagonal element of the full and all reduced density matrices, as it was proved in my studies and discussed in book [5], does not depend on the symmetry of the state and its dimensionality.

It can be expected that after integration we lose some information. In the case of electron system, the one-particle reduced density matrix must be used. This leads to loss of information connected with the two-particle correlations, which are described by the two-particle reduced density matrix.

What is not evident and deserves a special discussion: as it was proved by the author for an arbitrary many-electron system, the total spin S of the system in principle cannot be introduced in the DFT studies. This can be done at the two-particle reduced density matrix level. In the Section 2, we will discuss this problem in detail.

On the other hand, even at the framework of the two-particle reduced density matrix formalism, one cannot study the non-additive many-body effects, which determined by many-body forces, In this connection, I would like to mention that when I arrived from Moscow to Mexico by invitation of the Director of IF-UNAM Octavio Novaro and was working in his laboratory, we obtained a closed formula for the energy of N-body interactions [6].

Later on, in the author book [7], Chapter 4, different general cases for the many-body forces have been considered, see also paper [8]. In several publications, e.g., in Refs. [9, 10], it was conclusively demonstrated that the clusters built from closed-shell atoms (atoms without valence electrons) are stabilized by the three-body forces. The alkaline-earth clusters: Be_n, Mg_n, Ca_n, and etc. are the typical example of such clusters. They are stabilized by the three-body forces, It is also important to mention that for the stability of rare-gas clusters the three-body dispersion forces, which are known as the Axilrod-Teller-Muto forces, play a decisive role, see Section 4.3.3 in Ref. [7] and recent review by Johnson and co-authors [11].

It is instructive to discuss shortly the physical sense of the non-additivity and many-body forces concepts. As is well known, the interaction of charge particles is described by the Coulomb law

$$V = \sum_{a<b} \frac{q_a q_b}{r_{ab}}, \tag{2}$$

where charges q_i are considered as points. Eq. (2) contains only two-particle interactions, so, it is additive. However, in quantum mechanics, the charge particles are not points and they are not rigid. Atoms and molecules obey quantum-mechanical laws. If we consider the Coulomb interactions between charged atoms (or charged molecules), the additivity is lost. The interaction will depend on surrounding. In the case of three atoms, the third atom can polarize the electronic structure of two others, and this leads to three-body forces, since the interaction depends on three interatomic distances.

It should be noted that the possibility of application of DFT approaches to large systems, which were not available to be studied before, induced a euphoria in the DFT community. This euphoria led to wide using DFT methods without an analysis of the limitations following from quantum mechanics.

In many publications it was revealed that the applications of DFT method in some cases lead to incorrect results. First, it was recognized in the DFT studies of intermolecular interactions. The potential curves obtained by the early created DFT functionals for many stable in experiments dimers were repulsive, since in these DFT functionals the dispersion energy was not taken into account.

Then it became clear that DFT methods meet serious difficulties in studies of transition metals with nd electrons. These problems were analyzed by many authors, e.g., by Cramer and Truhlar [12]. I would like to stress that most of difficulties discussed in their review [12] are connected with the problem of spin in DFT approach and in principle cannot be resolved in the framework of the electron density, $\rho(r)$, which belong to the one-particle reduced density matrix, see Eq. (1). The spin problem in DFT approaches will be analyzed in detail in Sections 2 and 3.

Last years, many comparative studies of the relative precision of exchange-correlation (XC) functionals are published. Below I will discuss some important, from my viewpoint, DFT papers published on this topic in the last years. Certainly, the list of selected papers is only a little part of thousands DFT papers that are published each year.

Gillan et al. [13] analyzed different kind of XC functionals for liquid and ice water and water clusters. The conclusion was that many functionals are not satisfactory because do not describe correctly the dispersion. Let us stress that this situation takes place for such widely studied substance as water. The authors mentioned that after they included in XC functionals the non-local dispersion, the results still cannot be admitted as completely satisfactory.

In the Taylor et al. [14] paper, the precision of DFT calculations of intermolecular interactions with respect to highly accurate benchmarks for 10 dimers was analyzed. Their review is comprised 17 authors, among them are the well-known creators of XC functionals Angyán, Hirao, Scuseria, Truhlar and others.

The perspectives of DFT theory were discussed in the paper by Truhlar and collaborators [15]. The authors also analyzed recent Minnesota functionals. On the other hand, Mardirossian and Head-Gordon [16] benchmarked the Minnesota functionals using a very comprehensive database and came to conclusion that none of them are state-of-the-art for non-covalent interactions and isomerization energy.

In the article published by Medvedev et al. in collaboration with Perdue [17], 128 XC functionals created in period 1974–2015 were analyzed. The authors [17] made the comparison of the normalized error for different functionals for atoms and its ions, see **Figure 1**, using the CCSD approximation as a reference level. It is important to stress that according to their results the normalized error of $\rho(r)$ up to 2000 decreased and then increased due to the introduction of semiempirical functionals.

Kepp [18] made a comment on the conclusion by the authors [17] that after the early 2000 the functionals strayed from the path toward exactness. Kepp indicated that the "straying" is not chemically relevant to the studied systems. In their response [19], the authors performed a special calculation, from which follows that their results are valid also for heavier system, including molecules.

In the next paper published by Perdew with Medvedev et al. [20], the authors discuss the possibilities and pitfalls of statistical error analysis, taking as an example the ranking of approximate functionals by the accuracy of their self-consistent electron densities.

As mentioned above in our discussion of last publications, the DFT community is concerned on the still existing problems in some applications of DFT approaches and most of these problems are connected with the quality of XC functionals. I like and completely agree with the witty comment made by Savin [21]:

> *"The existing great number of different expressions for the XC functionals can be considered as evidence that we still have not satisfactory correct XC functionals".*

Among new publications, it is important to mention the very recent paper published by Perdue with collaborates [22], in which the problem of the symmetry breaking in DFT was discussed. The interesting paper was recently published by Bartlett [23]. Even the title of the paper "Adventures in DFT by a wavefunction theorist" looks quite intriguing. I also recommend readers the quite comprehensive and interesting review by Johnson and co-authors [11] published in 2021.

In my following discussions, I will analyze in detail the two problems:

The first problem is connected with the definition of spin in the KS-DFT framework. It will be shown that the concept of the total spin S of the state, in principle, cannot be defined in the frame of electron density formalism.

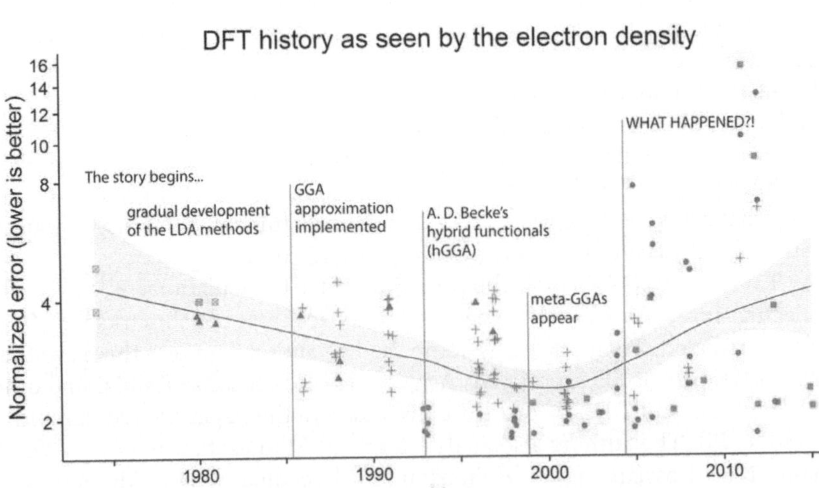

Figure 1.
The precision of calculating ρ using different functionals.

The second problem is related to the degenerate states in DFT. As follows from the general conceptions of quantum mechanics, at Born-Oppenheimer approximation in the case of degenerate states, the electronic and nuclear motions cannot be separated, they are mixt by so-called vibronic interactions. This problem will be discussed in Section 3.

In this chapter I will not discuss Ψ-versions of DFT that solves many problems in DFT but lost the simplicity of KS-DFT formulation. The Ψ-formalism was successfully combined with DFT approach by Győrling, Trickey, and some other investigators. Nevertheless, the existing problems in the applications of DFT approach are still remaining actual. I will discuss them in the next two sections. The discussion will be based on two theorems proved by the author in 2007 [24].

2. Electron density ρ and the total spin S

It was recognized sufficiently long ago that the concept of the total spin of the many-electron quantum state is quite difficult to base in the frame of DFT approaches. Studying the two-electron system, McWeeny [25] came to conclusion the electron density does not allow to identify the spin state. McWeeny formulated it as the following statement:

> *"Electron spin is in a certain sense extraneous to the DFT".*

In their analysis of DFT foundation, Weiner and Trickey [26] came to conclusion that

> *"... the way that the ρ-based XC potential takes account of spin is very obscure except in the simplest configurations".*

The statements of McWeeny and Weiner-Trickey are quite cautious. I would like to formulate it more definite:

> *"The conception of spin in principle cannot be defined in DFT at the level of the first reduced density matrix".*

To the best of my knowledge, the concept of spin was discussed in DFT community only for two electron systems. In my article [24], I proved the theorem that the electron density does not depend upon the total spin of the state for N-electron system. This proof was done applying the formalism of the permutation group.

I would like to mention that in the Ψ-formalism used by Slater [27], he presented the wave function as a linear combination of determinants, corresponding to a given value of the total spin S. It does not allow to obtain any conclusions about spin. On the other hand, as I noted above, in my articles, firstly in Ref. [24], I obtained the proof for N-particle system for the independency of the electron density from the total spin S of the state. It was achieved applying the permutation group apparatus. For understanding the following text, I presented in this section the short description of the formalism of permutation groups. For more detail description, see Appendix in the end of this chapter or chapter 2 in my book [28].

The permutation group is characterized by Young diagrams $[\lambda]$:

$$[\lambda] = [\lambda_1 \lambda_2 \dots \lambda_k],$$

$$\lambda_1 \geq \lambda_2 \geq \ldots \geq \lambda_k, \sum_{i=1}^{k} \lambda_i = N. \quad (3)$$

The rows in the Young diagram are represented in decreasing order. The presence of several rows with equal length λ_i is indicated by a power of λ_i. For example, $[\lambda] = [2^2 1^2]$ is depicted graphically as

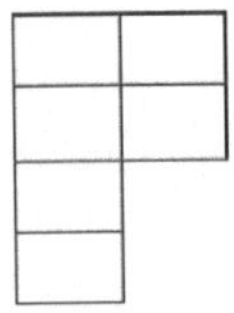

$[\lambda] = [2^2 1^2]$.

I would like to mention that in Appendix, I describe the unusual biography of Alfred Young who was not a professional mathematician but was a country clergyman.

Though the concept of spin has enabled to explain the nature of chemical bond, the electron spins are not involved directly in the formation of the latter. The interactions responsible for chemical bonding have a purely electrostatic nature. In non-relativistic approximation, the Hamiltonian does not depend on the spin, that means that the spin is saved, and we can operate with the value of the total spin S in the considered state.

According to PEP, the total electron wave function can be constructed as a sum of product of the spatial and spin wave functions symmetrized in respect to the irreducible representations $\Gamma^{[\lambda]}$ [28].

$$\Psi^{[1^N]} = \frac{1}{\sqrt{f_\lambda}} \sum_r \Phi_r^{[\lambda]} \Omega_{\tilde{r}}^{[\tilde{\lambda}]}. \quad (4)$$

In Eq. (4), $[\lambda]$ is the Young diagram and $\Gamma^{[\tilde{\lambda}]}$ denotes the representation conjugate to $\Gamma^{[\lambda]}$. Its matrix elements are

$$\Gamma_{\tilde{r}\tilde{t}}^{[\tilde{\lambda}]}(P) = (-1)^p \Gamma_{rt}^{[\lambda]}(P). \quad (5)$$

where p is the parity of permutation P. The spin Young diagram $[\tilde{\lambda}]$ is dual to $[\lambda]$, i.e., it is obtained from the latter by replacing rows by columns. For example,

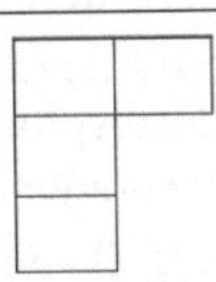

$[\lambda] = [21^2]$

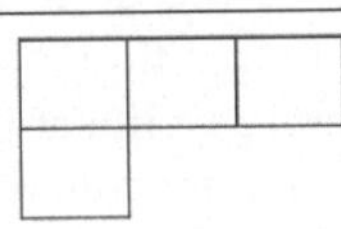

$[\tilde{\lambda}] = [31]$

Let us return to Eq. (4), where the sum is taken over all basis functions of the representation. The normalization of the total wave function is provided by the factor $1/\sqrt{f_\lambda}$. It should be mentioned that the electron spin has only two projections $s_z = \pm 1/2$, therefore the spin Young diagram $[\tilde{\lambda}]$ must have no more than two boxes per columns. In one box, the projection of spin $s_z = ½$ in the other box $s_z = -½$. It means that the total spin of this column equals 0. It is evident that the contribution to the total spin of the system of electrons will come only from uncoupled electron spins, that is, from the length of rows. The value of the total spin corresponding to spin Young diagram $[\tilde{\lambda}]$ is equal to

$$S = \frac{1}{2}\left(\tilde{\lambda}^{(1)} - \tilde{\lambda}^{(2)}\right). \tag{6}$$

Eq. (6) enables one to find easily the values of the spin S for each spin Young diagram. For example, the spin corresponding to the spin Young diagram $[\tilde{\lambda}] = [3\ 1]$, is equal to $S = 1$.

Let us mention that in the case of particles with $s > ½$, for a given Young diagram can correspond several values of S.If spin of the particle $s = 1$, to $[\lambda] = [31]$ can be attributed three values of $S = 1, 2,$ and 3, see Table 2 in Section C4 in Appendix C of my book on PEP [5].

As follows from experiment, the wave function of elementary particles can be only completely symmetric or antisymmetric. It allowed to generalize the PEP, primary formulated by Pauli only for electrons, for all elementary particles:

> *The only possible states of a system of identical particles possessing spin s are those for which the total wave function transforms upon interchange of any two particles as*
>
> $$P_{ij}\Psi(1, \ldots, i, \ldots j, \ldots, N) = (-1)^{2s}\Psi(1, \ldots, i, \ldots j, \ldots, N). \tag{7}$$
>
> *That is, it is symmetric for integer values of s (the Bose-Einstein statistics) and antisymmetric for half-integer values of s (the Fermi-Dirac statistics).*

According to Ehrenfest and Oppenheimer [29], this formulation is valid not only for elementary particles, but it is valid for different composite particles as well. As examples of composite particles, the authors considered atoms, molecules, and nuclei composed by electrons and protons (at that time the neutron had not been discovered). According to the presented above the general formulation of PEP for elementary particles, the wave-functions that described them can have only two types of symmetry: completely symmetric or antisymmetric, depending on their intrinsic value of spins.

The composite particles considered by Ehrenfest and Oppenheimer [29], were composed by fermions, that is, from particles with spin ½. So, the even number of particles leads to the Bose-Einstein statistics and odd number to the Fermi-Dirac statistics,

To the best of my knowledge, the scientists that had developed methods allowing to use the conception of spin in DFT calculations considered only two-particle systems. The general case of the N-electron system was considered firstly by the author in Ref. [24] where it was proved the theorem named by some authors as the Kaplan Theorem 2. This theorem was formulated in the following manner:

> *"The electron density of an arbitrary N-electron system, characterized by the N-electron wave function corresponding to the total spin S and constructed on some orthonormal orbital set, does not depend upon the total spin S and always preserves the same form as it is for a single-determinantal wave function."*

According to this theorem, for any permutation symmetry of the spatial wave function described by the Young diagram $[\lambda]$ that correspond to a definite value of spin S, the electron density is equal

$$\rho_t^{[\lambda]}(\boldsymbol{r}) = \sum_{n=1}^{N} |\varphi_n|^2. \tag{8}$$

It is a well-known expression of the electron density for the state described by the one-determinantal function with single-occupied orbitals. It can be shown that in the case of orbital configuration with arbitrary occupation numbers, the final expression (6) has not changed and will also correspond to the electron density for the one-determinantal function.

As follows from the discussion in the beginning of this section, at the first reduced density matrix approximation the concept of spin in principle cannot be introduced in the frame of traditional KS approach and at the gradient correction level as well. In more detail see the discussion based on the theory of permutation group in my paper, Ref. [24] or in some earlier papers.

From the analysis of the discussed above theorem follows that for different values of the total spin S, the expression for obtained electron density does not changed and have the same value as for wave function presented as a single Slater determinant. In this connection, it should be mentioned that about the ambiguity of the description by the electron density was known many years ago. I will cite two publications: more than 40 years ago Harriman [30] demonstrated that for each electron density ρ it can be constructed an arbitrary number of orthonormal orbitals, while in 2001 Cappelle and Vignale [31] showed that at the LSDA approximation it can be constructed different sets of potentials having the same ground state density.

Mean-while, different methods allowing taking into account the spin multiplet structure were developed, see Ref. [24] and references therein. In next section, I will discuss two groups of these methods.

3. Methods allowing take into account the spin multiplet structure within the DFT approach

It should be mentioned that the methods, taking into account the spin multiplet structure, are beyond the KS formalism. In most of these methods the Ψ-formalism is used. I will consider two widely used approaches:

1. One of the first publications, in which the spin multiplet structure was taken into account in the frame of DFT, was the paper by Ziegler et al. [32]. For each value of the total spin S, they built the appropriate combinations of the Slater determinants. The factor giving the value of spin was obtained by correcting the exchange energy, E_X. In publications [33–35], in which scientists applied the Ziegler et al. approach, only the exchange energy was considered. Mineva et al. [36] are stressed that the scientists [33–35] developing after Ziegler et al. the methods that allows to take into account the conception of spin in the DFT studies, are considered only the exchange energy, E_X, and did not consider the correlation energy, E_C. This led to the incorrect multiplet structure. To the best of my knowledge, this drawback has not been discussed in the DFT community.

   In some applications, the Ziegler et al. method was named as Multiplet Structure Method, or shortly MSM. In the following text, I will use this abbreviation. Usually the exchange-correlation functional, E_{XC}, is presented as a sum of exchange, E_X, and correlation E_C, energies:

$$E_{xc} = E_x + E_c. \quad (9)$$

In the studies based or developing MSM approach [32–35], the value of the total spin S was found using only the exchange energy E_X, and then applied to the total E_{xc}. It is evident that the contribution of the correlation energy E_X, is not the same as the exchange energy. Therefore, if E_c is multiplied by the same factor as it was found for E_x then it will give a wrong multiplet structure, because the exchange and correlations functionals should have a different dependence on S.

2. In the second group of methods [37–41], designated as restricted open-shell Kohn-Sham (ROKS) method, the open-shell theory of Roothaan [42] was used. In the first publication by Russo et al. [37], the Hamiltonian of Roothaan [42] was used, but the exchange term was replaced by the exchange-correlation functional. The authors [38–41], combined the ROKS methods with the MSM approach. As a result, the methods elaborated in publications [38–41] carried the same mistakes as the first ROKS method created by Russo et al. [37]. They do not provide the correct spin value for the correlation functional.

As stated above, both approaches MSM and ROKS, do not provide the correct total spin S for the correlation functionals.

According to calculations by Illas et al. [43], the ferromagnetic coupling is exaggerated, if the DFT method approaches are used. This agreed with our precise Mn_2 calculations performed with Mavridis group [44]. In most of calculations, Mn_2 had the ferromagnetic ground state with maximum value of the total spin, $S = 5$. I would like also to mention that in the following article [45] Illas and collaborators, using the Filatov-Shaik ROKS method [39, 40], failed to improve the agreement with experiment. The reason is that, as we noted above, in the ROKS calculation, the correlation energy was not considered.

The definition of the correlation energy was given by Löwdin [46] many years ago. According to it:

$$E_{corr} = E_{exact} + E_{HF}. \quad (10)$$

The exact quantum-mechanical calculations can be performed only for small electron systems, for larger systems, the "exact" energy will depend upon the method used for its calculation. Thus, the correlation energy is method dependent. It should be also mentioned that the correlation energy has not an analytical expression, which leads to some problems in its applications, see Ref. [24].

As it was noted in Introduction, the total spin S of the system can be introduced only at the two-particle reduced density matrix level. The modern state of the development of the two-particle reduced density matrix formalism was discussed in a large number of articles [47–55]. Unfortunately, the spin problem still has not been considered by the DFT community.

4. Symmetry properties of the density matrix; degenerate states

In an elegant proof, Hohenberg and Kohn [1] laid down the theoretical foundation of the DFT theory. In their fundamental paper, the degeneracy was not treated, since they considered the ground state, which very rare is degenerated, as it is in the case of O_2 molecule.

Very soon, in the DFT community it was accepted that the Levy-Lieb [56, 57] constraint search procedure allows to study the degenerate states in the DFT calculations. First, it was shown by Levy [56] in 1969 and then in 1983 by Lieb [57], who applied more abstract mathematical approach. I would like to mention that Bersuker [58] was the first who criticized the possibility of application of the DFT approach to degenerate states. Bersuker considered it on the special case of the Jahn-Teller effect. Let us mention that according to the following from quantum mechanics the Born-Oppenheimer approximation (the molecules can be calculated only at this approximation) the vibronic interaction mixed the electronic and nuclear motions, and the electron and nuclear densities may not be constructed. Thus, the Levy-Lieb [56, 57] constraint search procedure contradicts quantum mechanics.

In Section 2, we already discussed that the author proved, see Ref. [24], the theorem that the electron density of the arbitrary N-electron system, defined in Eq. (1), does not depend upon the total spin S and always preserves the same form as it is for a single-determinantal wave function. From this theorem follows that the wave function of N-electron system does not depend on the degeneracy of the state and on its symmetry as well. It was proved using the permutation group apparatus, described in detail in chapter 2 of my book [28], see also Appendix to this chapter.

In general, it can be two types of degenerate states, the spatial and the spin degeneracy. In the case of the spatial degeneracy, the particles are described by spatial wave functions, although when they are degenerated in the spin space, they are described by spin wave functions. For constructing the degenerate in space wave function, the point group formalism should be used. Let us consider a point group **G** with g elements. In the book [28], the author constructed the wave functions belonging to the f_α-dimensional representation $\Gamma^{(\alpha)}$ of an arbitrary point group **G** as:

$$\Psi_{ik}^{(\alpha)} = \frac{f_\alpha}{g}\sum_R \Gamma_{ik}^{(\alpha)}(R)^* R\Psi_0, \tag{11}$$

where $\Gamma_{ik}^{(\alpha)}(R)$ are the matrix elements of the representation $\Gamma^{(\alpha)}$ and the sum in Eq. (11) is taken over all g elements of the group **G**. The operations R of the group **G** are acting on some non-symmetrized product Ψ_0 of one-electron orbitals. If in $\Psi_{ik}^{(\alpha)}$ the second index k is fixed, then f_α function form a basis for the representation $\Gamma^{(\alpha)}$, each different indices k enumerates different bases.

If we have a f_α degenerate state, each of its f_α functions describe the system with the same probability and a pure state (the state described by wave function) cannot be selected. Therefore, the degenerate state must be considered as a mixed state, in which its basis functions enter the density with the same weight factors. The diagonal element of the density matrix in the case of degenerate state is written as:

$$D_k^{(\alpha)} = \frac{1}{f_\alpha}\sum_{i=1}^{f_\alpha}\left|\Psi_{ik}^{(\alpha)}\right|^2. \tag{12}$$

Using expression (12), I proved the theorem [24], according to which, the diagonal elements of the full density matrix is invariant for all operations of the group symmetry of the state, that is, it is a group invariant. It was proved that for every operation R of group G and all its irreducible representations $\Gamma^{(\alpha)}$

$$RD_k^\alpha = D_k^\alpha. \tag{13}$$

This means that the diagonal element of the full density matrix (and all reduced density matrices as well) transforms according to the totally symmetric one-dimensional representation A_1 of G regardless of the dimension of representation $\Gamma^{(\alpha)}$. It was proved for an arbitrary point group, but it is correct for any finite group. For the permutation group, this result was used in my publications [24, 59, 60] in analysis of the foundations of PEP. In these articles, I analyzed the case when PEP is not fulfilled and except of symmetrical and antisymmetrical states, an arbitrary permutation symmetry, including degenerate permutation states, are permitted. I have showed that if PEP is not fulfilled, this leads to contradictions with the concepts of particle independence and their identity. It was rigorously proved that the particles, described by wave functions with the permutation symmetry not allowed by PEP, may not exist in our Nature.

The arguments presented in Refs. [24, 59, 60], see also book [5], can be considered as a theoretical substantiation of PEP. They explained why in our Nature only completely symmetric or antisymmetric states, corresponding to one-dimensional representations of the permutation group, are realized. From this result, the important consequence follows

> *We may not expect that in future some unknown elementary particles can be discovered that are not fermions or bosons.*

On the other hand, according to the so-called *fractional* statistics, which is valid in the 2D-space, a continuum of intermedium cases between boson and fermion particles can exist, see subsection 5.4 in book [5]. As was showed by Leinaas and Myrheim [61] in their pioneer paper, in 2D-space can exist a continuum of states between boson and fermion symmetry. After Leinaas and Myrheim [61], Wilczek [62] introduced in 2D-space the *anyons*, which obey any statistics. However, we should take into account that anyons are quasiparticles defined in 2D-space. The real particles can exist only in 3D-space, and according to PEP, formulated for all elementary particles, see Eq. (7), the elementary particles can obey only the boson or fermion symmetry. It is important to stress that the discovery of the fractional statistics does not contradict PEP.

All experimental data, see my recent review [63], confirm the Pauli Exclusion Principle. Different very precise experiments did not show any Pauli-forbidden transitions.

This is confirmed also by very precise calculations of H_2 molecule [7], in which, certainly, PEP was taken into account. The quantum mechanical calculations of the H_2 dissociation energy and its first ionization potential [64, 65] are in a complete agreement with very precise experimental values, see Table 1.1 in [7]. From this follows not only an additional confirmation of PEP, but also a rather general conclusion that molecules obey the same quantum-mechanical laws that obey traditionally physical objects: atoms and solids; at nanoscale we should not distinguish between chemical and physical systems.

In the end, I would like to note that in some papers the authors claimed that they developed the non-Born-Oppenheimer DFT in the frame of the electron density approach. These publications were analyzed in my first paper on DFT limits [24], where it was shown that in spite of the authors claims, their formalisms must be attributed to the Born-Oppenheimer approximation.

Acknowledgements

I am grateful to Maestro Ronald Columbié-Leyva for new references and for the technical and software support.

Conflict of interest

No potential conflict of interest was reported by the author.

Appendix: Short necessary knowledge on the permutation group

The permutation symmetry is classified according to the irreducible representations of the permutation group π_N.[1] The latter are labeled by the Young diagrams

$$[\lambda] = [\lambda_1\lambda_2 \dots \lambda_k],$$
$$\lambda_1 \geq \lambda_2 \geq \dots \geq \lambda_k, \sum_{i=1}^{k} \lambda_i = N. \tag{14}$$

where λ_i is represented by a row of λ_i cells. The presence of several rows of equal length λ_i is convenient to indicate by a power of λ_i. For example,

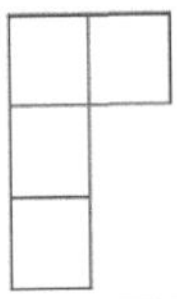

$[\lambda] = [21^2]$.

At present, the apparatus of permutations groups cannot be described without using the Young diagrams, I would like to note here some unusual details of the biography of Alfred Young. He was a country clergyman and has not any mathematical education. Young published studies were extending from 1900 to 1935, and in total he published 8 papers. The keystone of his studies was the reduction of the permutation groups to its irreducible representations in an explicit form. It is quite remarkable the gap of 25 years between his second paper in 1902 and the third in 1927. This gap will not be surprising, if we take into account that Young was a clergyman with numerous clerical duties.

It is obvious that one can form from two cells only two Young diagrams:

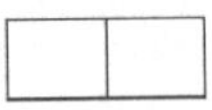

[2]

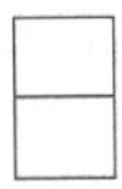

$[1^2]$.

For the permutation group of three elements, π_3, one can form from three cells three Young diagrams:

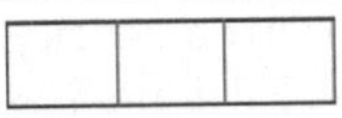

[3]

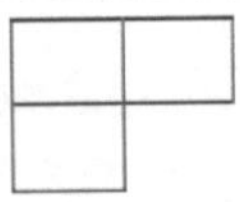

[2 1]

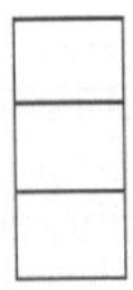

$[1^3]$

[1] For a more detailed treatise see books by Rutherford [66], Chapter 2 in book [28] or Appendix B in book [5].

The group π_4 has five Young diagrams:

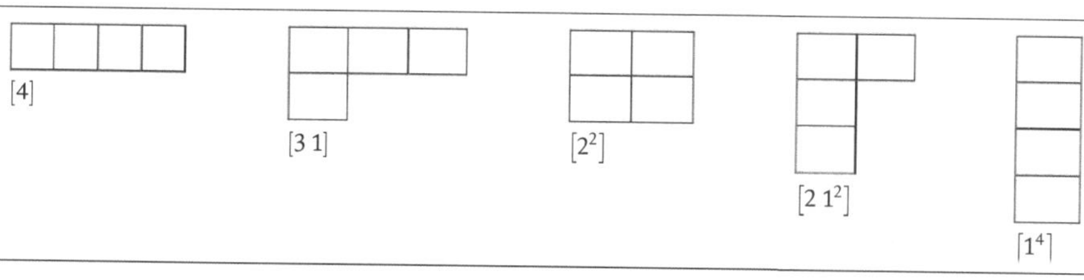

As we mentioned in the beginning of this Appendix, the representation of the permutation group π_N are labeled by the Young diagram $[\lambda]$. The rules how to find from a given Young diagram the matrices of representation of the permutation group were formulated by Young, but they were very complex. The Japanese mathematician Yamanouchi considerably simplified these rules. At present, it is called Young-Yamanouchi representation, which is described in detail in my book [28].
For each irreducible representation $\Gamma^{[\lambda]}$, the normalized basis functions can be easily constructed by the Young operator,

$$\omega_{rt}^{[\lambda]} = \sqrt{\frac{f_\lambda}{N!}} \sum_P \Gamma_{rt}^{[\lambda]}(P)P. \tag{15}$$

In Eq. (15), $\Gamma_{rt}^{[\lambda]}(P)$ are the matrix elements of the representation $\Gamma^{[\lambda]}$, f_λ is the dimension of $\Gamma^{[\lambda]}$ and the sum over operations P are taken for all $N!$ permutations of the group π_N. The normalized functions are obtained by acting the operator (15) on some non-symmetrized product of one-electron orbitals,

$$\Phi_0 = \varphi_1(1)\varphi_2(2)\ldots\varphi_N(N). \tag{16}$$

The normalized functions

$$\Phi_{rt}^{[\lambda]} = \omega_{rt}^{[\lambda]}\Phi_0 = \sqrt{\frac{f_\lambda}{N!}} \sum_P \Gamma_{rt}^{[\lambda]}(P)P\Phi_0, \tag{17}$$

are transformed in accordance with the representation $\Gamma^{[\lambda]}$. If in $\Phi_{rt}^{[\lambda]}$, Eq. (17), the second index t is fixed, then f_λ function form a basis for the representation $\Gamma^{[\lambda]}$, each different index k enumerates different bases. Let us prove this statement applying an arbitrary permutation Q of the group π_N to the function (17):

$$Q\Phi_{rt}^{[\lambda]} = \sqrt{\frac{f_\lambda}{N!}} \sum_P \Gamma_{rt}^{[\lambda]}(P)QP\Phi_0 = \sqrt{\frac{f_\lambda}{N!}} \sum_P \Gamma_{rt}^{[\lambda]}(P)\left(Q^{-1}R\right)R\Phi_0. \tag{18}$$

Using the invariance properties of a sum over all group elements and the property of orthogonal matrices, we obtain the matrix element of the product of permutations as a product of matrix elements.

$$\Gamma_{rt}^{[\lambda]}(P)\left(Q^{-1}R\right) = \sum_u \Gamma_{ru}^{[\lambda]}\left(Q^{-1}\right)\Gamma_{ut}^{[\lambda]}(R) = \sum_u \Gamma_{ur}^{[\lambda]}(Q)\Gamma_{ut}^{[\lambda]}(R). \tag{19}$$

In Eq. (19) we denoted the permutation QP by R. Substituting (19) in (18), we obtain finally

$$Q\Phi_{rt}^{[\lambda]} = \sqrt{\frac{f_\lambda}{N!}}\sum_{u}\Gamma_{ur}^{[\lambda]}(Q)\left(\sum_{R}\Gamma_{ut}^{[\lambda]}(R)R\Phi_0\right) = \sum_{u}\Gamma_{ur}^{[\lambda]}(Q)\Phi_{ut}^{[\lambda]}. \tag{20}$$

Thus, if the second index t is fixed, then f_λ functions form a basis for the representation $\Gamma^{[\lambda]}$, and each different index k enumerates different bases. It is also important to mention that in the function $\Phi_{rt}^{[\lambda]}$, index r characterizes the symmetry under permutation of the arguments, while index t characterizes the symmetry under permutation of the one-particle functions φ_a.

References

[1] Hohenberg P, Kohn W. Inhomogeneous electron gas. Physics Review. 1964;**136**:B864. DOI: 10.1103/PhysRev.136.B864

[2] Kohn W, Sham LJ. Self-consistent equations including exchange and correlation effects. Physics Review. 1965;**140**:A1133. DOI: 10.1103/PhysRev.140.A1133

[3] Zangwill A. A half century of density functional theory. Physics Today. 2015; **68**:34-39. DOI: 10.1063/PT.3.2846

[4] Arita M, Arapan S, Bowler DR, Miyazaki T. Large-scale DFT simulations with a linear-scaling DFT code CONQUEST on K-computer. Journal of Advanced Simulation in Science and Engineering. 2014;**1**:87-97. DOI: 10.15748/jasse.1.87

[5] Kaplan IG. The Pauli Exclusion Principle: Origins, Verifications and Applications. Chichester, UK: Wiley; 2017

[6] Kaplan IG, Santamaria R, Novaro O. Non-additive forces in atomic clusters. Molecular Physics. 1995;**84**:105-114. DOI: 10.1080/00268979500100071

[7] Kaplan IG. Intermolecular Interactions. Physical Picture, Computational Methods and Model Potential. Chichester: Wiley; 2006

[8] Kaplan IG. Symmetry properties of the electron density and following from it limits on the KS-DFT applications. Molecular Physics. 2018;**2018**(116): 658-665. DOI: 10.1080/00268976.2017.1393573

[9] Kaplan IG, Roszak S, Leszczinski J. Nature of binding in the alkaline–earth clusters: Be_3, Mg_3, and Ca_3. The Journal of Chemical Physics. 2000;**113**: 6245-6252. DOI: 10.1063/1.1287835

[10] Díaz-Torrejón CC, Kaplan IG. Many-body forces and stability of the alkaline-earth tetramers. Chemical Physics. 2011;**381**:67-71. DOI: 10.1016/j.chemphys.2011.01.011

[11] Price AJA, Bryenton KR, Johnson ER. Requirements for an accurate dispersion corrected density functional. The Journal of Chemical Physics. 2021;**154**:230902. DOI: 10.1063/5.0050993

[12] Cramer CJ, Truhlar DG. Density functional theory for transition metals and transition metal chemistry. Physical Chemistry Chemical Physics. 2009;**11**: 10757-10816. DOI: 10.1039/B907148B

[13] Gillan MJ, Alfè D, Michaelides A. Perspective: How good is DFT for water? The Journal of Chemical Physics. 2016; **144**:130901. DOI: 10.1063/1.4944633

[14] Taylor DE, Angyan JG, Galli G, Zhang C, Gygi F, Hirao K, et al. Blind test of density-functional-based methods on intermolecular interaction energies. The Journal of Chemical Physics. 2016;**145**:124105. DOI: 10.1063/1.4961095

[15] Yu HS, Li SL, Truhlar DG. Perspective: Kohn-Sham density functional theory descending a staircase. The Journal of Chemical Physics. 2016; **145**:130901. DOI: 10.1063/1.4963168

[16] Mardirossian N, Head-Gordon M. How accurate are the Minnesota density functionals for noncovalent interactions, isomerization energies, thermochemistry, and Barrier Heights involving molecules composed of main-group elements? Journal of Chemical Theory and Computation. 2016;**12**: 4303-4325. DOI: 10.1021/acs.jctc.6b00637

[17] Medvedev MG, Bushmarinov IS, Sun J, Perdue JP, Lyssenko K. Density functional theory is straying from the path toward the exact functional.

Science. 2017;**355**:49-52. DOI: 10.1126/science.aah5975

[18] Keep KP. Comment on "Density functional theory is straying from the path toward the exact functional". Science. 2017;**356**:496. DOI: 10.1126/science.aam9364

[19] Medvedev MG, Bushmarinov IS, Sun J, Perdue JP, Lyssenko K. Response to comment on "Density functional theory is straying from the path toward the exact functional". Science. 2017;**356**: 496. DOI: 10.1126/science.aam9550

[20] Marjewski AA, Medvedev MG, Gerasimova IS, Panovac MV, Perdew JP, Lysenko KA, et al. Interplay between test sets and statistical procedures in ranking DFT methods: The case of electron density studies. Mendeleev Communications. 2018;**28**:225-235. DOI: 10.1016/j.mencom.2018.05.001

[21] Savin A. Private Communication. 2021

[22] Perdew JP, Ruzsinszkya A, Sun J, Nepal NK, Kaplan AD. Interpretations of ground-state symmetry breaking and strong correlation in wavefunction and density functional theories. PNAS. 2021; **118**:1-6. DOI: 10.1073/pnas.2017850118

[23] Bartlett RJ. Adventures in DFT by a wavefunction theorist. The Journal of Chemical Physics. 2019;**151**:160901. DOI: 10.1063/1.5116338

[24] Kaplan IG. Problems in DFT with the total spin and degenerate states. International Journal of Quantum Chemistry. 2007;**107**:2595-2603. DOI: 10.1002/qua.21423

[25] McWeeny R. Density functions and density functionals. Philosophical Magazine B. 1994;**69**:727-735. DOI: 10.1080/01418639408240141

[26] Weiner B, Trickey SB. State energy functionals and variational equations in density functional theory. Journal of Molecular Structure. 2000;**501–502**: 65-83. DOI: 10.1016/S0166-1280(99) 00415-7

[27] Slater JC. Electron Structure of Molecules. Vol. 1. New York: McGraw-Hill; 1963

[28] Kaplan IG. Symmetry of Many-Electron Systems. New York: Academic Press; 1975

[29] Ehrenfest P, Oppenheimer JR. Note on the statistics of nuclei. Physics Review. 1931;**37**:333-338. DOI: 10.1103/PhysRev.37.333

[30] Harriman JE. Orthonormal orbitals for the representation of an arbitrary density. Physical Review A. 1981;**24**: 680-682. DOI: 10.1103/PhysRevA.24.680

[31] Capelle K, Vignale G. Nonuniqueness of the potentials of spin-density-functional theory. Physical Review Letters. 2001;**86**:5546-5549. DOI: 10.1103/PhysRevLett.86.5546

[32] Ziegler T, Rauk A, Baerends EJ. On the calculation of multiplet energies by the hartree-fock-slater method. Theoretica Chimica Acta. 1977;**43**: 261-271. DOI: 10.1007/BF00551551

[33] Noodleman L. Valence bond description of antiferromagnetic coupling in transition metal dimers. The Journal of Chemical Physics. 1981;**74**: 5737-5743. DOI: 10.1063/1.440939

[34] Dunlap BI. Explicit treatment of correlation within density-functional theories that use the kinetic-energy operator. Physical Review A. 1984;**29**: 2902-2905. DOI: 10.1103/PhysRevA.29.2902

[35] Daul C. Density functional theory applied to the excited states of coordination compounds. International Journal of Quantum Chemistry. 1994;**52**: 867-877. DOI: 10.1002/qua.560520414

[36] Mineva T, Goursot A, Daul C. Atomic multiplet energies from density functional calculations. Chemical Physics Letters. 2001;**350**:147-154. DOI: 10.1016/S0009-2614(01)01264-7

[37] Russo TV, Martin RL, Hay PJ. Density functional calculations on first-row transition metals. The Journal of Chemical Physics. 1994;**101**:7729-7737. DOI: 10.1063/1.468265

[38] Frank I, Hutter J, Marx D, Parrinello M. Molecular dynamics in low-spin excited states. The Journal of Chemical Physics. 1998;**108**:4060-4069. DOI: 10.1063/1.475804

[39] Filatov M, Shaik S. Spin-restricted density functional approach to the open-shell problem. Chemical Physics Letters. 1998;**288**:689-697. DOI: 10.1016/S0009-2614(98)00364-9

[40] Filatov M, Shaik S. Application of spin-restricted open-shell Kohn–Sham method to atomic and molecular multiplet states. The Journal of Chemical Physics. 1999;**110**:116-**125**. DOI: 10.1063/1.477941

[41] Grimm S, Nonnenberg C, Frank I. Restricted open-shell Kohn–Sham theory for π–π^* transitions. I. Polyenes, cyanines, and protonated imines. The Journal of Chemical Physics. 2003; **119**:11574-11584. DOI: 10.1063/1.1623742

[42] Roothaan CCJ. Self-consistent field theory for open shells of electronic systems. Reviews of Modern Physics. 1960;**32**:179-185. DOI: 10.1103/RevModPhys.32.179

[43] Illas F, Moreira I d PF, Bofill JM, Filatov M. Spin symmetry requirements in density functional theory: The proper way to predict magnetic coupling constants in molecules and solids. Theoretica Chimica Acta. 2006;**116**: 587-597. DOI: 10.1007/s00214-006-0104-6

[44] Tzeli D, Miranda U, Kaplan IG, Mavridis A. First principles study of the electronic structure and bonding of Mn_2. The Journal of Chemical Physics. 2008;**129**:154310. DOI: 10.1063/1.2993750

[45] Moreira I, Moreira PF, Costa R, Filatov M, Illas F. Restricted ensemble-referenced Kohn–Sham versus broken symmetry approaches in density functional theory: Magnetic coupling in Cu binuclear complexes. Journal of Chemical Theory and Computation. 2007;**3**:764-774. DOI: 10.1021/ct7000057

[46] Löwdin PO. Correlation problem in many-electron quantum mechanics I. Review of different approaches and discussion of some current ideas. In: Prigogine I, editor. Advances in Chemical Physics. Vol. 2. London, UK: Interscience Publishers LTD.; 1959. pp. 207-322. DOI: 10.1002/9780470143483.ch7

[47] Ziesche P. Pair density functional theory—A generalized density functional theory. Physics Letters A. 1994;**195**:213-220. DOI: 10.1016/0375-9601(94)90155-4

[48] Ziesche P. Attempts toward a pair density functional theory. International Journal of Quantum Chemistry/Symposium. 1996;**60**:1361-1374. DOI: 10.1002/(SICI)1097-461X(1996)60:7%3C1361::AID-QUA18%3E3.0.CO;2-7

[49] Gonis A, Schulthesis IC, van Ek J, Turchi PE. A general minimum principle for correlated densities in quantum many-particle systems. Physical Review Letters. 1996;**77**: 2981-2984. DOI: 10.1103/PhysRevLett.77.2981

[50] Levy M, Ziesche P. The pair density functional of the kinetic energy and its simple scaling property. The Journal of Chemical Physics. 2001;**115**:9110-9112. DOI: 10.1063/1.1413976

[51] Furche F. Towards a practical pair density-functional theory for many-electron systems. Physical Review A. 2004;**70**:022514. DOI: 10.1103/PhysRevA.70.022514

[52] Buijse MA, Baerends EJ, Snijders JG. Analysis of correlation in terms of exact local potentials: Applications to two-electron systems. Physical Review A. 1989;**40**:4190-4202. DOI: 10.1103/PhysRevA.40.4190

[53] Gritsenko O, van Leeuwen R, Baerends EJ. Analysis of electron interaction and atomic shell structure in terms of local potentials. The Journal of Chemical Physics. 1994;**101**:8955-8963. DOI: 10.1063/1.468024

[54] Gritsenko OV, Baerends EJ. Effect of molecular dissociation on the exchange-correlation Kohn-Sham potential. Physical Review A. 1996;**54**:1957-1972. DOI: 10.1103/PhysRevA.54.1957

[55] Cuevas-Saavedra R, Staroverov VN. Exact expressions for the Kohn–Sham exchange-correlation potential in terms of wave-function-based quantities. Molecular Physics. 2016;**114**:1050-1058. DOI: 10.1080/00268976.2015.1131861

[56] Levy M. Universal variational functionals of electron densities, first-order density matrices, and natural spin-orbitals and solution of the v-representability problem. Proceedings of the National Academy of Sciences of the United States of America. 1979;**76**:6062-6065. DOI: 10.1073/pnas.76.12.6062

[57] Lieb EH. Density functionals for Coulomb systems. International Journal of Quantum Chemistry. 1983;**24**:243-277. DOI: 10.1002/qua.560240302

[58] Bersuker IB. Limitations of density functional theory in application to degenerate states. Journal of Computational Chemistry. 1997;**18**:260-267. DOI: 10.1002/(SICI)1096-987X(19970130)18:2%3C260::AID-JCC10%3E3.0.CO;2-M

[59] Kaplan IG. The Pauli exclusion principle. Can it be proved? Foundations of Physics. 2013;**43**:1233-1251. DOI: 10.1007/s10701-013-9742-4

[60] Kaplan IG. Modern state of the Pauli exclusion principle and the problems of its theoretical foundation. Symmetry. 2021;**13**:1-21. DOI: 10.3390/sym13010021

[61] Leinaas JM, Myrheim J. On the theory of identical particles. Nuovo Cimento. 1977;**37B**:1-23. DOI: 10.1007/BF02727953

[62] Wilczek F. Magnetic flux, angular momentum, and statistics. Physical Review Letters. 1982;**48**:1144-1146. DOI: 10.1103/PhysRevLett.48.1144

[63] Kaplan IG. The Pauli exclusion principle and the problems of its experimental verification. Symmetry. 2020;**12**:1-15. DOI: 10.3390/sym12020320

[64] Kolos W, Rychlewski J. Improved theoretical dissociation energy and ionization potential for the ground state of the hydrogen molecule. The Journal of Chemical Physics. 1993;**98**:3960-3967. DOI: 10.1063/1.464023

[65] Wolniewicz L. Nonadiabatic energies of the ground state of the hydrogen molecule. The Journal of Chemical Physics. 1995;**103**:1792-1799. DOI: 10.1063/1.469753

[66] Rutherford DE. Substitutional Analysis. London, UK: Hafner Publishing Co.; 1968

Chapter 3

The Density Functional Theory and Beyond: Example and Applications

Abstract

Density Functional Theory is one of the most widely used methods in quantum calculations of the electronic structure of matter in both condensed matter physics and quantum chemistry. Despite the importance of the density functional theory to find the correlation-exchange energy, but this quantity remains inaccurate. So we have to go beyond DFT to correct this quantity. In this framework, the random phase approximation has gained importance far beyond its initial field of application, condensed matter physics, materials science, and quantum chemistry. RPA is an approach to accurately calculate the electron correlation energy.

Keywords: DFT, LDA, GGA, RPA, Schrödinger equation

1. Introduction

The study of the microscopic properties of a physical system in the condensed matter branch requires the solution of the Schrödinger equation. When the studied system is composed of a large number of interacting atoms, the analytical solution of the Schrödinger equation becomes impossible. However, certain numerical calculation methods provide access to a solution to this fundamental equation for increasingly large systems. The calculation methods, called ab-initio like the density functional theory (DFT), propose to solve the Schrödinger equation without adjustable parameters. The density functional theory quickly established itself as a relatively fast and reliable way to simulate electronic and structural properties for all of the elements of the periodic table ranging from molecules to crystals. In this chapter, we recall the principle of this theory which considers electron density as a fundamental variable and that all physical properties can be expressed as a function of it.

2. Schrödinger equation

It is a fundamental equation to be solved to describe the electronic structure of a system with several nuclei and electrons and for a non-relativistic quantum description of a molecular or crystalline system and which is written:

$$H\Psi = \left(-\sum_{i}^{n} \frac{\hbar^2 \nabla_i^2}{2m} - \sum_{I}^{N} \frac{\hbar^2 \nabla_I^2}{2M} - \sum_{i,I} \frac{Z_I e^2}{|\overleftarrow{r_i} - \overleftarrow{R_I}|} + \sum_{i<j} \frac{e^2}{|\overleftarrow{r_i} - \overleftarrow{r_j}|} + \sum_{I<J} \frac{Z_I Z_J e^2}{|\overleftarrow{R_I} - \overleftarrow{R_J}|} \right) \Psi, \tag{1}$$

where H is the molecular Hamiltonian and Ψ is the wave function. It is therefore a question of seeking the solutions of this equation. We can write the Hamiltonian in the form:

$$H = T_e + T_n + V_{e-e} + V_{n-n} + V_{n-e}. \tag{2}$$

We give the definition for each term:
$T_e = -\sum_i^n \frac{\hbar^2 \nabla_i^2}{2m}$: The kinetic energy of n electrons of mass m.
$T_n = -\sum_I^N \frac{\hbar^2 \nabla^2}{2M}$: The kinetic energy of N nuclei of mass M.
$V_{e-e} = \sum_{i<j} \frac{e^2}{|\overleftarrow{r_i} - \overleftarrow{r_j}|}$: The electron–electron repulsive potential energy.
$V_{e-n} = -\sum_{i,I} \frac{Z_I e^2}{|\overleftarrow{r_i} - \overleftarrow{R_I}|}$: The attractive potential energy nucleus-electron.
$V_{n-n} = \sum_{I<J} \frac{Z_I Z_J e^2}{|\overleftarrow{R_I} - \overleftarrow{R_J}|}$: The nucleus-nucleus repulsive potential energy.

For a system of N nuclei and n electrons, Schrödinger equation is too complex to be able to be solved analytically. The exact solution of this equation is only possible for the hydrogen atom and hydrogenoid systems. In order to simplify the solution of this equation, Max Born and Robert Oppenheimer [1] have proposed an approximation aiming to simplify it.

3. The Born-Oppenheimer approximation

We consider that we can decouple the movement of electrons from that of nuclei, by considering that their movement of nuclei is much slower than that of electrons: we consider them as fixed in the study of the movement of the electrons of the molecule. The inter-nuclear distances are then treated as parameters. It has an immediate computational consequence, called an adiabatic hypothesis. It is in fact the same approximation and since the Oppenheimer approximation is still used in quantum chemistry, during chemical reactions or molecular vibrations, we can consider according to the classical Born-Oppenheimer approximation that the distribution of electrons (adapts) almost instantaneously, when from the relative motions of nuclei to the resulting Hamiltonian variation. This is due to the lower inertia of the electrons $M = 1800m_e$ then the electron wave function can therefore be calculated when we consider that the nuclei are immobile, from where

$$T_n = 0; V_{n-n} = constant, \tag{3}$$

and so the Hamiltonian becomes

$$H = T_e + V_{e-e} + V_{e-n} + V_{n-n}. \tag{4}$$

$$H = H_{ele} + V_{n-n}, \tag{5}$$

with H_{ele}: electronic Hamiltonian which is equal to:

$$H_{ele} = T_e + V_{e-e} + V_{e-n}. \tag{6}$$

Therefore the Born Oppenheimer approximation gives us:

$$H = T_e + V_{e-e} + V_{e-n}. \tag{7}$$

We use another notation to simplify the calculations

$$H = T + V_{ext} + U. \tag{8}$$

$$T = T_e; U = V_{e-e} = V_H; V_{ext} = V_{e-n}. \tag{9}$$

The Born-Oppenheimer approximation results in the Eq. (7) which keeps a very complex form: it always involves a wave function with several electrons. This approximation significantly reduces the degree of complexity but also the new wave function of the system depends on N bodies while other additional approximations are required to be able to effectively solve this equation. The remainder of this chapter will deal with approximations allowing to arrive at a solution of this equation within the framework of the density functional theory (DFT) and the random phase approximation (RPA).

4. Density Functional theory (DFT)

Density Functional Theory is one of the most widely used methods for calculating the electronic structure of matter in both condensed matter physics and quantum chemistry. The DFT has become, over the last decades, a theoretical tool which has taken a very important place among the methods used for the description and the analysis of the physical and chemical properties for the complex systems, particularly for the systems containing a large number electrons. DFT is a reformulation of the N-body quantum problem and as the name suggests, it is a theory that only uses electron density as the fundamental function instead of the wave function as is the case in the method by Hartree and Hartree-Fock. The principle within the framework of the DFT is to replace the function of the multielectronic wave with the electronic density as a base quantity for the calculations. The formalism of the DFT is based on the two theorems of P. Hohenberg and W. Kohn [2].

4.1 Hohenberg and Kohn theorems

Hohenberg-Kohn (HK) reformulated the Schrödinger equation no longer in terms of wave functions but employing electron density, which can be defined for an N-electron system by:

$$n = 2N \int dr_1 \int dr_2 \ldots \int dr_{n-1} \Psi^* (r_1, r_2, \ldots \ldots r_{n-1}, r) \Psi(r_1, r_2, \ldots \ldots r_{n-1}, r), \tag{10}$$

this equation depends only on the three position parameters r = (x, y, z), position vector of a given point in space. This approach is based on two theorems demonstrated by Hohenberg and Kohn.

Theorem 1: For any system of interacting particles in an external potential $V_{ext}\left(\vec{r}\right)$, the potential $V_{ext}\left(\vec{r}\right)$ is only determined, except for an additive constant, by the electron density $n_0\left(\vec{r}\right)$ in its ground state.

The first HK Theorem can be demonstrated very simply by using reasoning by the absurd. Suppose there can be two different external potentials $V_{ext}^{(1)}$ and $V_{ext}^{(2)}$ associated with the ground state density $n\left(\overleftarrow{r}\right)$. These two potentials will lead to two different Hamiltonians $H^{(1)}$ and $H^{(2)}$ whose wave functions $\psi^{(1)}$ and $\psi^{(2)}$ describing the ground state are different. As described by the ground state of $H^{(1)}$ we can therefore write that:

$$E^{(1)} = \left\langle \psi^{(1)} | H^{(1)} | \psi^{(1)} \right\rangle < \left\langle \psi^{(2)} | H^{(1)} | \psi^{(2)} \right\rangle. \tag{11}$$

This strict inequality is valid if the ground state is not degenerate which is supposed in the case of the approach of HK. The last term of the preceding expression can be written:

$$\left\langle \psi^{(2)} | H^{(1)} | \psi^{(2)} \right\rangle = \left\langle \psi^{(2)} | H^{(2)} | \psi^{(2)} \right\rangle + \left\langle \psi^{(2)} | H^{(1)} - H^{(2)} | \psi^{(2)} \right\rangle, \tag{12}$$

$$\left\langle \psi^{(2)} | H^{(1)} | \psi^{(2)} \right\rangle = E^{(2)} + \int \left[V_{ext}^{(1)}\left(\overleftarrow{r}\right) - V_{ext}^{(2)}\left(\overleftarrow{r}\right) \right] n_0 \left(\overleftarrow{r}\right) d^3 r, \tag{13}$$

$$E^{(1)} < E^{(2)} + \int \left[V_{ext}^{(1)}\left(\overleftarrow{r}\right) - V_{ext}^{(2)}\left(\overleftarrow{r}\right) \right] n_0 d^3 r. \tag{14}$$

It will also the same reasoning can be achieved by considering $E^{(2)}$ instead of $E^{(1)}$. We then obtain the same equation as before, the symbols (1) and (2) being inverted:

$$E^{(2)} = \left\langle \psi^{(2)} | H^{(2)} | \psi^{(2)} \right\rangle < \left\langle \psi^{(1)} | H^{(2)} | \psi^{(1)} \right\rangle, \tag{15}$$

$$\left\langle \psi^{(1)} | H^{(2)} | \psi^{(1)} \right\rangle = \left\langle \psi^{(1)} | H^{(1)} | \psi^{(1)} \right\rangle + \left\langle \psi^{(1)} | H^{(2)} - H^{(1)} | \psi^{(1)} \right\rangle, \tag{16}$$

$$\left\langle \psi^{(1)} | H^{(2)} | \psi^{(1)} \right\rangle = E^{(1)} + \int \left[V_{ext}^{(2)}\left(\overleftarrow{r}\right) - V_{ext}^{(1)}\left(\overleftarrow{r}\right) \right] n_0 \left(\overleftarrow{r}\right) d^3 r. \tag{17}$$

$$E^{(2)} < E^{(1)} + \int \left[V_{ext}^{(2)} - V_{ext}^{(1)} \right] n_0 d^3 r, \tag{18}$$

we obtain the following contradictory equality:

$$E^{(1)} + E^{(2)} < E^{(1)} + E^{(2)}. \tag{19}$$

The initial hypothesis is therefore false; there cannot exist two external potentials differing by more than one constant leading at the same density of a non-degenerate ground state. This completes the demonstration.

$\Rightarrow$ the external potential of the ground state is a density functional.

> **Theorem 2:** The previous theorem only exposes the possibility of studying the system via density. It only allows knowledge of the density associated with the studied system. The Hohenberg-Kohn variational principle partially answers this problem:
>
> a universal functional for the energy $E[n]$ can be defined in terms of the density. The exact ground state is the overall minimum value of this functional.

Since the fundamental energy of the system is uniquely determined by its density, then energy can be written as a density functional. By following reasoning similar to that of the first part we show that the minimum of the functional corresponds to the energy of the ground state, indeed, the total energy can be written:

$$E_{HK}[n] = \int n\left(\overleftarrow{r}\right) V_{ext}\left(\overleftarrow{r}\right) d^3r + F_{HK}[n], \tag{20}$$

$F[n]$ is a universal functional of n(r):

$$F_{HK}[n] = T[n] + U[n]. \tag{21}$$

And the number of particles:

$$N = \int n\left(\overleftarrow{r}\right) dr. \tag{22}$$

Thus, we see that by minimizing the energy of the system with respect to the density we will obtain the energy and the density of the ground state. Despite all the efforts made to evaluate this functional E[n], it is important to note that no exact functional is yet known.

4.2 Ansatz of Kohn-Sham

Since the kinetic energy of a gas of interacting electrons being unknown, in this sense, Walter Kohn and Lu Sham [3] (KS) proposed in 1965 an ansatz which consists in replacing the system of electrons in interaction, impossible to solve analytically, by a problem of independent electrons evolving in an external potential. In the case of a system without interaction, the functional E[n] is reduced to kinetic energy and the interest of the reformulation introduced by Kohn and Sham is that we can now define a monoelectronic Hamiltonian and write the equations monoelectronic Kohn-Sham. According to KS the energy is written in the following form:

$$E_{HK}[n] = T_s[n] + \int V_{ext}(r) n(r) d^3r + E_{hartree}[n] + E_{xc}[n], \tag{23}$$

with the functional:

$$F_{HK}[n] = E_{HK}[n] - \int V_{ext}\left(\overleftarrow{r}\right) n\left(\overleftarrow{r}\right) d^3r. \tag{24}$$

$$F_{HK}[n] = T_s[n] + E_c[n] + E_{hartree}[n] + E_x[n] = T_s[n] + E_{hartree}[n] + E_{xc}[n]. \tag{25}$$

$T_s[n]$: representing the kinetic energy of a fictitious gas of non-interacting electrons but of the same density is given by:

$$T_s[n] = \sum_i \int dr \Psi_i^*\left(\overleftarrow{r}\right) \frac{-\nabla^2}{2} \Psi_i\left(\overleftarrow{r}\right). \tag{26}$$

$$E_{hartree}[n] = \frac{e^2}{8\pi\varepsilon_0} \iint \frac{n\left(\overleftarrow{r}\right) n\left(\overleftarrow{r'}\right)}{|\overleftarrow{r} - \overleftarrow{r'}|} d^3r d^3r'. \tag{27}$$

$T_s[n]$: the kinetic energy without interaction.
$E_{xc}[n]$: the exchange-correlation energy.
$E_{hartree}[n]$: the electron–electron potential energy.

$$E_{xc}[n] = E_{HK}[n] - \int V_{ext}(\overleftarrow{r})n(\overleftarrow{r})d^3r - T_s[n] - E_{hartree}[n]. \tag{28}$$

$$E_{xc}[n] = F[n] - T_s[n] - E_{hartree}[n]. \tag{29}$$

Based on the second Hohenberg-Kohn theorem, which shows that the electron density of the ground state corresponds to the minimum of the total energy and on the condition of conservation of the number of particles

$$\delta N\left[n(\overleftarrow{r})\right] = \int \delta n(\overleftarrow{r})dr = 0, \tag{30}$$

So we have:

$$\delta\left\{E_{HK}[n] - \mu\left(\int n(\overleftarrow{r})d^3r - N\right)\right\} = 0. \tag{31}$$

$$\frac{\delta E_{HK}[n]}{\delta n(\overleftarrow{r})} = \mu, \tag{32}$$

$$\frac{\delta T_s[n]}{\delta n(\overleftarrow{r})} + v^{eff}(\overleftarrow{r}) = \mu, \tag{33}$$

$$v^{eff}(\overleftarrow{r}) = V_{ext}(\overleftarrow{r}) + \frac{e^2}{8\pi\varepsilon_0}\int \frac{n(\overleftarrow{r'})}{|\overleftarrow{r} - \overleftarrow{r'}|}d^3r' + \frac{\delta E_{xc}[n]}{\delta n(\overleftarrow{r})}, \tag{34}$$

therefore, the kinetic energy without interaction $T_s[n]$ is determined by:

$$\frac{\delta T_s[n]}{\delta n(\overleftarrow{r})} = (3\pi^2 n)^{\frac{5}{3}}\frac{\hbar^2}{2m} = \frac{\hbar^2}{2m}k_F^2. \tag{35}$$

Finally, the mono-electronic Hamiltonian of Kohn-sham in atomic unit is put in the form:

$$H = \frac{-\nabla^2}{2} + v^{eff}. \tag{36}$$

The Hamiltonian is iteratively computed, the self-consistency of a loop is reached when the variation of the calculated quantity is lower than the fixed convergence criterion. The wave functions are calculated by a conjugate gradient method (or equivalent). The density is built from the wave functions, convergence is reached when the density is sufficiently close to the density of the previous step. When seeking to optimize the atomic structure of the system, an additional loop is added. With each iteration of this loop, the atomic positions are changed. It is said that the system is minimized when the forces are lower than the convergence criterion on the amplitude of the forces.

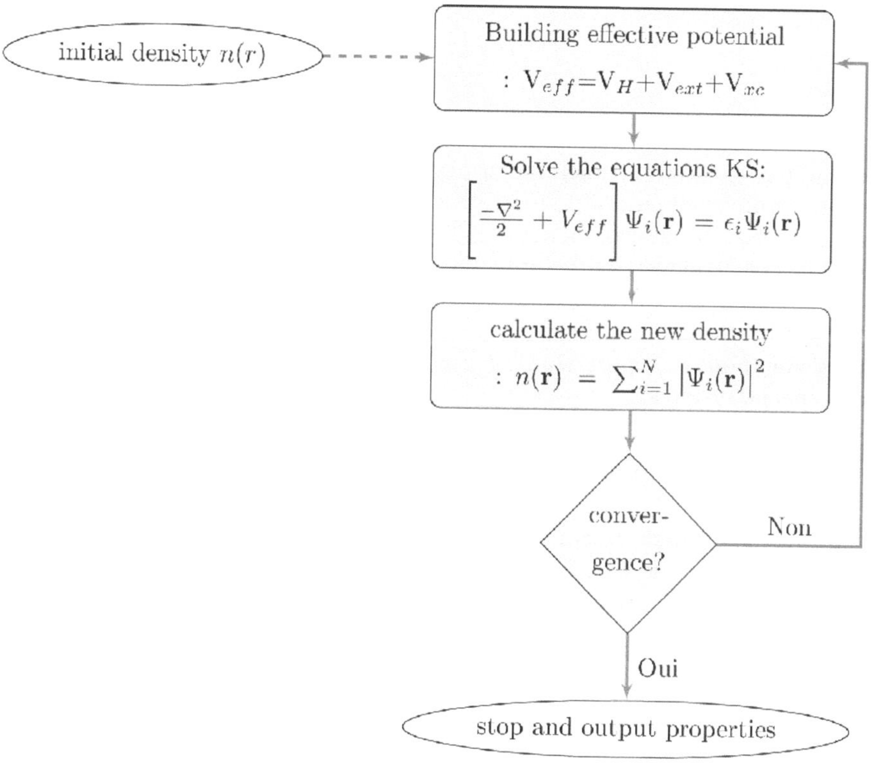

4.3 Expression of the exchange and correlation term

As described above, DFT is at the stage of Kohn-Sham equations, a perfectly correct theory insofar as the electron density which minimizes the total energy is exactly the density of the system of N interacting electrons. However, DFT remains inapplicable because the exchange-correlation potential remains unknown. It is therefore necessary to approximate this exchange-correlation potential. Two types of approximations exist the local density approximation or LDA and the generalized gradient approximation or GGA as well as the derived methods which are based on a non-local approach.

4.3.1 Local density approximation (LDA)

In only one model case, that of the uniform gas of electrons (corresponding quite well to the electrons of the conduction band of a metal), we know the exact expressions or with an excellent approximation of the terms of exchange and correlation respectively. In this LDA (Local Density Approximation), the electron density is assumed to be locally uniform and the exchange-correlation functional is of the form:

$$E_{xc}^{LDA} = \int n\left(\overleftarrow{r}\right)\varepsilon_{xc}^{homo}\left(n\left(\overleftarrow{r}\right)\right)dr. \tag{37}$$

$$\varepsilon_{xc}^{homo}\left(n\left(\overleftarrow{r}\right)\right) = \varepsilon_x[n] + \varepsilon_c[n]. \tag{38}$$

The function of $\varepsilon_{xc}^{homo}\left(n\left(\overleftarrow{r}\right)\right)$ is determined from a quantum computation parameterization for a constant electron density $n\left(\overleftarrow{r}\right) = n$;

$$\varepsilon_{xc}^{homo}\left(n\left(\overleftarrow{r}\right)\right)=\frac{3}{4}\left(\frac{3n(r)}{\pi}\right)^{\frac{1}{3}}. \tag{39}$$

Ceperley-Alder [4] numerically determined the contribution of the correlations. The search for analytical functions that come as close as possible to these results leads to the development of various functionalities with varying degrees of success. In general, the LDA approximation gives good results in describing the structural properties, i.e. it allows to determine the energy variations with the crystalline structure although it overestimates the cohesion energy, also concerning the mesh parameter for the majority of solids and good values of elastic constants like the isotropic modulus of compressibility. But this model remains insufficient in inhomogeneous systems.

4.3.2 Generalized gradient approximation (GGA)

To overcome the shortcomings of the LDA method, the generalized gradient approximation considers exchange-correlation functions depending not only on the density at each point, but also on its gradient [5], of the general form.

$$E_{xc}^{GGA}\left[n_\alpha,n_\beta\right]=\int n\left(\overleftarrow{r}\right)\varepsilon_{xc}\left[n_\alpha,n_\beta,\nabla_{n_\alpha(r)}\nabla_{n_\beta\left(\overleftarrow{r}\right)}\right]dr, \tag{40}$$

α and β are spins, in this case again, a large number of expressions have been proposed for this factor ε_{xc} leading to so many functionals. In general, the GGA improves compared to the LDA a certain number of properties such as the total energy or the energy of cohesion, but does not lead to a precise description of all the properties of a semiconductor material namely its electronic properties.

4.3.3 Functional hybrid HSE

The functions of DFT have been proved to be quite useful in explaining a wide range of molecular characteristics. The long-term nature of the exchange interaction, and the resulting huge processing needs, are a key disadvantage for periodic systems. This is especially true for metallic systems that necessitate BZ sampling. A new hybrid functionality, recently proposed by Heyd et *al.* [6], addresses this problem by separating the description of the exchange and the interaction into a short and long part. The expression of the exchange-correlation energy in HSE03 is given by:

$$E_{xc}^{HSE03}=\frac{1}{4}E_x^{sr,\mu}+\frac{3}{4}E_x^{PBE,sr,\mu}+E_x^{PBE,lr,\mu}+E_c^{PBE}. \tag{41}$$

As can be seen from the Eq. (41) only the exchange component of the electron–electron interaction is split into a short and long (lr) range (sr) part. The full electron correlation is represented by the standard correlation portion of the density of the GGA functional. Note that the term hybrid refers to the combined use of the exact exchange energy of the Hartree-Fock model and the exchange-correlation energy at the DFT level. The construction of hybrid functionals has been a good advancement in the field of exchange-correlation energy processing by allowing an explicit incorporation of the nonlocal character through the use of the exact term of exchange energy.

5. Random phase approximation

Despite the DFT is relevance in determining the exchange-correlation energy, it is still insufficient to characterize elastic characteristics. To rectify this amount, we must go beyond DFT. This is due to the random phase approximation as a method for calculating the electronic correlation energy accurately. RPA appears in the 1950s' [7–10] as a method of solving the N-body problem and arises from the desire to describe better (i.e. better than in a mean-field approximation) the physics of uniform electron gas, where the correlation between the positions of long-range electrons is important. In fact, collective oscillations (called plasma oscillations) are observed in an electron gas, which is the direct consequence of the long-range correlation between the electrons. Bohm and Pines, who introduced RPA, propose to place these collective oscillations at the center of solving the N-body problem, hoping that a good description of one will provide a good understanding of the other. RPA has been used with some success in the literature to describe systems containing Van der Waals interactions and in particular involving [11] scattering forces, which are known to be difficult to process. RPA introduced within the framework of DFT via the fluctuation-dissipation theorem with adiabatic connection (ACFDT).

6 Adiabatic-connection fluctuation-dissipation theory

The adiabatic connection fluctuation-dissipation (AC-FDT) [12] technique will be explained in order to discover the exact exchange-correlation energy in RPA. It will serve as the starting point for introducing the random phase approximation because it provides a general formulation for the exact correlation energy.

6.1 Adiabatic-connection (AC)

The adiabatic connection (AC) is a way to express the exact exchange-correlation energy function. The central idea in this approach is to build an interpolation Hamiltonian, which connects a Hamiltonian of an independent particle (reference Hamiltonian) $\hat{H}_0 = \hat{H}(\lambda = 0)$ and the physical Hamiltonians (multi-body Hamiltonian) $\hat{H} = \hat{H}(\lambda = 1)$, with λ being a connection parameter. The AC technique can be used to derive the total energy of the ground state of a Hamiltonian of multiple interacting bodies, in which a continuous set of Hamiltonians dependent on the coupling force (λ) is introduced by:

$$\begin{aligned}\hat{H}(\lambda) &= \hat{H}_0 + \lambda\hat{H}_1(\lambda) = \hat{T} + V_{ne} + \hat{V}(\lambda) + \lambda W_{ee} \\ &= \sum_{i=1}^{N}\left[\frac{-1}{2}\nabla_i^2 + v_\lambda^{ext(i)}\right] + \sum_{i>j=1}^{N}\frac{\lambda}{|r_i - r_j|}.\end{aligned} \quad (42)$$

With N being the number of electrons, v_λ^{ext} is an external potential with $v_{\lambda=1}^{ext}(r) = v^{ext}(r)$, being the external physical potential of the fully interactive system. Additionally, v_λ^{ext} can be spatially non-local for $\lambda \neq 1$. Following that, the reference Hamiltonian, or the Hamiltonian for an independent particle specified by the Eq. (42) for $\lambda = 0$, is of the mean field type, or is known in English as (Mean-field (MF)), i.e., a simple synthesis on a single-particle Hamiltonian:

$$\hat{H}_0 = \sum_{i=1}^{N}\left[\frac{-1}{2}\nabla_i^2 + v_{\lambda=0}^{ext}(r_i)\right] = \sum_{i=1}^{N}\left[\frac{-1}{2}\nabla_i^2 + v^{ext}(r_i) + v^{MF}(r_i)\right]. \quad (43)$$

With v^{MF} is an average field potential resulting from the electron–electron interaction. It can be the Hartree-Fock (HF) potential (v^{HF}) or the Hartree-plus correlation-exchange potential (v^{Hxc}) in the DFT. According to the two Eqs. (42) and (43), the perturbative Hamiltonian becomes:

$$\begin{aligned}\hat{H}_1(\lambda) &= \sum_{i>j=1}^{N} \frac{1}{|r_i - r_j|} + \frac{1}{\lambda} \sum_{i=1}^{N} \left[v_{\lambda}^{ext}(r_i) - v_{\lambda=0}^{ext}(r_i) \right] \\ &= \sum_{i>j=1}^{N} \frac{1}{|r_i - r_j|} + \frac{1}{\lambda} \sum_{i=1}^{N} \left[v_{\lambda}^{ext}(r_i) - v^{ext}(r_i) - v^{MF}(r_i) \right]. \end{aligned} \tag{44}$$

In the construction of the total energy, the ground state wave function $|\Psi_\lambda\rangle$ is introduced for the system λ, such that

$$H(\lambda)|\Psi_\lambda\rangle = E(\lambda)|\Psi_\lambda\rangle. \tag{45}$$

Adopt the normalization condition, $\langle\Psi_\lambda|\Psi_\lambda\rangle = 1$, the total interacting energy of the ground state can then be obtained using the theorem of Hellmann-Feynman [13]

$$E(\lambda = 1) = E_0 + \int_0^1 d\lambda \times \left\langle \Psi_\lambda \middle| \left(\hat{H}_1(\lambda) + \lambda \frac{d\hat{H}_1(\lambda)}{d\lambda} \right) \middle| \Psi_\lambda \right\rangle, \tag{46}$$

The energy of order zero is E_0. It should be noted that the adiabatic connecting path chosen in Eq. (46) is not unique. In DFT, the path is chosen so that the electron density remains constant throughout the journey. This suggests a λ -dependency that $\hat{H}_1(\lambda)$ is not aware of.

6.2 The random phase approximation in the framework of adiabatic-connection fluctuation-dissipation theory

We will quickly discuss the concept of RPA in the context of DFT, which has served as the foundation for current RPA computations. The total ground state energy for an interacting N electron system is a (implicit) function of the electron density n(r) in the Kohn-Sham approximation (KS-DFT) and can be divided into four terms:

$$E[n(r)] = T_s[\psi(r)] + E_{ext}[n(r)] + E_H[n(r)] + E_{xc}[\psi_i(r)]. \tag{47}$$

In the KS framework, the electron density is obtained from the single particle $\psi_i(r)$ orbitals via $n(r) = \sum_i^{occ} |\psi_i(r)|^2$. Among the four terms of the Eq. (47) only $E_{ext}[n(r)]$ and $E_H[n(r)]$ are explicit functions of n(r). T_s is treated exactly in KS-DFT in terms of single particle $\psi_i(r)$ orbitals which are themselves functional of n(r). The unknown correlation-exchange (XC) energy term, which is approximated as an explicit functionality of n(r) (and its local gradients) in conventional functional functions (LDA and GGA) and as a function of $\psi_i(r)$ in more advanced functions, contains the complete complexity of many bodies (hybrid density functions, RPA, etc.). In DFT, several existing approximations of E_{xc} can be categorized using a hierarchical approach called Jacob's scale [14]. But what if we wish to improve the accuracy of E_xc in a larger number of systems? To that purpose, starting with the technically accurate manner of generating E_xc using the AC technique mentioned above is instructive. As previously stated, the AC path is used in KS-DFT in order to maintain the correct electron density. Reducing the Eq. (46) for the total energy of the exact ground state E = E (λ = 1) to:

$$E = E_0 + \int_0^1 d\lambda \left\langle \Psi_\lambda \Big| \frac{1}{2} \sum_{i \neq j=1}^{N} \frac{1}{|r_i - r_j|} \Big| \Psi_\lambda \right\rangle + \int_0^1 d\lambda \left\langle \Psi_\lambda \Big| \sum_{i=1}^{N} \frac{d}{d\lambda} v_\lambda^{ext}(r_i) \Big| \Psi_\lambda \right\rangle$$

$$= E_0 + \frac{1}{2}\int_0^1 d\lambda \int\int drdr' \times \left\langle \Psi_\lambda | \hat{n}(r) \left[\frac{\hat{n}(r') - \delta(r - r')}{|r - r'|} \right] | \Psi_\lambda \right\rangle + \int drn(r) \left[v_\lambda^{ext}(r) - v_{\lambda=0}^{ext}(r) \right]. \tag{48}$$

$$\hat{n}(r) = \sum_{i=1}^{N} \delta(r - r_i), \tag{49}$$

$\hat{n}(r)$ is the electron density operator and $n(r) = \langle \Psi_\lambda | \hat{n}(r) | \Psi_\lambda \rangle$, for any $0 \leqslant \lambda \leqslant 1$. For the reference state $|\Psi_0\rangle$ of KS (given by the slater determinant of orbitals $\Psi_i(r)$ occupied by a single particle, we get

$$E_0 = \left\langle \Psi_0 \Big| \sum_{i=1}^{N} \left[-\frac{1}{2} \nabla^2 + v_{\lambda=0}^{ext}(r_i) \right] \Big| \Psi_0 \right\rangle = T_s[\Psi_i(r)] + \int drn(r) v_{\lambda=0}^{ext}(r). \tag{50}$$

$$E = T_s[\Psi_i(r)] + \int drn(r) v_{\lambda=1}^{ext} + \frac{1}{2}\int_0^1 d\lambda \int\int drdr' \langle \Psi_\lambda | \frac{\hat{n}(r)[\hat{n}(r') - \delta(r - r')]}{|r - r'|} | \Psi_\lambda. \tag{51}$$

From the Eqs. (47) and (51), we obtained:

$$E_H[n(r)] = \frac{1}{2}\int drdr' \frac{n(r)n(r')}{|r - r'|}. \tag{52}$$

$$E_{ext}[n(r)] = \int drn(r) v_{\lambda=1}^{ext}(r). \tag{53}$$

We get the formally exact correlation-exchange energy expression XC;

$$E_{xc} = \frac{1}{2}\int d\lambda \int\int drdr' \frac{n_{xc}^{\lambda}(r, r')n(r)}{|r - r'|}, \tag{54}$$

with $n_{xc}^{\lambda}(r, r')$ is defined by

$$n_{xc}^{\lambda}(r, r') = \frac{\langle \Psi_\lambda | \delta\hat{n}(r) \delta\hat{n}(r') | \Psi_\lambda \rangle}{n(r)} - \delta(r - r'). \tag{55}$$

The mathematical expression for the so-called XC-hole is 55, with $\delta\hat{n} = \hat{n}(r) - n(r)$ denoting the fluctuation of the density operator $\hat{n}(r)$ around its expectation value n(r). The hole (XC) is also related to the density-density correlation function, as shown by the Eq. (55). It illustrates how the presence of an electron at point r reduces the density of all other electrons at point r' in physical terms. The temperature fluctuation-dissipation (FDT) theorem is used to relate the density-density correlations (fluctuations) in the Eq. (55) to the response (dissipation) features of the system in the second step. In statistical physics, FDT is a powerful approach. It shows that the reaction of a system in thermodynamic equilibrium to a tiny external disturbance is the same as the response to spontaneous internal fluctuations in the absence of disturbance [15]. FDT is applicable to both thermal and quantum mechanical fluctuations and shows itself in a variety of physical phenomena. A good example of the latter is the dielectric formulation of the many-body problem by Nozières and Pines [16]. The FDT at zero temperature performed at [16] is relevant in this situation.

$$\langle\Psi_\lambda|\delta\hat{n}(r)\delta\hat{n}(r')|\Psi_\lambda\rangle = -\frac{1}{\pi}\int_0^\infty d\omega Im\chi^\lambda(r,r',\omega), \quad (56)$$

with $\chi^\lambda(r,r',\omega)$, is the linear density-response function of the system. Using the Eqs. (54) and (55) and $v(r,r') = \frac{1}{|rr'|}$, we arrive at the renamed ACFD expression for XC energy in DFT

$$E_{xc} = \frac{1}{2}\int_0^1 d\lambda \int\int drdr'v(r,r') \times \left[-\frac{1}{\pi}\int_0^\infty d\omega Im\chi^\lambda(r,r',\omega) - \delta(r-r')n(r)\right] = \frac{1}{2\pi}\int_0^1 d\lambda \int\int drdr' \times v(r,r') \times \left[-\frac{1}{\pi}\int_0^\infty d\omega^\lambda(r,r',i\omega) - \delta(r-r')n(r)\right]. \quad (57)$$

The analytical structure of $\chi^\lambda(r,r',\omega)$ and the fact that it becomes real on the imaginary axis are the reasons why the above frequency integration can be conducted along the imaginary axis. The problem of computing the energy XC on one of the response functions of a succession of fictional systems along the path AC is transformed by the expression ACFD in the Eq. (57), which must also be tackled in practice. RPA is a particularly basic approximation of the response function in this context:

$$\chi^\lambda_{RPA}(r,r',i\omega) = \chi^0(r,r',i\omega) + \int dr_1 dr_2 \chi^0(r,r_1,i\omega) \times \lambda v(r_1 - r_2)\chi^\lambda_{RPA}(r_2,r',\omega), \quad (58)$$

$\chi^0(r,r_1,i\omega)$, is the response function of independent particles of KS of the reference system $\lambda = 0$ and is known explicitly in terms of orbitals $\psi_i(r)$ single particle (KS), orbital energies ε_i and occupancy factors f_i:

$$\chi^0(r,r',i\omega) = \sum_{ij}\frac{\left(f_i - f_j\right)\psi_i^*(r)\psi_j(r)\psi_j^*(r')\psi_i(r')}{\varepsilon_i - \varepsilon_j - i\omega}. \quad (59)$$

From the Eqs. (57) and (58), the energy XC in RPA can be split into an exchange-exact (EX) and the correlation term RPA:

$$E_{xc}^{RPA} = E_x^{EX} + E_{RPA}^c. \quad (60)$$

$$E_x^{EX} = -\sum_{ij} f_i f_j \int\int drdr'\psi_i^*(r)\psi_j(r)v(r,r')\psi_j^*(r')\psi_i(r'). \quad (61)$$

$$E_c^{RPA} = \frac{1}{2\pi}\int_0^\infty d\omega Tr\left[ln\left(1-\chi^0(i\omega)v\right) + \chi^0(i\omega)v\right]. \quad (62)$$

7. Approximation of pseudo-potentials

The goal is to study the ground state of a system made up of nuclei, core electrons and valence electrons. The heart electrons are often closely linked to

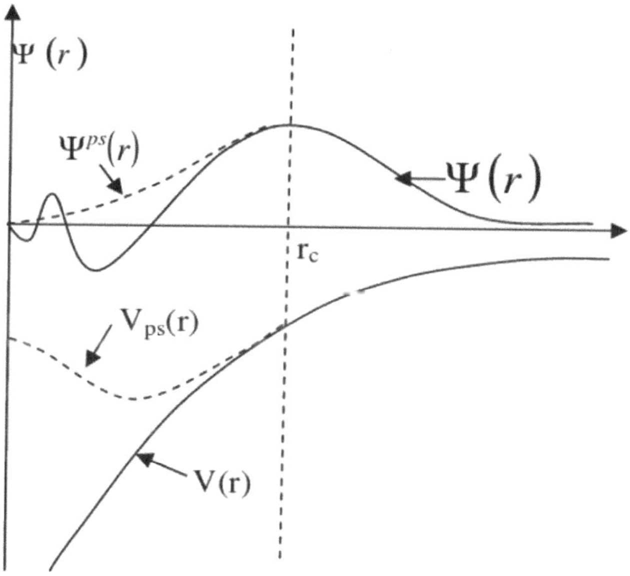

Figure 1.
Schematic illustration of all-electron potential (solid lines) and pseudo-potential (broken lines) and their corresponding wave functions.

nuclei, they are considered (frozen). This approximation makes it possible to develop the valence wave functions on a reduced number of plane waves having a kinetic energy lower than the energy of the cut-off (E $_{cut}$ $E_{cut} \geqslant \frac{\hbar^2}{2m}|K+G|^2$), which allows correct treatment of the problem depends on the pseudo-potential used and the system studied. It consists in replacing the ionic potential $V_{el,nu}$ by a pseudopotential V^{ps} (see **Figure 1**) which acts on a set of wave pseudo-functions instead and places true wave functions and having the same eigenstates in the atomic Schrdinger equation. This idea has been developing since the end of the 1950s. This potential is constructed so as to reproduce the scattering properties for the true valence wave functions, while ensuring that the pseudo-wave function does not have a node in the core region defined by a cutoff radius r_c which is optimized for each orbital. Beyond the core region, the pseudopotential is reduced to the ionic potential so that the pseudo-wave function is equal to the true wave function. The use of a pseudo-potential reduces on the one hand the number of electrons considered in the problem by taking into account only the valence electrons and on the other hand makes it possible to restrict the base of plane waves for the electrons of valence by eliminating most of the oscillations of the wave functions in the heart region.

8. Projection-augmented plane wave method (PAW)

We are always looking for the precision of the computation and thus to minimize the reasonable computation time as much as possible, we advise to use the least hard potentials possible, that is to say requiring the fewest plane waves to ensure convergence. The method of plane waves augmented by projection (PAW) (Projector Augmented Waves) [17] best meets this requirement, which explains their use in this thesis. The projection augmented plane wave method is an approach developed by P. Blochl, which models heart states from wave functions for an isolated atom; it assumes that these states are not modified when the atom is placed in a compound (approximation of frozen hearts). The PAW approach consists of using two kinds of basic functions: one uses partial atomic wave functions inside cores and the other uses a set of functions. Wave planes in the interstitial region. There is then a separation of space into two parts. The cause of this separation is the behavior of the effective potential seen by the valence shell electrons: in the interstitial zone, the total electron density is low. The wave functions oscillate rapidly which makes it possible to give strong variations in the amplitude of the total electron density. Because of these variations, the Fourier decomposition of the potential created by this charge distribution has components large wave vectors, which disadvantages the treatment in plane waves. We are therefore led to write in a different way the wave functions of the valence electrons in these two regions of space:

- In the interstitial region, plane waves are used.
- Inside the hearts, a partial wave decomposition solutions of the Schrödinger equation for the isolated atom which are much better adapted to the potential which prevails there.

9. Application

In spite of the significance of diamond and silicon, there's still a need for solid hypothetical and exploratory information on the elastic constants of these materials, in specific on the versatile constants of the third-order. This lack of experimental data limits the capacity of researchers to create modern materials with a focus on mechanical reactions. Besides, this lack triggered interest in other theoretical calculations. M. Barhoumi et al. [18] have proposed to calculate the elastic properties with different approximations of DFT and beyond with ACFDT in RPA, since the RPA has significant advantages, especially for those interested in functional density theory. It correctly describes the dispersion and van der Waals interactions. In this direction, they have found that the results obtained with RPA are in good agreement with the previous published [19–33]. Also, it should be noted that from the calculated elastic constants, other structural properties such as elastic modulus, shear modulus, Young's modulus and Poisson's ratio can be derived.

10. Conclusion

In this chapter, we have introduced the general method of calculating the ground state energy of a crystalline solid by application of DFT. We have just described how it is possible to determine the energy of the ground state of a solid by studying a fictitious system of independent particles giving rise to the same density as the real electronic system. On the other hand, we have highlighted the approximations necessary to be able to apply this theory. Despite the importance of the DFT to find the exchange-correlation energy, but this quantity remains inaccurate to describe the elastic properties. So we have to go beyond DFT with RPA to correct this handicap. In this direction, we have shown that RPA is a good description of electronic correlation energy.

References

[1] M. Born et R. Oppenheimer. Annalen der Physik, 84, (1927) 457-484.

[2] P. Hohenberg. Phys. Rev. 136, (1964) B864-B864.

[3] W. Kohn et L. J. Sham, Phys. Rev. 50, (1965) A1133-A1138.

[4] D. M. Ceperley et B. J. Alder. Phys. Rev. Lett., 45, (1980) 566-569.

[5] J. P. Perdew, K. Burke et M. Ernzehof. Phys. Rev. Lett., 78, (1997) 385.

[6] HSE, J. Heyd, G. E. Scuseria, and M. Ernzerhof, J. Chem. Phys. 118, 8207 (2003)

[7] D. Bohm, and D. Pines. Phys. Rev. 82 (1951) 625.

[8] D. Pines, and D. Bohm. Phys. Rev. 85 (1952).

[9] D. Bohm, and D. Pines. Phys. Rev. 92 (1953) 609.

[10] M. Gell-Mann, and K. A. Brueckner. Phys. Rev. 106 (1957) 364.

[11] H. Rydberg, M Dion, N. Jacobson, E. Schroeder, P Hyldgaard, S. I. Simak, D. C. Langreth, and B. I. Lundqvist. Phys. Rev. Lett. 91 (2003).

[12] O. Gunnarsson et B. Lundqvist. Phys. Rev. B 13, (1976) 4274.

[13] H. Hellmann, Einfiihrung in die Quantenchemie (Deuticke, Leipzig,1937); R.P.Feynman, Phys. Rev. 56, (1939) 340.

[14] J. Perdew et K. Schmidt. In: Van Doren V, Van Alsenoy C, Geerlings P (eds). AIP, Melville, New York (2001).

[15] R. Kubo. Rep. Prog. Phys. 29, (1966) 255.

[16] D. P. Nozières, and D. Pines, A dielectric formulation of the many body problem: Application to the free electron gas. Nuovo Cim 9, 470-490 (1958).

[17] P. E. Blochl. Phys. Rev. B 50, (1994) 17953-17979.

[18] M.Barhoumi, D.Rocca, M.Said, and S.Lebgueb, Solid State Communications 324, (2021) 114136.

[19] A. Hmiel, J. M. Winey, and Y. M. Gupta, Phys. Rev B 93, (2016) 174113.

[20] Z. J. Fu, G. F. Ji, X. R. Chen, and Q. Q. Gou, Commun. Theor. Phys. 51, (2009) 1129.

[21] H. J. McSkimin, and J. P. Andreatch, J. Appl. Phys. 43, (1972) 2944.

[22] M. H. Grimsditch, and A. K. Ramdas, Phys. Rev. B 11, (1975) 3139.

[23] A. Migliori, H. Ledbetter, R. G. Leisure, C. Pantea, and J. B. Betts, J. Appl. Phys. 104, (2008) 053512.

[24] M. H. Grimsditch, E. Anastassakis, and M. Cardona, Phys. Rev. B 18, (1978) 901.

[25] E. Anastassakis, A. Cantarero, and M. Cardona, Phys. Rev. B 41, 7529 (1990).

[26] J. Winey, A. Hmiel, and Y. Gupta, J. Phys. Chem. Solids 93, (2016) 118.

[27] C. S. G. Cousins, Phys. Rev. B 67, (2003) 024107.

[28] O. H. Nielsen, and R. M. Martin, Phys. Rev. B 32, (1985) 3792.

[29] J. Zhao, J. M. Winey, and Y. M. Gupta Phys. Rev. B 75, (2007) 094105.

[30] M. Å opuzyski, and J. A. Majewski, Phys. Rev B 76, (2007) 045202.

[31] J. J. Hall, Phys. Rev. 161, (1967) 756-761.

[32] H. J. McSkimin, and P. Andreatch, J. Appl. Phys. 35, (1964) 3312.

[33] J. Philip, and M. Breazeale, J. Appl. Phys. 52, (1981) 3383.

Chapter 4

Boltzmann Populations of the Fluxional $Be_6B_{11}^-$ and Chiral Be_4B_8 Clusters at Finite Temperatures Computed by DFT and Statistical Thermodynamics

Abstract

Total energy computations using density functional theory are typically carried out at a zero temperature; thus, entropic and thermic contributions to the total energy are neglected, even though functional materials work at finite temperatures. This book chapter investigates the Boltzmann populations of the fluxional $Be_6B_{11}^-$ and chiral Be_4B_8 isomers at finite temperature estimated within the framework of density functional theory, CCSD(T), and statistical thermodynamics. A couple of steps are taken into account to compute the Boltzmann populations. First, to identify a list of all possible low-energy chiral and achiral structures, an exhaustive and efficient exploration of the potential/free energy surfaces is carried out using a multi-level and multi-step global hybrid genetic algorithm search coupled with Gaussian code. Second, the thermal or so-called Boltzmann populations were computed in the framework of statistical thermodynamics for temperatures ranging from 20 to 1500 K at DFT and CCSD(T) theoretical levels. The results show the effects of temperature on the distribution of isomers define the putative global minimum at finite temperature due to the minimization of the Gibbs free energy and maximization of entropy. Additionally, we found that the fluxional $Be_6B_{11}^-$ cluster is strongly dominant at hot temperatures, whereas the chiral Be_4B_8 cluster is dominant at room temperature. The methodology and results show the thermal effects in the relative population hence molecular properties.

Keywords: Global minimum, infrared spectrum, DFT, boron cluster, fluxional, density functional theory, temperature, Boltzmann, Gibbs free energy, entropy, CCSDT, statistical thermodynamics

1. Introduction

Boron is the smallest and lightest semi-metal atom [1, 2] and a neighbor of carbon in the periodic table. Moreover, it has high ionization energy of 344.2 kJ/mol [3], and an affinity for oxygen atoms, which is the basis of borates [3, 4]. In recent years, the pure boron clusters, the metal, and non-metal doped boron clusters, have attracted considerable attention [1, 5–13] due to their unpredictable chemistry [14, 15] and high potential to form novel structures [16]. The potential of boron atoms to form stable molecular networks [17] lies in the fact that they have three valence electrons and four available orbitals, which implies they are electron-deficient. Boron electron deficiency gives origin to a vast number of allotropic forms and uncommon geometries [6, 16] such as nanotubes [13, 18], borospherenes [19], borophene [16], cages [13, 20], planar [21], quasi planar [22], rings [23, 24], chiral [22, 25–28], boron-based helix clusters [25, 29], and fluxional boron clusters [10, 29–32] that have recently attracted the interest of experimental and theoretical researchers. Aromaticity, antiaromaticity, and conflicting aromaticity dominate the chemical bonding in boron-based clusters [25, 33–35]. The two most-used indices for quantifying aromaticity are the harmonic oscillator model of aromaticity, based on the geometric structure, and the nucleus-independent chemical shift, based on the magnetic response. Aromaticity is not observable, cannot be directly measured [36], and correlates with electronic delocalization [37]. The fluxionality in boron and boron-doped-based molecular systems is highly relevant in terms of its catalytic activity [38] and is due to electronic delocalization [25]. Moreover, in boron-based nanoscale rotors, electronic localization o delocalization contributes significantly to stability, magnetic properties, and chemical reactivity [36], and it is a function of the atomic structure, size, bonding, charge, and temperature [39]. So far, doping a boron cluster with non-metals [40] dramatically affects its structure, stability, and reactivity, like shut-down the fluxionality of the boron-doped anion B_{19}. In contrast, doping a boron cluster with metals [7, 9, 24, 41–43] like beryllium-doped boron clusters, exhibit remarkable properties such as fluxionality [16, 29, 32, 44–46], aromaticity [29, 47], and characteristics similar to borophene [1]. Furthermore, previous theoretical studies showed that the boron fullerenes B_{60} and B_{80} can be stabilized by surrounding the boron clusters with beryllium atoms [48, 49], which effectively compensates for boron electronic deficiency [49]. These effects make beryllium-doped boron clusters interesting, joined with the fact, nowadays, dynamic structural fluxionality in boron nanoclusters is a topic of interest in nanotechnology [19, 50]. Particularly attractive are the chiral helices $Be_6B_{11}^-$, reported by Guo [29], and Buelna-Garcia et al. [39] as one of the low-lying and fluxional isomers. Later, a chemical bonding and mechanism of formation study of the beryllium-doped boron chiral cluster $Be_6B_{10}^{-2}$ and coaxial triple-layered anionic Be_6B_{11} sandwich structures were reported [25, 46]. In these structures, the chirality arises due to the formation of a boron helix. Particularly, the chirality of nanoclusters has attracted attention due to their chiroptical properties, potential application in efficient chiral discrimination [51, 52], nonlinear optics [53] and chiral materials with interesting properties [22, 54], and of course, not to mention that chiral structures play a decisive role in biological activity [55]. Previous theoretical studies joint with experimental photoelectron spectroscopy reported the first pure boron chiral B_{30} structure as the putative global minimum [22] at T = 0. In these pair of planar enantiomers, the chirality arises due to the hexagonal hole and its position. In the past years, the lowest energy structures of the B_{39} borospherene were reported as chiral due to their hexagonal and pentagonal holes [26]. Similarly, the B_{44} cluster was reported as a chiral structure due to its nonagonal holes [28]. That is, in these clusters, holes in the structure cause chirality. So far, the chirality depends on the geometry; In contrast, fluxionality strongly depends on temperature. A boron molecular Wankel motor [56–58] and sub nanoscale tank treads have been reported [59, 60]; however, the temperature have not been considered. Nevertheless, most theoretical density functional studies assume that the temperature is zero and neglect temperature-dependent and entropic contributions; consequently, their finite temperature properties remain unexplored [61, 62]. Experimental studies are carried out in non-zero temperatures, then it is necessary to understand the effect of the temperature on the cluster properties and the lowest energy structure's determination [61–63]. Herein, we investigate the effect of temperature-entropy term on the Boltzmann population, which needs the elucidation of the putative global minimum and its low-energy isomers [39, 64–68]. The properties observed in a molecule are statistical averages over the ensemble of geometrical conformations that are ruled by the Boltzmann distributions of isomers. So we need an efficiently sampling of the free energy surface to know the distribution of isomers at different temperatures [39, 68–71]. A considerable change in the isomer distribution and the energetic separation among them is the first notable effect of temperature [39]. Useful materials work at finite temperatures; in that conditions, Gibbs free energy is minimized whereas, the entropy of the atomic cluster is maximized [39, 72]. and determines the putative global minimum at a finite temperature [39]. Although in the mid 1960's, Mermin et al. [73] studied the thermal properties of the inhomogeneous electron gas, most DFT calculations are typically performed at zero temperature. Recently, over again, DFT was extended to finite temperature [74–76], but nowadays, as far as we know, it is not implemented in any public software and practical calculations are not possible. Taking temperature into account requires dealing with small systems' thermodynamics; The Gibbs free energy of classical thermodynamics also applies for small systems, known as thermodynamics of small systems [77–79]. The thermodynamics of clusters have been studied by various theoretical and simulation tools [61, 68, 77, 80–86] like molecular-dynamics simulations. Previous reports investigated the behavior of $Al_{12}C$ cluster at finite temperature employing Car-Pirinello molecular dynamics [61], and dynamical behavior of Borospherene in the framework of Born-Oppenheimer Molecular Dynamics [5, 10]. Under the harmonic superposition approximation, the temperature-entropy term can be computed with the vibrational frequencies on hand. The entropy and thermal effects have been considered for gold, copper, water, and sodium clusters [71, 87–95]. Franco-Perez et al. [96] reported the thermal corrections to the chemical reactivity at finite temperature, their piece of work validates the usage of reactivity indexes calculated at zero temperature to infer chemical behavior at room temperature. Gazquez et al. [97] presented a unified view of the temperature-dependent approach to the DFT of chemical reactivity. Recently, the effect of temperature was considered by Castillo-Quevedo et al. reported the reaction rate and the lowest energy structure of copper Cu_{13} clusters at finite temperature [98, 99]. Dzib et al. reported Eyringpy; A Python code able to compute the rate constants for reactions in the gas phase and in solution [100], Vargas-Caamal et al. computed the temperature-dependent dipole moments for the $HCl(H_2O)_n$ clusters [101], Shkrebtii et al. computed the temperature-dependent linear optical properties of the Si(100) surface [102], several authors take into account the temperature in gold clusters [87–89], and thermochemical behavior study of the sorghum molecule [103], and more recently, Buelna-Garcia et al. [104] employing density functional theory and nanothermodynamics reported the lowest energy structure of neutral chiral Be_4B_8 at a finite temperature. and reported that the fluxionality of the anionic Be_6B_{11} clusters depends strongly on temperature [39]. In this work, we employed density functional theory, statistical thermodynamics, and CCSD(T) to compute the Gibbs free energy and the Boltzmann population at absolute temperature T for each neutral chiral Be_4B_8 and anionic Be_6B_{11} isomers. We think that this provides useful information about which isomers will be dominant at hot temperatures. No work has previously been attempted to investigate entropy-driven isomers in the fluxional $Be_6B_{11}^-$ and chiral Be_4B_8 cluster at CCSDT level of theory to the best of our knowledge. The remainder of the manuscript is organized as follows: Section 2 gives the computational details and a brief overview of the theory and algorithms used. The results and discussion are presented in Section 3. We discuss the effect of the symmetry in the energetic ordering and clarify the origin of the 0.41 kcal/mol difference in energy between two structures with symmetries C_2 and C_1 appear when we compute the Gibbs free energy. A comparison among the energies computed at a single point CCSD(T) against the DFT levels of theory and the T_1 diagnostic is presented. Conclusions are given in Section 4.

2. Theoretical methods and computational details

2.1 Global minimum search

Despite advances in computing power, the minimum global search in molecular and atomic clusters remains a complicated task due to several factors. The exploration should be systematic and unbiased [68, 105]; a molecule's degrees of freedom increase with the number of atoms [68, 106–109]; a molecule composed of N number of atoms possesses 3 N degrees of freedom (i.e., a linear molecule has three degrees of translation, two of rotation, and [3 N-6] of vibrational modes); and, as a consequence, the potential/free energy surface depends on a large number of variables. The number of local minima increases exponentially as a function of the number of atoms in the molecule. Moreover, the total energy computation requires a quantum mechanical methodology to produce a realistic value for energy. In addition to that, there should be many initial structures. It is essential to sample a large region of the configuration space to ensure that we are not missing structures, making an incomplete sampling of the configurational space and introducing a significant problem to calculating the thermodynamic properties [64]. A complete sampling of the potential/free energy surface is impossible, but a systematic exploration of the potential energy surface is extremely useful. Although searching for a global minimum in molecular systems is challenging, the design and use of algorithms dedicated to the search for global minima, such as simulated annealing, [110–115] kick method [116, 117], genetic algorithms [118–120], Gradient Embedded Genetic Algorithm [121–123] and basin hopping [124, 125], has been accomplished over the years. In the past few years, one of us designed and employed genetic algorithms [12, 13, 29, 39, 98, 99, 104, 126, 127] and kick methodology [101, 127–133] coupled with density functional theory to explore atomic and molecular clusters' potential energy surfaces. They have led us to solve the minimum global search in a targeted way. In this chapter, our computational procedure to elucidate the low-energy structures employs a recently developed nature-inspired hybrid strategy that combines a *Cuckoo* search [134] and genetic algorithms coupled to density functional theory that has been implemented in the GALGOSON code v1.0. Nature-inspired metaheuristic algorithms have been applied in almost all areas of science, engineering, and industry, work remarkably efficiently, and have many advantages over deterministic methods [135]. GALGOSON systematically and efficiently explores potential/free energy surfaces (PES/FES) of the atomic clusters to find the minimum energy structure. The methodology consists of a three-step search strategy where, in the first and second steps, we explore the PES, and in the third step, we explore the FES. First, the code builds a generation of random initial structures with an initial population of two hundred individuals per atom in the Be-B clusters using a kick methodology. The process to make 1D, 2D, and 3D structures is similar to that used in previous work [12] and are restricted by two conditions [12] that can be summarized as follows: (a) All the atoms are confined inside a sphere with a radius determined by adding all atoms' covalent radii and multiplied by a factor established by the user, typically 0.9. (b) The bond length between any two atoms is the sum of their covalent radii, modulated by a scale factor established by the user, typically close to 1.0; this allows us to compress/expand the bond length. These conditions avoid the high-energy local minima generated by poorly connected structures (too compact/ loose). Then, structures are optimized at the PBE0/3-21G level of theory employing Gaussian 09 code. As the second step, all energy structures lying in the energy range of 20 kcal/mol were re-optimized at the PBE0-GD3/LANL2DZ level of theory and joints with previously reported global minimum structures. Those structures comprised the initial population for the genetic algorithm. The optimization in this stage was at the PBE0-D3/LANL2DZ level of theory. The criterion to stop the generation is if the lowest energy structure persists for 10 generations. In the third step, structures lying in 10 kcal/mol found in the previous step comprised the initial population for the genetic algorithm that uses Gibbs free energy extracted from the local optimizations at the PBE0-D3/def2-TZVP, taking into account the zero-point energy (ZPE) corrections. The criterion to stop is similar to that used in the previous stage. In the final step, the lowest energy structures are evaluated at a single point energy at the CCSD(T)/def2-TZVP// PBE0-D3/def2-TZVP level of theory. All the calculations were done employing the Gaussian 09 code [136].

2.2 Thermochemistry properties

All the information about a quantum system is contained in the wave function; similarly, the partition function provides all the information need to compute the thermodynamic properties and it indicates the states accessible to the system at temperature T. Previous theoretical studies used the partition function to compute temperature-dependent entropic contributions [137] on $[Fe(pmea)(NCS)_2]$ complex, infrared spectroscopy on anionic Be_6B_{11} cluster [39], and rate constant [100]. In this work, the thermodynamic functions are calculated using the temperature-dependent partition function Q shown in Eq. (1).

$$Q(T) = \sum_i g_i\, e^{-\Delta E_i/k_B T} \tag{1}$$

In Eq. (1), the g_i is the degeneracy or multiplicity, using degeneracy numbers is equivalent to take into account all degenerate states and the sum runs overall energy levels, and k_B is the Boltzmann constant, T is the temperature and ΔE_i is the total energy of a molecule [100, 138]. At high temperatures, all thermal states are accessible due to the term $-\Delta E_i/k_B T$ tends to zero, and the partition tends to infinity. An exact calculation of Q could be complicated due to the coupling of the internal modes, a way to decouple the electronic and nuclei modes is through the use of Born-Oppenheimer approximation. (BOA) This approach says that the electron movement is faster than the nuclei and assumes that the molecular wave function is the electronic and nuclear wavefunction product $\psi = \psi_e \psi_n$. The vibrations change the momentum of inertia as a consequence, affect the rotations; this fact tightly couple the vibrational and rotational degrees of freedom; The separation of rotational and vibrational modes is called the rigid rotor, harmonic oscillator (RRHO) approximation, under this approximation, the molecule is treated rigidly, this is generally good when vibrations are of small amplitude. Here the vibration will be modeled in terms of harmonic oscillator and rotations in terms of the rigid rotor. Within BOA and RRHO approximations, the partition function is factorized into electronic, translational, vibrational, and rotational energies. Consequently, the partition function, Q, can be given in Eq. (2) as a product of the corresponding contributions [100, 139], and under the rigid rotor, harmonic oscillator, Born-Oppenheimer, ideal gas, and a particle-in-a-box approximations.

$$Q = q_{trans} q_{rot} q_{vib} q_{elec}. \tag{2}$$

Table 1 shows the contributions of electronic, translational, vibrational, and rotational to the partition function.

We computed all partition functions at temperature T and a standard pressure of 1 atm. The equations are equivalent to those given in the Ref. [100], and any standard text of thermodynamics [138, 139] and they apply for an ideal gas. The implemented translational partition function in the Gaussian code [136] is the partition function, $q = q_{trans}$, given in **Table 1**. In this study, the $q = q_{trans}$ is computed as a function of T and is used to calculate the translational entropy. In addition to using vibrational modes to identify true lowest energy structures from transition states, we also used them to compute the vibrational partition function. In this study is considered vibrational modes, ν, under the harmonic oscillator approximation, and total vibrational energy consists of the sum of the energies of each vibrational mode. In computing the electronic partition, we considered that the energy gap between the first and higher excited states is more considerable than $k_B T$, as a consequence electronic partition function, $q = q_{elect}$, is given by $q_{elect} = \omega_0$, q_{rot}, q_{rot}^{nl}, $q = q_{trans}$ are used to compute the entropy contributions given in **Table 2**.

The vibrational frequencies are calculated employing the Gaussian code, and all the information needed to compute the total partition function is collected from the output. The Gibbs free energy and the enthalpy are computed employing the Eqs. (3) and (4).

$$H = U + nRT, \tag{3}$$

$$G = H - TS. \tag{4}$$

2.3 Boltzmann population

The properties observed in a molecule are statistical averages over the ensemble of geometrical conformations or isomers accessible to the cluster [140]. So, the molecular properties are ruled by the Boltzmann distributions of isomers that can change due to temperature-entropic term [23, 71, 101], and the soft vibrational modes that clusters possess make primary importance contributions to the entropy [93]. The Boltzmann populations of the low-energy isomers of the cluster $Be_6B_{11}^-$ and Be_4B_8 are computed through the probabilities defined in Eq. (5)

Contribution	Partition function
Translational	$q_{trans} = \left(\frac{2\pi m k_B T}{h^2}\right)^{\frac{3}{2}} \frac{k_B T}{P}$
Rotational linear	$q_{rot}^{l} = \frac{T}{\sigma \Theta_{rot}}, \Theta_{rot} = \frac{\hbar^2}{2 I k_B}$
Rotational nonlinear	$q_{rot}^{nl} = \frac{\pi^{1/2}}{\sigma}\left[\frac{T^{3/2}}{(\Theta_{rotA}\Theta_{rotB}\Theta_{rotC})^{1/2}}\right], \Theta_{rotj} = \frac{\hbar^2}{2 I_j k_B}, j = A, B, C$
Vibrational	$q_{vib}^{pol} = \prod_{i=1}^{n_{\upsilon[a]}} \frac{e^{-\Theta_{vib_i}/2T}}{1-e^{-\Theta_{vib_i}/T}}, \Theta_{vib_i} = \frac{h\nu_i}{k_B}$
Electronic	$q_{elec} = \omega_0$

Table 1.
Contributions to the partition function.

	Internal energy	Entropy
Translational	$U_{trans} = \frac{3}{2}RT$	$S_{trans} = R(\ln q_{trans} + \frac{5}{2})$
Rotational linear	$U^{l}_{rot} = RT$	$S^{l}_{rot} = R(\ln q^{l}_{rot} + 1)$
Rotational nonlinear	$U^{nl}_{rot} = \frac{3}{2}RT$	$S^{nl}_{rot} = R(\ln q^{nl}_{rot} + \frac{3}{2})$
Vibrational	$U^{pol}_{vib} = R\sum_{i}^{n_{\upsilon\|a\|}} \Theta_{vib_i}\left(\frac{1}{2} + \frac{1}{e^{\Theta_{vib_i}/T}-1}\right)$	$S^{pol}_{vib} = R\sum_{i}^{n\upsilon}\left[\frac{\Theta_{vib_i}/T}{e^{\Theta_{vib_i}/T}-1} - \ln\left(1 - e^{-\Theta_{vib_i}/T}\right)\right]$
	$\Theta_{vib_i} = \frac{h\nu_i}{k_B}$	
Electronic	$U_{elec} = 0$	$S_{elec} = R\ln q_{elec}$

Table 2.
Contributions to internal energy and entropy.

$$P(T) = \frac{e^{-\beta\Delta G^K}}{\sum e^{-\beta\Delta G^K}} \tag{5}$$

where $\beta = 1/k_B T$, and k_B is the Boltzmann constant, T is the temperature in Kelvin, ΔG is the Gibbs free energy of the kth isomer. Eq. (5) establishes that the distribution of molecules will be among energy levels as a function of the energy and temperature. Eq. (5) is restricted so that the sum of all probabilities of occurrence, at fixed temperature T, Pi (T) is equal to 1 and given by Eq. (6)

$$\sum_{i=1}^{n} P(T) = 1 \tag{6}$$

It is worth mentioning that the energy difference among isomers is determinant in the computation of the solid–solid transition, T_{ss} point. T_{ss} occurs when two competing structures are energetically equaled, and there is simultaneous coexistence of isomers at T. in other words, the T_{ss} point is a function of the energy difference between two isomers and the relative energy ΔG that the cluster possesses. Boltzmann distribution finds a lot of applications as like simulated method annealing applied to the search of structures of minimum energy, rate of chemical reaction [100], among others. For the calculation of the Boltzmann populations, we used a homemade Python/Fortran code called BOFA (Boltzmann-Optics-Full-Ader).

2.4 Computational details

The global exploration of the potential and free energy surfaces of the $Be_6B_{11}^-$ and Be_4 B_8 clusters were done with a hybrid Cuckoo-genetic algorithm written in Python. All local geometry optimization and vibrational frequencies were carried out employing the density functional theory (DFT) as implemented in the Gaussian 09 [136] suite of programs, and no restrictions in the optimizations were imposed. Final equilibrium geometries and relative energies are reported at PBE0 [141] /def2-TZVP [142] level of theory, taking into account the D3 version of Grimme's dispersion corrections [143] and including the zero-point (ZPE) energy corrections. As Pan et al. [144] reported, the computed relative energies with PBE0 functional are very close to the CCSD(T) values in B_9^- boron cluster. The def2-TZVP basis from the Ahlrichs can improve computations accuracy and describe the Be-B clusters [29]. To gain insight into its energetics, we evaluated the single point energy CCSD (T)/def2TZVP//PBEO-D3/def2-TZVP level of theory for the putative global minima and the low-energy $Be_6B_{11}^-$ isomers, and employing Orca code at the DLPNO-CCSD(T) theoretical level for the low-energy isomers of Be_4B_8 cluster. Boltzmann-Optics-Full-Ader, (BOFA) is employed in the computation of the Boltzmann populations. The code is available with the corresponding author.

3. Results and discussion

3.1 The lowest energy structures and energetics

Figure 1 shows the lowest energy structure of $Be_6B_{11}^-$ clusters and seven low-energy competing isomers computed at the PBE0-D3/def2-TZVP basis set. For the putative global minimum at the PBE0-D3/def2-TZVP, the optimized average B-B bond length is 1.64 Å. In contrast, the optimized B-Be bond length is 2.01 Å. At the PBE0-D3/def2-TZVP and temperature of 298.15 K, the putative global minimum with 54% of the relative population has C_1 symmetry with a singlet electronic state ^1A. It is a distorted, oblate spheroid with three berylliums in one face and two in the other face. Nine-boron and one-beryllium atoms are forming a ring located around the spheroid's principal axes and the remaining two boron atoms are located close to the boron ring in one of its faces. The second higher energy structure, at 298.15 K, lies only 0.61 kcal/mol Gibbs free energy above the putative global minima, and it has C_1 symmetry with a singlet electronic state ^1A. It is a prolate spheroid with 19% of the relative population at a temperature of 298.15 K. The next two higher energy isomers, at 298.15 K, lies at 0.85 and 1.23 kcal/mol Gibbs energy above the putative global minimum. They are prolate, coaxial Triple-Layered structures with C_s, and C_{2v} symmetries with singlet electronic states, ^1A, respectively. This clearly, shows that the low-symmetry structure C_1 become more energetically preferred than the C_{2v} symmetry by Gibbs free energy difference of 0.38 kcal/mol at 298.15 K, due to entropic effects and in agreement with a similar result found in Au_{32} [105]. Indeed,

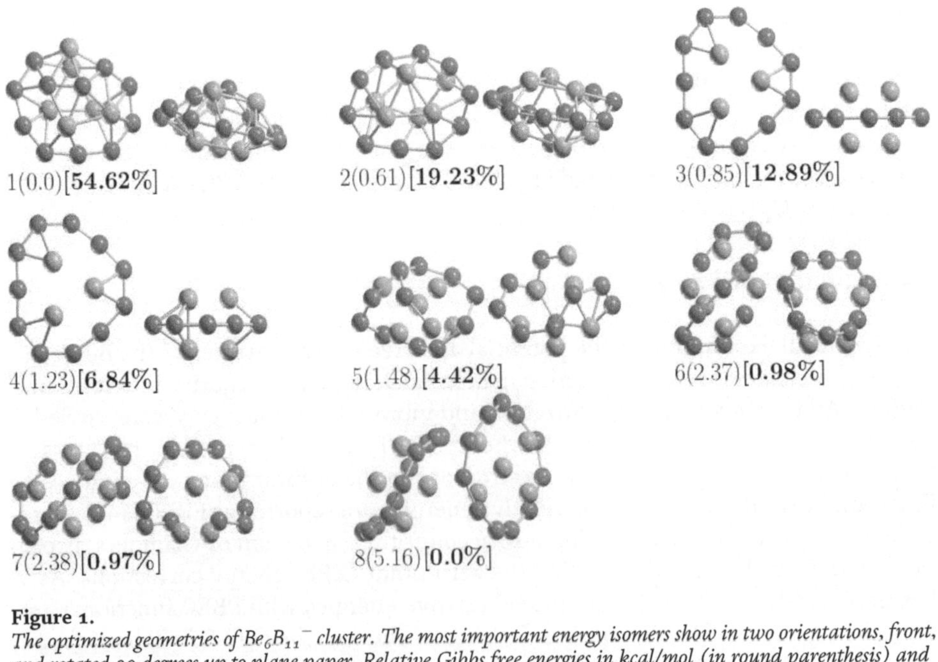

Figure 1.
The optimized geometries of $Be_6B_{11}^-$ cluster. The most important energy isomers show in two orientations, front, and rotated 90 degrees up to plane paper. Relative Gibbs free energies in kcal/mol (in round parenthesis) and the relative population [in square parenthesis], at PBE0-D3/Def2-TZVP level of theory. The criterium to plot them is until the probability occupation is zero. The pink- and yellow-colored spheres represent the boron and beryllium atoms, respectively.

and according to our computations, those structures are strongly dominating at temperatures higher than 377 K. The next structure is shown in **Figure 1**(5), is located at 1.48 kcal/mol above the global minimum; it is close to a spherical shape and correspond to a prolate structure with C_1 symmetry, and a singlet electronic state ^1A; this structure only has 4.4% of the relative population at 298.15 K. The next two structures, located at 2.37 kcal/mol Gibbs free energy above the global minimum, are the chiral helix-type structures, reported by Guo [29] as minimum global. They are prolate structures with C_{2v} symmetries, and their relative population is around only 1%. We must point out that those chiral-helix structures never become the lowest energy structures in all ranges of temperature. The relative population is zero for structures located at higher relative Gibbs free energy than 5.1 kcal/mol, and at 298.15 K, there is no contribution of these isomers to any total molecular property. A full understanding of the molecular properties requires the search of global minimum and all its closest low-energy structures [64]. The separation among isomers by energy-difference is an important and critical characteristic that influences the relative population and, consequently, the total molecular properties. We computed the global minima and the first seven low-energy to gain insight into how the energy-gap among isomers change and how the energy-ordering of the low-energy structures is affected at a single point CCSD(T)/def2-TZVP level of theory corrected with the zero-point energy computed at the PBE0-D3/def2-TZVP level of theory. At the CCSD(T) level of theory, the global minima, the seven lowest energy isomers, and the energy order agree with previous work [39], as seen in the first row of **Table 3**. The second row of **Table 3** shows the corrected CCSDT+E_{ZPE} energy. Interestingly, the energetic ordering does not change when we take into account the ZPE energy. Nevertheless, the energy difference among isomers was reduced drastically. we can deduce that the ZPE energy inclusion is essential in the isomers' energy ordering and molecular properties. The third row of **Table 3** shows the energy-order considering the Gibbs free energy computed at 298.15 K; at this temperature, the isomers energy-ordering is changed, the second isomers take the putative global minima place, and the first isomers take the fifth place. Interestingly, this energy-ordering is at 298.15 K. This energy-ordering is a complete function of the temperature that we will discuss later in the relative population section. The fourth row in **Table 3** shows the electronic energy taking into account the ZPE energy. It follows the same trend in energy-ordering when considering the Gibbs free energy, and it is the same putative global minima. The fifth row in **Table 3** is just electronic energy. It almost follows the CCSD(T) energies trend, except the

Be_6B_{11}	Level	i_1	i_2	i_3	i_4	i_5	i_6	i_7	i_8
	$CCSD(T)$	0.0	1.75	1.84	1.84	4.10	4.13	2.64	2.42
	$CCSD(T) + \varepsilon_{ZPE}$	0.0	0.58	0.85	0.86	1.19	1.23	1.68	1.81
	ΔG	0.0	−1.48	0.89	0.88	−0.63	−0.25	4.14	−0.87
Be_6B_{11}	$\varepsilon_0 + \varepsilon_{ZPE}$	0.0	−0.29	1.51	1.52	2.41	2.42	5.0	−0.08
	ε_0	0.0	0.87	2.50	2.50	5.32	5.32	5.96	0.52
	Point Group Symmetry	C_1	C_1	C_2	C_2	C_S	C_{2v}	C_1	C_1
	Electronic ground state	^1A	^1A	^1A	^1A	$^1A'$	$^1A_1$	^1A	^1A
	Frequencies (cm^{-1})	230	119	102	100	46	43	161	151

Table 3.
The relative energies in kcal·mol^{-1}, coupled-cluster single-double and perturbative triple, CCSD(T), CCSD(T) with zero-point energy (ε_{ZPE}), ($CCSD(T) + \varepsilon_{ZPE}$), Gibbs free energy ($\Delta G$) at 298.15 K, electronic energy with ε_{ZPE} (%mcalEε_o), electronic energy (ε_o), point group symmetry, electronic ground state, and the lowest frequency in cm^{-1} for eight low-energy isomers.

isomers number eight that take the second place located at 0.52 kcal/mol above the putative global minima. The sixth, seventh and eighth rows on **Table 3** show the point group symmetry, electronic ground state, and the lowest vibrational frequency of each isomer. When we take the Gibbs free energy to energy-ordering structures, the second isomers interchange to the first place, becoming the lowest energy structure; The energy ordering change drastically, whereas the electronic energy almost follows the same trend CCSD(T) energy-ordering. This shows us that the level of theory and the inclusion of entropy and temperature change the energy-ordering; therefore, the total molecular properties.

3.2 Boltzmann population of $Be_6B_{11}^-$ cluster

Figure 2a shows the most important and strongly dominating $T_{ss1\text{-}g}$ point that is located at 377 K temperature scale with a relative population of 33%. For temperatures ranging from 10 to 377 K, the relative population is strongly dominated by the putative global minima isomer distorted oblate spheroid with C_1 symmetry and this relative population is similar to $-T^{-3}$ function with one point of inflection located at 180 K. After decreases monotonically up to 377 K. At the $T_{ss1\text{-}g}$ point, the distorted oblate spheroid with C_1 symmetry co-exist and compete with the coaxial Triple-Layered structures with C_s symmetry; This implies that the distorted oblate spheroid will be replaced with the coaxial Triple-Layered structures. Above temperature 377 K, the relative population is strongly dominated by the coaxial Triple-Layered structures with C_s symmetry, located at 0.85 kcal/mol above the global minima at temperature 298.15 K. This relative population depicted in blue-solid line in panel (a) has behavior as a sigmoid function, from temperatures ranging from 377 to 600 K, it grows rapidly and from temperatures ranging from 600 to 1500 K, it almost keeps constant with 60%. The second $T_{ss2\text{-}g}$ point is located at temperature 424 K with a relative population of 22.9%, and this point the global minima distorted oblate spheroid with C_1 symmetry co-exist, and compete with the coaxial Triple-Layered structures with C_{2v} symmetry, located at 1.23 kcal/mol above the global minima at 298.15 K. The relative population of the coaxial Triple-Layered C_{2v} symmetry depicted in green-solid line in panel (a) also has a behavior of a sigmoid function and up to 600 K it keeps constant with 32% of relative population. The $T_{ss3\text{-}g}$, and $T_{ss4\text{-}g}$ points are located at 316.7 K, and 349 K axis temperature with

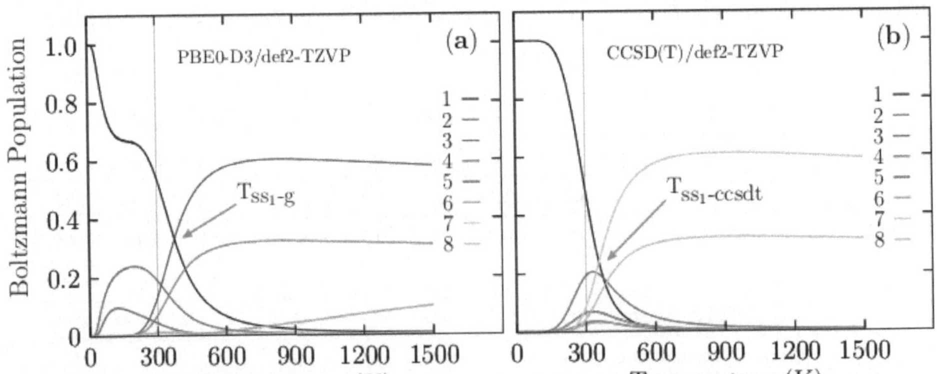

Figure 2.
Panel (a) shows the Boltzmann population of the $Be_6B_{11}^-$ (ensemble at thermal equilibrium) for the temperatures ranging from 10 to 1500 K computed at the PBE0-D3/def2-TZVP level of theory. Panel (b) shows the Boltzmann population for the temperatures ranging from 10 to 1500 K computed at the CCSDT/def2-TZVP//PBE0-GD3/def2-TZVP level of theory. At the temperature of 350 K, four structures co-exist with 20% of probability.

relative populations 14% and 17%, respectively. These relative populations correspond to the second isomer located just 0.61 kcal/mol at 298.15 K above the global minima, and co-existing at the temperatures 316.7 K and 349 K with the coaxial Triple-Layered structures with C_s, and C_{2v} symmetries, respectively. At low temperatures range, this isomer's relative population depicted in red-solid line of **Figure 2a** is around only 20%, and up to room temperature, it decreases exponentially to zero. At temperatures up to 600 K, the relative population is zero; hence, at high temperatures these isomers do not contribute to the molecular properties. The relative population lower than 10%, depicted in violet-solid line shows in **Figure 2a**, correspond to the isomers located at 1.48 kcal/mol above global minima at 298.15 K. Interesting, this structure is the putative minimum global when the CCSD(T) energy is employed in the ordering energetic, Despite that, this structure's relative population clearly shows that this structure does not contribute to molecular properties in all ranges of temperatures.

3.3 The lowest energy structures of Be_4B_8 clusters

Figure 3 shows the low-energy configurations of Be_4B_8 clusters optimized at PBE0-D3/def2-TZVP level of theory taking into account ZPE energy correction. The optimized average B-B bond length of the putative chiral global minimum is 1.5867 Å, in good agreement with an experimental bond length of 1.57–1.59 Å [145, 146], and also within agreement with others previous DFT calculations [39]. The most recurring motif within the lower energy isomers of B_8Be_4 is a sandwich structure, (SSh) in which the boron atoms form a hollow distorted ellipsoid ring

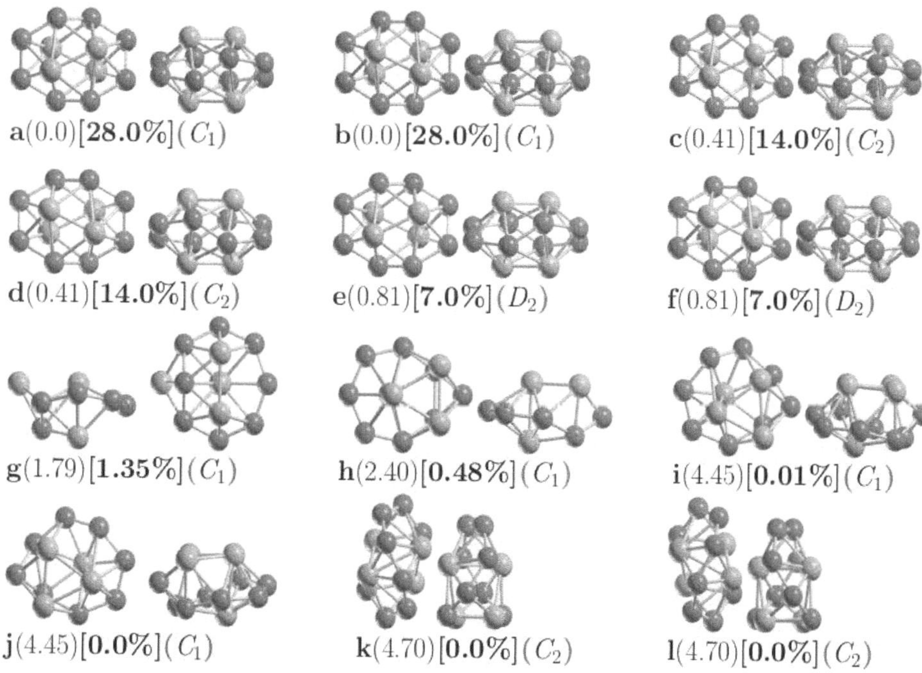

Figure 3.
Optimized geometries of a neutral Be_4B_8 cluster at the PBE0-D3/def2TZVP level of theory with zero-point correction energy. These are shown in front and side views. The first letter is the isomer label, the relative Gibbs free energies in kcal·mol^{-1} (in round parenthesis) at 298.15 K, the relative population (in square brackets), and the group symmetry point (in red round parenthesis). The structures with labels (a and b), (c and d), (e and f), (i and j), (k and l) and (h) are chiral. The purple- and yellow-colored spheres represent the boron and beryllium atoms, respectively.

with each of the Be-Be dimers capping the top and bottom with C_1 point group symmetry. Isomers 1 and 2 are also listed as i_1 and i_2 in **Table 4**, are enantiomers differing in the orientation of the Be-Be dimers with respect to the boron skeleton. The Be-Be bond length for the six lowest energy enantiomers is 1.9874, 1.9876, and 1.9881 Å for symmetries C_1, C_2, and D_2, respectively, in good agreement with the bond length of the Be-Be in Be_2B_8 cluster 1.910 Å [44]. To gain insight into the energy hierarchy of isomers and validate our DFT calculations, relative energies were computed at different levels of theory, and differences between them are shown in **Table 4**. Energy computed at different methods yield different energies due mainly to the functional and basis-set employed, [39, 147], so the energetic ordering change; consequently, the probability of occurrence and the molecular properties will change. The first line of **Table 4** shows the relative Gibbs free energy computed at PBE0-D3/def2-TZVP and room temperature. The small relative Gibbs free energies (0.41, and 0.81 kcal/mol) differences among the six enantiomer structures i_1 to i_6 in **Table 4** are caused by the rotational entropy being a function of the symmetry number that in turn depends on the point group symmetry. An increase/decrease in the value of rotational entropy changes the Gibbs free energy. The Gibbs free energy computed with and without symmetry will differ by a factor $RT\ln(\sigma)$. Here, R is the universal gas constant, T, the temperature, and σ is the symmetry number. The computed factor at room temperature with $\sigma = 2$ is $RT\ln(\sigma) = 0.41$ kcal/mol, and it is $RT\ln(\sigma) = 0.81$ kcal/mol with $\sigma = 4$, in agreement with the values shown in the first line of **Table 4**. As the temperature increases, the energy differences between the factors $RT\ln(\sigma)$ become larger. These small relative Gibbs free energies are responsible for different values of probability of occurrence at low temperatures for the similar isomers with different point group symmetry. This strongly suggests that there must be atomic clusters with low and high symmetries in the Boltzmann ensemble to compute the molecular properties correctly. The second line in **Table 4** shows single point (SP) relative energies computed at the CCSD(T) [148], the energetic ordering of isomers listed in the first line of **Table 4** follows almost the trend of energetic ordering at SP CCSD(T) level, notice that just the achiral isomers label i_7 to i_8 in **Table 4** are interchanged in energetic ordering. The third line **Table 4** shows single point relative energies computed at

Level	i_1	i_2	i_3	i_4	i_5	i_6	i_6	i_8	i_9	i_{10}
ΔG	0.0	0.0	0.41	0.41	0.81	0.81	1.79	2.40	4.45	4.45
CCSD(T)	0.0	0.0	0.0	0.0	0.0	0.0	3.61	3.38	5.38	5.38
CCSD(T) + ε_{ZPE}	0.0	0.0	0.0	0.0	0.0	0.0	2.71	2.51	4.51	4.51
DLPNO-CCSD(T)	0.0	0.0	0.0	0.0	0.0	0.0	0.75	1.37	5.0	5.0
DLPNO-CCSD(T) + ε_{ZPE}	0.0	0.0	0.0	0.0	0.0	0.0	−0.20	0.50	4.10	4.10
$\varepsilon_0 + \varepsilon_{ZPE}$	0.0	0.0	0.0	0.0	0.0	0.0	2.38	2.80	5.03	5.03
ε_0	0.0	0.0	0.0	0.0	0.0	0.0	3.28	3.68	5.90	3.28
Point Group Symmetry	C_1	C_1	C_2	C_2	C_1	C_1	C_1	C_1	C_2	C_2
T_1	0.019	0.019	0.019	0.019	0.019	0.019	0.019	0.019	0.019	0.019

Table 4.
Single-point relative energy calculations of the low-energy structures from i_1 to i_{10} at different levels of theory: coupled cluster single-double and perturbative triple (CCSD(T)), CCSD(T) with zero-point energy (CCSD(T) + ε_{ZPE}, CCSD(T)) employing the domain-based local pair natural orbital coupled-cluster theory (DLPNO-CCSD(T)), with TightPNO setting, and with ε_{ZPE} (DLPNO-CCSD(T) + ε_{ZPE}), Gibbs free energy (ΔG) at 298.15 K, electronic energy with ε_{ZPE} ($\varepsilon_0 + \varepsilon_{ZPE}$), electronic energy ($\varepsilon_0$), point group symmetry, and T_1 diagnostic. All relative energies are given in kcal·mol^{-1}.

the CCSD(T) [148]/def2-TZVP//PBE0-D3/def2-TZVP; the energetic ordering is similar to pure CCSD(T) energy. DLPNO-CCSD(T) relative energies, with and without ZPE correction, are shown in lines four and five of **Table 4**, the first follows the trend of pure CCSD(T) energy, and the second, the ZPE value, interchange the isomers, label i_7 in **Table 4**, to be the putative global minimum. Here we can say that the ZPE energy inclusion is essential in distributing isomers and molecular properties. The sixth and seventh lines of **Table 4** show the electronic energy with and without ZPE correction, and both of them follow the trend of the Gibbs free energy given in line number one. Line number 8 in **Table 4** shows the point group symmetry for each isomer. The T_1 diagnostic for each isomer is shown in line nine of **Table 4**, all of them are lower than the recommended value 0.02 [148] so the systems are appropriately characterized.

3.4 Boltzmann population of Be_4B_8 clusters

As we mentioned earlier, the determination of the structure is the first step to study any property of a material. Moreover, we have to consider that an observed molecular property in a Boltzmann ensemble is a weighted sum of all individual contributions of each isomer that form the ensemble. At temperature 0 K, the electronic energy plus zero-point energy determine the putative global minimum and all nearby low-energy structures, whereas, at temperatures larger than 0 K, the Gibbs free energy defines the putative global minimum. **Figure 4a** shows the probability of occurrence computed at PBE0-D3/def2-TZVP level of theory for each particular chiral and achiral Be_4B_8 isomers for temperatures ranging from 20 to 1900 K. **Figure 4b** shows the probability of occurrence computed at CCSD(T) level of theory. Notice, there is not a significant difference in the probabilities of occurrence between the two panels, thus the computation of probabilities at DFT level of theory is very similar to those computed at CCSDT level of theory. A closer examination of the panel (b) shown that in the temperature ranging from 20 to 300 K, all molecular properties are dominated by the chiral structure depicted in **Figure 3a** because its probability of occurrence is almost constant. We point out that in this range of temperature, the C_1, C_2, and D_2 symmetries strongly dominate with different probabilities of occurrence of 28, 14 y 7% respectively. At this point, there is

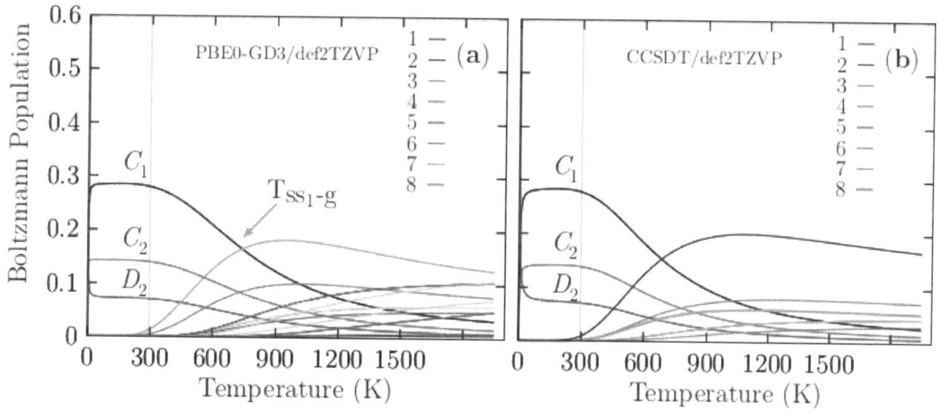

Figure 4.
Panel (a) shows the probability of occurrence for temperatures ranging from 20 to 1900 K at the PBE0-D3/def2-TZVP level of theory. Panel (b) shows the probability of occurrence for temperatures ranging from 20 to 1900 K at the CCSDT/def2-TZVP level of theory. In panel (a), the transition solid–solid point ($T_{ss1\text{-}g}$) is located at 739 K with 16.6% of probability, while in panel (b) the $T_{ss1\text{-}g}$ is located at 780 K with 15% probability.

a co-existence of chiral structures and achiral structures, shown in **Figure 3**, above this point the achiral structure (**Figure 3g**) becomes dominant. The second transformation solid–solid point located at 1017 K and 10% of probability also coexist the chiral putative global minimum with symmetry C_1 and achiral structure (**Figure 3h**) located at 2.51 kcal/mol CCSDT energy at above the putative global minimum. The Boltzmann population computed at PBE0-D3/def2-TZVP level of theory follows the trend of the Boltzamnn population computed at CCSD(T) level of theory.

4. Conclusions

We computed the Boltzmann population of anionic Be_6B_{11} and neutral Be_4B_8 cluster at the SP CCSDT and DFT levels of theory. If one increases the system's temperature, entropic effects start to play an important role, and Gibbs free energy is minimized, and entropy is maximized. The fluxionality of the $Be_6B_{11}^-$ cluster is strongly dependent on temperature that is shown by its Boltzmann population. At the CCSDT level of theory, the Boltzmann population of the $Be_6B_{11}^-$ cluster indicate there are four competing structures, so a mixture of isomers co exist at a specific temperature, so we expect that around a temperature of 350 K, four structures could be observed. The observed properties in a molecule are statistical averages over the ensemble of isomers. The molecular properties at cold temperatures are due to the lowest energy structure $Be_6B_{11}^-$ at CCSD(T) level of theory and zero temperature whereas at hot temperatures, the molecular properties are due to the coaxial Triple-Layered structure with C_1 symmetry. At room temperature the molecular properties are due to a mixture of spectra of the three systems that coexist at 350 K. Regarding Be_4B_8 cluster, all molecular properties at cold and room temperatures are dominated by pair of enantiomers putative global minima. The computed Boltzmann populations at PBE0-D3/def2-TZVP level of theory is similar at the computed Boltzmann populations at CCSDT/def2-TZVP level of theory, so at the DFT level, the Boltzmann populations, hence the molecular properties are well calculated. As future work, the inclusion of anharmonic effects should be taken into account.

References

[1] Dongliang, K.; Weiguo, S.; Hongxiao, S.; Cheng, L.; Xiaoyu. Probing the structure and electronic properties of beryllium doped boron clusters: A planar BeB_{16} cluster motif for metallo-borophene. Sci. Rep. 2019, 9, 14367–14387. doi:10.1038/s41598-019-50905-7.

[2] Mannix, A.J.; Zhang, Z.; Guisinger, N.P.; Yakobson, B.I.; Hersam, M.C. Borophene as a prototype for synthetic 2D materials development. Nature Nanotechnology 2018, 13, 444–450. doi: 10.1038/s41565-018-0157-4.

[3] Kabay, N.; Bryjak, M.; Hilal, N. Boron Separation Processes; Elsevier Science, 2015.

[4] DeFrancesco, H.; Dudley, J.; Coca, A., Boron Chemistry: An Overview. In Boron Reagents in Synthesis; ACS Publications, 2016; chapter 1, pp. 1–25, doi:10.1021/bk-2016-1236.ch001.

[5] Jalife, S.; Liu, L.; Pan, S.; Cabellos, J. L.; Osorio, E.; Lu, C.; Heine, T.; Donald, K.J.; Merino, G. Dynamical behavior of boron clusters. Nanoscale 2016, 8, 17639–17644, doi:10.1039/C6NR06383G.

[6] Zhai, H.J.; Zhao, Y.F.; Li, W.L.; Chen, Q.; Bai, H.; Hu, H.S.; Piazza, Z.A.; Tian, W.J.; Lu, H.G.; Wu, Y.B.; Mu, Y. W.; Wei, G.F.; Liu, Z.P.; Li, J.; Li, S.D.; Wang, L.S. Observation of an all-boron fullerene. Nature Chemistry 2014, pp. 727–731. doi:10.1038/nchem.1999.

[7] Sun, W.; Kang, D.; Chen, B.; Kuang, X.; Ding, K.; Lu, C. Tuning of Structure Evolution and Electronic Properties through Palladium-Doped Boron Clusters: PdB_{16} as a Motif for Boron-Based Nanotubes. The Journal of Physical Chemistry A 2020, 1, 123–134, doi:10.1021/acs.jpca.0c05197.

[8] Jian, T.; Chen, X.; Li, S.D.; Boldyrev, A.I.; Li, J.; Wang, L.S. Probing the structures and bonding of size-selected boron and doped-boron clusters. Chem. Soc. Rev. 2019, 48, 3550–3591, doi: 10.1039/C9CS00233B.

[9] Chen, T.T.; Li, W.L.; Chen, W.J.; Li, J.; Wang, L.S. $La_3\ B_{1-\ 4}$: an inverse triple-decker lanthanide boron cluster. Chem. Commun. 2019, 55, 7864–7867. doi:10.1039/C9CC03807H.

[10] Martínez-Guajardo, G.; Cabellos, J. L.; Díaz-Celaya, A.; Pan, S.; Islas, R.; Chattaraj, P.K.; Heine, T.; Merino, G. Dynamical behavior of Borospherene: A Nanobubble. Sci. Report 2015, 22, 11287–11297. doi:https://www.nature.com/articles/srep11287.

[11] Li, P.; Du, X.; Wang, J.J.; Lu, C.; Chen, H. Probing the Structural Evolution and Stabilities of Medium-Sized MoBn0/− Clusters. The Journal of Physical Chemistry C 2018, 122, 20000–20005, doi:10.1021/acs.jpcc.8b05759.

[12] Grande-Aztatzi, R.; Martínez-Alanis, P.R.; Cabellos, J.L.; Osorio, E.; Martínez, A.; Merino, G. Structural evolution of small gold clusters doped by one and two boron atoms. Journal of Computational Chemistry 2014, 35, 2288–2296, doi:10.1002/jcc.23748.

[13] Dong, X.; Jalife, S.; Vásquez-Espinal, A.; Ravell, E.; Pan, S.; Cabellos, J.L.; Liang, W.y.; Cui, Z.h.; Merino, G. $Li_2\ B_{12}$ and $Li_3\ B_{12}$: Prediction of the Smallest Tubular and Cage-like Boron Structures. Angewandte Chemie International Edition 2018, 57, 4627–4631, doi:10.1002/anie.201800976.

[14] Brothers, P.J. Boron complexes of porphyrins and related polypyrrole ligands: unexpected chemistry for both boron and the porphyrin. Chem. Commun. 2008, pp. 2090–2102. doi: 10.1039/B714894A.

[15] Axtell, J.C.; Saleh, L.M.A.; Qian, E. A.; Wixtrom, A.I.; Spokoyny, A.M.

Synthesis and Applications of Perfunctionalized Boron Clusters. Inorganic Chemistry 2018, 57, 2333–2350, doi:10.1021/acs.inorgchem.7b02912.

[16] Piazza, Z.A.; Hu, H.S.; Li, W.L.; Zhao, Y.F.; Li, J.; Wang, L.S. From planar boron clusters to borophenes and metalloborophenes. Nature Reviews Chemistry 2017, 1, 0071, doi:10.1038/s41570-017-0071.

[17] Kondo, T. Recent progress in boron nanomaterials. Science and technology of advanced materials 2017, 18, 780–804. doi:10.1080/14686996.2017.1379856.

[18] Kiran, B.; Bulusu, S.; Zhai, H.J.; Yoo, S.; Zeng, X.C.; Wang, L.S. Planar-to-tubular structural transition in boron clusters: B20 as the embryo of single-walled boron nanotubes. Proceedings of the National Academy of Sciences 2005, 102, 961–964, doi:10.1073/pnas.0408132102.

[19] Wang, Y.J.; Zhao, Y.F.; Li, W.L.; Jian, T.; Chen, Q.; You, X.R.; Ou, T.; Zhao, X.Y.; Zhai, H.J.; Li, S.D.; Li, J.; Wang, L.S. Observation and characterization of the smallest borospherene, B_{28} and B_{28}. The Journal of Chemical Physics 2016, 144, 064307, doi:10.1063/1.4941380.

[20] Lv, J.; Wang, Y.; Zhang, L.; Lin, H.; Zhao, J.; Ma, Y. Stabilization of fullerene-like boron cages by transition metal encapsulation. Nanoscale 2015, 7, 10482–10489. doi:10.1039/C5NR01659B.

[21] Piazza, Z.A.; Hu, H.S.; Li, W.L.; Zhao, Y.F.; Li, J.; Wang, L.S. Planar hexagonal B36 as a potential basis for extended single-atom layer boron sheets. Nature Communications 2014, 1, 3113, doi:10.1038/ncomms4113.

[22] Li, W.L.; Zhao, Y.F.; Hu, H.S.; Li, J.; Wang, L.S. $[B_{30}]^-$: A Quasiplanar Chiral Boron Cluster. Angewandte Chemie International Edition 2014, 53, 5540–5545, doi:10.1002/anie.201402488.

[23] An, W.; Bulusu, S.; Gao, Y.; Zeng, X.C. Relative stability of planar versus double-ring tubular isomers of neutral and anionic boron cluster B20 and B20-. The Journal of Chemical Physics 2006, 124, 154310, doi:10.1063/1.2187003.

[24] Dong, X.; Jalife, S.; Vásquez-Espinal, A.; Barroso, J.; Orozco-Ic, M.; Ravell, E.; Cabellos, J.L.; Liang, W.y.; Cui, Z.h.; Merino, G. Li_2B_{24}: the simplest combination for a three-ring boron tube. Nanoscale 2019, 11, 2143–2147. doi:10.1039/C8NR09173K.

[25] Feng, L.Y.; Guo, J.C.; Li, P.F.; Zhai, H.J. Boron-Based Chiral Helix Be_6B_{10} and Be_6B_{11} Clusters: Structures, Chemical Bonding, and Formation Mechanism. Chemistry - An Asian Journal 2020, 15, 1094–1104, doi: 10.1002/asia.201901640.

[26] Chen, Q.; Li, W.L.; Zhao, Y.F.; Zhang, S.Y.; Hu, H.S.; Bai, H.; Li, H.R.; Tian, W.J.; Lu, H.G.; Zhai, H.J.; Li, S.D.; Li, J.; Wang, L.S. Experimental and Theoretical Evidence of an Axially Chiral Borospherene. ACS Nano 2015, 9, 754–760, doi:10.1021/nn506262c.

[27] Chen, Q.; Chen, T.T.; Li, H.R.; Zhao, X.Y.; Chen, W.J.; Zhai, H.J.; Li, S.D.; Wang, L.S. B_{31} and B_{32}: chiralquasi-planar boron clusters. Nanoscale 2019, 11, 9698–9704. doi:10.1039/C9NR01524H.

[28] Tai, T.B.; Nguyen, M.T. A new chiral boron cluster B_{44} containing nonagonal holes. Chem. Commun. 2016, 52, 1653–1656. doi:10.1039/C5CC09111J.

[29] Guo, J.C.; Feng, L.Y.; Wang, Y.J.; Jalife, S.; Vásquez-Espinal, A.; Cabellos, J.L.; Pan, S.; Merino, G.; Zhai, H.J. Coaxial Triple-Layered versus Helical $Be_6B_{11}^-$ Clusters: Dual Structural Fluxionality and Multifold Aromaticity. Angewandte Chemie International

Edition 2017, 56, 10174–10177, doi: 10.1002/anie.201703979.

[30] Wang, Y.J.; Feng, L.Y.; Zhai, H.J. Starting a subnanoscale tank tread: dynamic fluxionality of boron-based $B_{10}Ca$ alloy cluster. Nanoscale Adv. 2019, 1, 735–745. doi:10.1039/C8NA00256H.

[31] Gu, F.L.; Yang, X.; Tang, A.C.; Jiao, H.; von R. Schleyer, P. Structure and stability of B+13 clusters. Journal of Computational Chemistry 1998, 19, 203–214, doi:https://doi.org/10.1002/(SICI)1096-987X(19980130)19:2<203::AID-JCC13>3.0.CO;2-I.

[32] Yu, X.; Xu, C.; Cheng, L. Theoretical investigation on anti-sandwich beryllium-boron clusters $Be_{2m}B_n$ (m=1-3): Fluxionality and multi-aromaticity. Computational and Theoretical Chemistry 2020, 1188, 112949. doi:https://doi.org/10.1016/j.comptc.2020.112949.

[33] Oña, O.B.; Torres-Vega, J.J.; Torre, A.; Lain, L.; Alcoba, D.R.; Vásquez-Espinal, A.; Tiznado, W. Chemical bonding analysis in boron clusters by means of localized orbitals according to the electron localization function topology. Theoretical Chemistry Accounts 2015, 134, 28–37, doi:10.1007/s00214-015-1627-5.

[34] Alexandrova, A.N.; Boldyrev, A.I.; Zhai, H.J.; Wang, L.S. All-boron aromatic clusters as potential new inorganic ligands and building blocks in chemistry. Coordination Chemistry Reviews 2006, 250, 2811 – 2866. 18th Main Group Chemistry, doi:https://doi.org/10.1016/j.ccr.2006.03.032.

[35] Zubarev, D.Y.; Boldyrev, A.I. Comprehensive analysis of chemical bonding in boron clusters. Journal of Computational Chemistry 2007, 28, 251–268, doi:10.1002/jcc.20518.

[36] Poater, J.; Duran, M.; Solà, M.; Silvi, B. Theoretical Evaluation of Electron Delocalization in Aromatic Molecules by Means of Atoms in Molecules (AIM) and Electron Localization Function (ELF) Topological. Chemical Reviews 2005, 105, 3911–3947, doi:10.1021/cr030085x.

[37] Mandado, M.; González-Moa, M.J.; Mosquera, R.A. QTAIM n-center delocalization indices as descriptors of aromaticity in mono and poly heterocycles. Journal of Computational Chemistry 2007, 28, 127–136, doi: 10.1002/jcc.20468.

[38] Zhai, H.; Alexandrova, A.N. Fluxionality of Catalytic Clusters: When It Matters and How to Address It. ACS Catalysis 2017, 7, 1905–1911, doi: 10.1021/acscatal.6b03243.

[39] Buelna-Garcia, C.E.; Cabellos, J.L.; Quiroz-Castillo, J.M.; Martinez-Guajardo, G.; Castillo-Quevedo, C.; de Leon-Flores, A.; Anzueto-Sanchez, G.; Martin-del Campo-Solis, M.F. Exploration of Free Energy Surface and Thermal Effects on Relative Population and Infrared Spectrum of the Be_6B_{11}Fluxional Cluster. Materials 2021, 14. doi:10.3390/ma14010112.

[40] Van Duong, L.; Tho Nguyen, M. Silicon doped boron clusters: how to make stable ribbons? Phys. Chem. Chem. Phys. 2017, 19, 14913–14918. doi: 10.1039/C7CP01740E.

[41] Romanescu, C.; Galeev, T.R.; Li, W. L.; Boldyrev, A.I.; Wang, L.S. Transition-Metal-Centered Monocyclic Boron Wheel Clusters (M c Bn): A New Class of Aromatic Borometallic Compounds. Accounts of Chemical Research 2013, 46, 350–358, doi: 10.1021/ar300149a.

[42] Chen, T.T.; Li, W.L.; Bai, H.; Chen, W.J.; Dong, X.R.; Li, J.; Wang, L.S. ReB8 and ReB9− : New Members of the Transition-Metal-Centered Borometallic Molecular Wheel Family. The Journal of Physical Chemistry A 2019, 123, 5317–5324, doi:10.1021/acs.jpca.9b03942.

[43] Popov, I.A.; Jian, T.; Lopez, G.V.; Boldyrev, A.I.; Wang, L.S. Cobalt-centred boron molecular drums with the highest coordination number in the CoB_{16} cluster. Nature Communications 2015, 6, 8654. doi: 10.1038/ncomms9654.

[44] Cui, Z.h.; Yang, W.s.; Zhao, L.; Ding, Y.h.; Frenking, G. Unusually Short Be-Be Distances with and without a Bond in Be_2F_2 and in the Molecular Discuses Be_2B_8 and Be_2B_{7}–. Angewandte Chemie International Edition 2016, 55, 7841–7846, doi:https://doi.org/10.1002/anie.201601890.

[45] Hermann, A.; Ashcroft, N.W.; Hoffmann, R. Binary Compounds of Boron and Beryllium: A Rich Structural Arena with Space for Predictions. Chemistry – A European Journal 2013, 19, 4184–4197, doi:https://doi.org/10.1002/chem.201203890.

[46] Feng, L.Y.; Guo, J.C.; Li, P.F.; Zhai, H.J. Boron-based binary Be_6B_{210-} cluster: three-layered aromatic sandwich, electronic transmutation, and dynamic structural fluxionality. Phys. Chem. Chem. Phys. 2018, 20, 22719–22729. doi:10.1039/C8CP04332A.

[47] Han, L.H.; Wang, Y.J.; Zhai, H.J. Boron-based Be_2B_5 +0- alloy clusters: inverse sandwiches with pentagonal boron ring and reduction-induced structural transformation to molecular wheel structure. New J. Chem. 2021, 45, 4675–4682. doi:10.1039/D0NJ05961G.

[48] Gribanova, T.N.; Minyaev, R.M.; Minkin, V.I. Stabilization of non-typical forms of boron clusters by beryllium doping. Chemical Physics 2019, 522, 44–54. doi:https://doi.org/10.1016/j.chemphys.2019.02.008.

[49] Gribanova, T.N.; Minyaev, R.M.; Minkin, V.I.; Boldyrev, A.I. Novel architectures of boron. Structural Chemistry 2020, 31, 2105–2128. doi: 10.1007/s11224-020-01606-9.

[50] Wang, Y.J.; Feng, L.Y.; Guo, J.C.; Zhai, H.J. Dynamic Mg2 B8 Cluster: A Nanoscale Compass. Chemistry An Asian Journal 2017, 12, 2899–2903, doi: 10.1002/asia.201701310.

[51] Ayuso, D.; Neufeld, O.; Ordonez, A. F.; Decleva, P.; Lerner, G.; Cohen, O.; Ivanov, M.; Smirnova, O. Synthetic chiral light for efficient control of chiral light-matter interaction. Nature Photonics 2019, 13, 866–871. doi: 10.1038/s41566-019-0531-2.

[52] Ayuso, D.; Ordonez, A.; Decleva, P.; Ivanov, M.; Smirnova, O. Polarization of chirality, 2020, [arXiv:physics.optics/2004.05191].

[53] Guo, P.; Yang, B.; Zhang, L.; Zhao, L. Temperature dependent chiroptical response of sigmoidalgold clusters: probing the stability of chiral metal clusters. Chem. Sci. 2018, 9, 5614–5622. doi:10.1039/C8SC00344K.

[54] Barroso, J.; Cabellos, J.L.; Pan, S.; Murillo, F.; Zarate, X.; Fernandez-Herrera, M.A.; Merino, G. Revisiting racemization mechanism of helicenes. Chem. Commun. 2018, 54, 188–191. doi: 10.1039/C7CC08191J.

[55] Ebeling, D.; Šekutor, M.; Stiefermann, M.; Tschakert, J.; Dah, J.E. P.; Carlson, R.M.K.; Schirmeisen, A.; Schreiner, P.R. Assigning the absolute configuration of single aliphatic molecules by visual inspection. Nature Communications 2018, 9, 2420. doi: 10.1038/s41467-018-04843-z.

[56] Tai, T.B.; Ceulemans, A.; Nguyen, M.T. Disk Aromaticity of the Planar and Fluxional Anionic Boron Clusters B_{20}-/2-. Chemistry - A European Journal 2012, 18, 4510–4512, doi: 10.1002/chem.201104064.

[57] Fagiani, M.R.; Song, X.; Petkov, P.; Debnath, S.; Gewinner, S.; Schöllkopf, W.; Heine, T.; Fielicke, A.; Asmis, K.R. Structure and Fluxionality of B13+

Probed by Infrared Photodissociation Spectroscopy. Angewandte Chemie International Edition 2017, 56, 501–504, doi:10.1002/anie.201609766.

[58] Yang, Y.; Jia, D.; Wang, Y.J.; Zhai, H.J.; Man, Y.; Li, S.D. A universal mechanism of the planar boron rotors, B_{13+}, B_{15+}, and B_{19-}: inner wheels rotating in pseudo-rotating outer bearings. Nanoscale 2017, 9, 1443–1448. doi:10.1039/C6NR09074E.

[59] Wang, Y.J.; Zhao, X.Y.; Chen, Q.; Zhai, H.J.; Li, S.D. B11-: a moving subnanoscale tank tread. Nanoscale 2015, 7, 16054–16060. doi: 10.1039/C5NR03732H.

[60] Wang, Y.J.; Guo, J.C.; Zhai, H.J. Why nanoscale tank treads move? Structures, chemical bonding, and molecular dynamics of a doped boron cluster B10C. Nanoscale 2017, 9, 9310–9316. doi:10.1039/C7NR03193A.

[61] Seitsonen, A.P.; Laasonen, K.; Nieminen, R.M.; Klein, M.L. Structure of CAl12. The Journal of Chemical Physics 1995, 103, 8075–8080, doi: 10.1063/1.470172.

[62] Chandrachud, P.; Joshi, K.; Kanhere, D.G. Thermodynamics of carbon-doped Al and Ga clusters: Ab initio molecular dynamics simulations. Phys. Rev. B 2007, 76, 235423. doi: 10.1103/PhysRevB.76.235423.

[63] Doye, J.P.K.; Calvo, F. Entropic effects on the structure of Lennard-Jones clusters. The Journal of Chemical Physics 2002, 116, 8307–8317, doi: 10.1063/1.1469616.

[64] Li, Z.H.; Jasper, A.W.; Truhlar, D.G. Structures, Rugged Energetic Landscapes, and Nanothermodynamics of Aln (2, <n< 65) Particles. Journal of the American Chemical Society 2007, 129, 14899–14910, doi:10.1021/ja073129i.

[65] Darby, S.; Mortimer-Jones, T.V.; Johnston, R.L.; Roberts, C. Theoretical study of Cu–Au nanoalloy clusters using a genetic algorithm. The Journal of Chemical Physics 2002, 116, 1536–1550, doi:10.1063/1.1429658.

[66] P. K. Doye, J.; J. Wales, D. Global minima for transition metal clusters described by Sutton-Chen potentials. New J. Chem. 1998, 22, 733–744. doi: 10.1039/A709249K.

[67] Ohno, K.; Maeda, S. Global Reaction Route Mapping on Potential Energy Surfaces of Formaldehyde, Formic, and Their Metal-Substituted Analogues. The Journal of Physical Chemistry A 2006, 110, 8933–8941, doi:10.1021/jp061149l.

[68] Baletto, F.; Ferrando, R. Structural properties of nanoclusters: Energetic, thermodynamic, and kinetic effects. Rev. Mod. Phys. 2005, 77, 371–423. doi: 10.1103/RevModPhys.77.371.

[69] Li, Z.H.; Truhlar, D.G. Nanothermodynamics of metal nanoparticles. Chem. Sci. 2014, 5, 2605–2624. doi:10.1039/C4SC00052H.

[70] Li, Z.H.; Jasper, A.W.; Truhlar, D.G. Structures, Rugged Energetic Landscapes, and Nanothermodynamics of Al_n (2<n<) Particles. Journal of the American Chemical Society 2007, 129, 14899–14910, doi:10.1021/ja073129i.

[71] Grigoryan, V.G.; Springborg, M. Temperature and isomeric effects in nanoclusters. Phys. Chem. Chem. Phys. 2019, 21, 5646–5654. doi:10.1039/C9CP00123A.

[72] Sutton, C.; Levchenko, S.V. First-Principles Atomistic Thermodynamics and Configurational Entropy. Frontiers in Chemistry 2020, 8, 757. doi:10.3389/fchem.2020.00757.

[73] Mermin, N.D. Thermal Properties of the Inhomogeneous Electron Gas. Phys. Rev. 1965, 137, A1441–A1443. doi: 10.1103/PhysRev.137.A1441.

[74] Pittalis, S.; Proetto, C.R.; Floris, A.; Sanna, A.; Bersier, C.; Burke, K.; Gross, E.K.U. Exact Conditions in Finite-Temperature Density-Functional Theory. Phys. Rev. Lett. 2011, 107, 163001. doi:10.1103/PhysRevLett.107.163001.

[75] Gonis, A.; Däne, M. Extension of the Kohn-Sham formulation of density functional theory to finite temperature. Journal of Physics and Chemistry of Solids 2018, 116, 86–99. doi: https://doi.org/10.1016/j.jpcs.2017.12.021.

[76] Eschrig, H. T> 0 ensemble-state density functional theory via Legendre transform. Phys. Rev. B 2010, 82, 205120. doi:10.1103/PhysRevB.82.205120.

[77] Strøm, B.A.; Simon, J.M.; Schnell, S. K.; Kjelstrup, S.; He, J.; Bedeaux, D. Size and shape effects on the thermodynamic properties of nanoscale volumes of water. Phys. Chem. Chem. Phys. 2017, 19, 9016–9027. doi:10.1039/C7CP00874K.

[78] Hill, T.L. Extension of Nanothermodynamics to Include a One-Dimensional Surface Excess. Nano Letters 2001, 1, 159–160, [https://doi.org/10.1021/nl010009e]. doi:10.1021/nl010009e.

[79] Gibbs, J. Thermodynamics; Scientific Papers, Dover Publications, 1961.

[80] Hill, T.L. Thermodynamics of Small Systems. The Journal of Chemical Physics 1962, 36, 3182–3197, doi:10.1063/1.1732447.

[81] Calvo, F. Thermodynamics of nanoalloys. Phys. Chem. Chem. Phys. 2015, 17, 27922–27939. doi:10.1039/C5CP00274E.

[82] Bixon, M.; Jortner, J. Energetic and thermodynamic size effects in molecular clusters. The Journal of Chemical Physics 1989, 91, 1631–1642, doi:10.1063/1.457123.

[83] Kristensen, W.D.; Jensen, E.J.; Cotterill, R.M.J. Thermodynamics of small clusters of atoms: A molecular dynamics simulation. The Journal of Chemical Physics 1974, 60, 4161–4169, doi:10.1063/1.1680883.

[84] Wales, D.J. Structure, Dynamics, and Thermodynamics of Clusters: Tales from Topographic Potential Surfaces. Science 1996, 271, 925–929, doi:10.1126/science.271.5251.925.

[85] Jena, P.; Khanna, S.; Rao, B. Physics and Chemistry of Finite Systems: From Clusters to Crystals; Number v.2 in NATO ASI Series : advanced science institutes series: Series C, Mathematical and physical sciences, Kluwer Academic Publishers, 1992.

[86] Fox, H.; Horsfield, A.P.; Gillan, M.J. Density functional calculations of surface free energies. The Journal of Chemical Physics 2006, 124, 134709, doi:10.1063/1.2184313.

[87] Ghiringhelli, L.M.; Gruene, P.; Lyon, J.T.; Rayner, D.M.; Meijer, G.; Fielicke, A.; Scheffler, M. Not so loosely bound rare gas atoms: finite-temperature vibrational fingerprints of neutral gold-cluster complexes. New Journal of Physics 2013, 15, 083003. doi:10.1088/1367-2630/15/8/083003.

[88] Schebarchov, D.; Baletto, F.; Wales, D.J. Structure, thermodynamics, and rearrangement mechanisms in gold clusters—insights from the energy landscapes framework. Nanoscale 2018, 10, 2004–2016. doi:10.1039/C7NR07123J.

[89] Goldsmith, B.R.; Florian, J.; Liu, J. X.; Gruene, P.; Lyon, J.T.; Rayner, D.M.; Fielicke, A.; Scheffler, M.; Ghiringhelli, L.M. Two-to-three dimensional transition in neutral gold clusters: The crucial role of van der Waals

interactions and temperature. Phys. Rev. Materials 2019, 3, 016002. doi: 10.1103/PhysRevMaterials.3.016002.

[90] Beret, E.C.; Ghiringhelli, L.M.; Scheffler, M. Free gold clusters: beyond the static, monostructure description. Faraday Discuss. 2011, 152, 153–167. doi: 10.1039/C1FD00027F.

[91] Lv, Z.L.; Xu, K.; Cheng, Y.; Chen, X. R.; Cai, L.C. Ab initio investigation of the lower energy candidate structures for $(H_2O)_5$+ water cluster. The Journal of Chemical Physics 2014, 141, 054309, doi:10.1063/1.4891721.

[92] Malloum, A.; Fifen, J.J.; Dhaouadi, Z.; Engo, S.G.N.; Jaidane, N.E. Structures and relative stabilities of ammonia clusters at different temperatures: DFT vs. ab initio. Phys. Chem. Chem. Phys. 2015, 17, 29226–29242. doi:10.1039/C5CP03374H.

[93] Malloum, A.; Fifen, J.J.; Dhaouadi, Z.; Nana Engo, S.G.; Jaidane, N.E. Structures and spectroscopy of medium size protonated ammonia clusters at different temperatures, $H^+(NH_3)_{10-16}$. The Journal of Chemical Physics 2017, 146, 044305, doi:10.1063/1.4974179.

[94] Malloum, A.; Fifen, J.J.; Conradie, J. Structures and infrared spectroscopy of large sized protonated ammonia clusters. The Journal of Chemical Physics 2018, 149, 244301, doi:10.1063/1.5053172.

[95] Fifen, J.J.; Agmon, N. Structure and Spectroscopy of Hydrated Sodium Ions at Different Temperatures the Cluster Stability Rules. Journal of Chemical Theory and Computation 2016, 12, 1656–1673, doi:10.1021/acs.jctc.6b00038.

[96] Franco-Perez, M.; Gazquez, J.L.; Vela, A. Electronic chemical response indexes at finite temperature in the canonical ensemble. The Journal of Chemical Physics 2015, 143, 024112, doi: 10.1063/1.4923260.

[97] Gazquez, J.L.; Franco-Perez, M.; Ayers, P.W.; Vela, A. Temperature-dependent approach to chemical reactivity concepts in density functional theory. International Journal of Quantum Chemistry 2019, 119, e25797, doi: https://doi.org/10.1002/qua.25797.

[98] Castillo-Quevedo, C.; Buelna-Garcia, C.E.; Paredes-Sotelo, E.; Robles-Chaparro, E.; Martinez-Guajardo, G.; Quiroz-Castillo, J.M.; de Leon-Flores, A.; Gaxiola, T.; Castillo, S.J.; Vasquez-Espinal, A.; Cabellos, J.L. Relative abundances and enantiomerization energy of the chiral Cu13 cluster at finite temperature, 2021, [arXiv:cond-mat.mtrl-sci/2109.03981].

[99] Castillo-Quevedo, C.; Buelna-Garcia, C.E.; Paredes-Sotelo, E.; Robles-Chaparro, E.; Zamora-Gonzalez, E.; Martin-del Campo-Solis, M.F.; Quiroz-Castillo, J.M.; del Castillo-Castro, T.; Martínez-Guajardo, G.; de Leon-Flores, A.; Cortez-Valadez, M.; Ortiz-Chi, F.; Gaxiola, T.; Castillo, S.J.; Vásquez-Espinal, A.; Pan, S.; Cabellos, J.L. Effects of Temperature on Enantiomerization Energy and Distribution of Isomers in the Chiral Cu13 Cluster. Molecules 2021, 26. doi:10.3390/molecules26185710.

[100] Dzib, E.; Cabellos, J.L.; Ortíz-Chi, F.; Pan, S.; Galano, A.; Merino, G. Eyringpy: A program for computing rate constants in the gas phase and in solution. International Journal of Quantum Chemistry 2019, 119, e25686, doi:10.1002/qua.25686.

[101] Vargas-Caamal, A.; Cabellos, J.L.; Ortiz-Chi, F.; Rzepa, H.S.; Restrepo, A.; Merino, G. How Many Water Molecules Does it Take to Dissociate HCl? Chemistry – A European Journal 2016, 22, 2812–2818, doi:10.1002/chem.201504016.

[102] Shkrebtii, A.; Heron, J.; Cabellos, J.; Witkowski, N.; Pluchery, O.; Mendoza, B.; Borensztein, Y. Temperature Dependent Optical

Response of Si (100): Theory vs. Experiment. MRS Proceedings 2011, 1370, mrss11-1370-yy05-10. doi: 10.1557/opl.2011.1039.

[103] Mendoza-Wilson, A.M.; Balandrán-Quintana, R.R.; Cabellos, J.L. Thermochemical behavior of sorghum procyanidin trimers with C_4-C_8 and C_4-C_6 interflavan bonds in the reaction with superoxide anion radical and H_2O_2-forming NADH-oxidase flavoenzyme. Computational and Theoretical Chemistry 2020, 1186, 112912. doi:https://doi.org/10.1016/j.comptc.2020.112912.

[104] Buelna-Garcia, C.E.; Robles-Chaparro, E.; Parra-Arellano, T.; Quiroz-Castillo, J.M.; del Castillo-Castro, T.; Martínez-Guajardo, G.; Castillo-Quevedo, C.; de León-Flores, A.; Anzueto-Sánchez, G.; Martin-del Campo-Solis, M.F.; Mendoza-Wilson, A.M.; Vásquez-Espinal, A.; Cabellos, J.L. Theoretical Prediction of Structures, Vibrational Circular Dichroism, and Infrared Spectra of Chiral Be_4B_8 Cluster at Different Temperatures. Molecules 2021, 26. doi: 10.3390/molecules26133953.

[105] Ji, M.; Gu, X.; Li, X.; Gong, X.; Li, J.; Wang, L.S. Experimental and Theoretical Investigation of the Electronic and Geometrical Structures of the Au_{32} Cluster. Angewandte Chemie International Edition 2005, 44, 7119–7123, doi:10.1002/anie.200502795.

[106] Wille, L.T.; Vennik, J. Computational complexity of the ground-state determination of atomic clusters. Journal of Physics A: Mathematical and General 1985, 18, L419–L422. doi:10.1088/0305-4470/18/8/003.

[107] Xu, S.G.; Zhao, Y.J.; Liao, J.H.; Yang, X.B. Understanding the stable boron clusters: A bond model and first-principles calculations based on high-throughput screening. The Journal of Chemical Physics 2015, 142, 214307, doi: 10.1063/1.4922059.

[108] Rossi, G.; Ferrando, R. Searching for low-energy structures of nanoparticles: a comparison of different methods and algorithms. Journal of Physics: Condensed Matter 2009, 21, 084208. doi:10.1088/0953-8984/21/8/084208.

[109] Cheng, L.; Feng, Y.; Yang, J.; Yang, J. Funnel hopping: Searching the cluster potential energy surface over the funnels. The Journal of Chemical Physics 2009, 130, 214112, doi:10.1063/1.3152121.

[110] Kirkpatrick, S.; Gelatt, C.D.; Vecchi, M.P. Optimization by Simulated Annealing. Science 1983, 220, 671–680. doi:10.1126/science.220.4598.671.

[111] Metropolis, N.; Rosenbluth, A.W.; Rosenbluth, M.N.; Teller, A.H.; Teller, E. Equation of State Calculations by Fast Computing Machines. J. Chem. Phys. 1953, 21, 1087–1092. doi:http://dx.doi.org/10.1063/1.1699114.

[112] Xiang, Y.; Gong, X.G. Efficiency of generalized simulated annealing. Phys. Rev. E 2000, 62, 4473–4476. doi: 10.1103/PhysRevE.62.4473.

[113] Xiang, Y.; Gubian, S.; Suomela, B.; Hoeng, J. Generalized Simulated Annealing for Global Optimization: the GenSA Package for R. The R Journal 2013, 5, 13–29.

[114] Vlachos, D.; Schmidt, L.; Aris, R. Comparison of small metal clusters: Ni, Pd, Pt, Cu, Ag, Au. Z. Phys. D Atom Mol. Cl. 1993, 26, 156–158. doi:10.1007/BF01425649.

[115] Granville, V.; Krivanek, M.; Rasson, J.P. Simulated annealing: a proof of convergence. IEEE Trans. Pattern Anal. Mach. Intell. 1994, 16, 652–656. doi:10.1109/34.295910.

[116] Saunders, M. Stochastic search for isomers on a quantum mechanical surface. Journal of Computational

Chemistry 2004, 25, 621–626, doi: 10.1002/jcc.10407.

[117] Saunders, M. Stochastic exploration of molecular mechanics energy surfaces. Hunting for the global minimum. Journal of the American Chemical Society 1987, 109, 3150–3152, doi: 10.1021/ja00244a051.

[118] Hsu, P.J.; Lai, S.K. Structures of bimetallic clusters. J. Chem. Phys. 2006, 124, 044711–0. doi: http://dx.doi.org/ 10.1063/1.2147159.

[119] Qin, W.; Lu, W.C.; Zhao, L.Z.; Zang, Q.J.; Wang, C.Z.; Ho, K.M. Stabilities and fragmentation energies of Si_n clusters (n = 2-33). J. Phys.: Condens. Matter 2009, 21, 455501.

[120] Goldberg, D.E. Genetic Algorithms in Search, Optimization and Machine Learning, 1st ed.; Addison-Wesley Longman Publishing Co., Inc.: Boston, MA, USA, 1989.

[121] Alexandrova, A.N.; Boldyrev, A.I. Search for the $Li_n(0/+1/-1)$ (n = 5-7) Lowest-Energy Structures Using the ab Initio Gradient Embedded Genetic Algorithm (GEGA). Elucidation of the Chemical Bonding in the Lithium Clusters. J. Chem. Theory Comput. 2005, 1, 566–580, doi:10.1021/ct050093g.

[122] Alexandrova, A.N.; Boldyrev, A.I.; Fu, Y.J.; Yang, X.; Wang, X.B.; Wang, L. S. Structure of the Na_xCl_x (x=1,4) clusters via ab-initio genetic algorithm and photoelectron spectroscopy. J. Chem. Phys. 2004, 121, 5709–5719. doi: 10.1063/1.1783276.

[123] Alexandrova, A.N. $H{\cdot}(H_2O)n$ Clusters: Microsolvation of the Hydrogen Atom via Molecular ab Initio Gradient Embedded Genetic Algorithm (GEGA). J. Phys. Chem. A 2010, 114, 12591–12599, doi:10.1021/jp1092543.

[124] Harding, D.; Mackenzie, S.R.; Walsh, T.R. Structural Isomers and Reactivity for Rh_6 and Rh_{6+}. J. Phys. Chem. B 2006, 110, 18272–18277, doi: 10.1021/jp062603o.

[125] Wales, D.J.; Doye, J.P.K. Global Optimization by Basin-Hopping and the Lowest Energy Structures of Lennard-Jones Clusters Containing up to 110 Atoms. J. Phys. Chem. A 1997, 101, 5111–5116, doi:10.1021/jp970984n.

[126] Mondal, S.; Cabellos, J.L.; Pan, S.; Osorio, E.; Torres-Vega, J.J.; Tiznado, W.; Restrepo, A.; Merino, G. 10-π-Electron arenes a la carte: structure and bonding of the $[E(C_nH_n)\text{-}E]_{n\text{-}6}$ (E = Ca, Sr, Ba; n = 6-8) complexes. Phys. Chem. Chem. Phys. 2016, 18, 11909–11918. doi: 10.1039/C6CP00671J.

[127] Ravell, E.; Jalife, S.; Barroso, J.; Orozco-Ic, M.; Hernandez-Juarez, G.; Ortiz-Chi, F.; Pan, S.; Cabellos, J.L.; Merino, G. Structure and Bonding in CE_5 - (E=Al-Tl) Clusters: Planar Tetracoordinate Carbon versus Pentacoordinate Carbon. Chemistry - An Asian Journal 2018, 13, 1467–1473, doi:10.1002/asia.201800261.

[128] Pan, S.; Moreno, D.; Cabellos, J.L.; Romero, J.; Reyes, A.; Merino, G.; Chattaraj, P.K. In Quest of Strong Be-Ng Bonds among the Neutral Ng-Be Complexes. The Journal of Physical Chemistry A 2014, 118, 487–494, doi: 10.1021/jp409941v.

[129] Cui, Z.h.; Ding, Y.h.; Cabellos, J.L.; Osorio, E.; Islas, R.; Restrepo, A.; Merino, G. Planar tetracoordinate carbons with a double bond in CAl3E clusters. Phys. Chem. Chem. Phys. 2015, 17, 8769–8775. doi:10.1039/ C4CP05707D.

[130] Vargas-Caamal, A.; Pan, S.; Ortiz-Chi, F.; Cabellos, J.L.; Boto, R.A.; Contreras-Garcia, J.; Restrepo, A.; Chattaraj, P.K.; Merino, G. How strong are the metallocene metallocene interactions? Cases of ferrocene, ruthenocene, and osmocene. Phys.

Chem. Chem. Phys. 2016, 18, 550–556. doi:10.1039/C5CP05956A.

[131] Cui, Z.h.; Vassilev-Galindo, V.; Luis Cabellos, J.; Osorio, E.; Orozco, M.; Pan, S.; Ding, Y.h.; Merino, G. Planar pentacoordinate carbon atoms embedded in a metallocene framework. Chem. Commun. 2017, 53, 138–141. doi:10.1039/C6CC08273D.

[132] Vargas-Caamal, A.; Ortiz-Chi, F.; Moreno, D.; Restrepo, A.; Merino, G.; Cabellos, J.L. The rich and complex potential energy surface of the ethanol dimer. Theoretical Chemistry Accounts 2015, 134, 16, doi:10.1007/s00214-015-1615-9.

[133] Flórez, E.; Acelas, N.; Ibargüen, C.; Mondal, S.; Cabellos, J.L.; Merino, G.; Restrepo, A. Microsolvation of NO_3^-: structural exploration and bonding analysis. RSC Adv. 2016, 6, 71913–71923. doi:10.1039/C6RA15059D.

[134] Joshi, A.; Kulkarni, O.; Kakandikar, G.; Nandedkar, V. Cuckoo Search Optimization- A Review. Materials Today: Proceedings 2017, 4, 7262–7269. International Conference on Advancements in Aeromechanical Materials for Manufacturing (ICAAMM-2016): Organized by MLR Institute of Technology, Hyderabad, Telangana, India, doi:https://doi.org/10.1016/j.matpr.2017.07.055.

[135] Bharti, A.; Prerna.; Banerjee, T. Applicability of Cuckoo Search Algorithm for the Prediction of Multicomponent Liquid–Liquid Equilibria for Imidazolium and Phosphonium Based Ionic Liquids. Industrial & Engineering Chemistry Research 2015, 54, 12393–12407, doi:10.1021/acs.iecr.5b03449.

[136] Frisch, M.J.; Trucks, G.W.; Schlegel, H.B.; Scuseria, G.E.; Robb, M.A.; Cheeseman, J.R.; Scalmani, G.; Barone, V.; Mennucci, B.; Petersson, G.A.; Nakatsuji, H.; Caricato, M.; Li, X.; Hratchian, H.P.; Izmaylov, A.F.; Bloino, J.; Zheng, G.; Sonnenberg, J.L.; Hada, M.; Ehara, M.; Toyota, K.; Fukuda, R.; Hasegawa, J.; Ishida, M.; Nakajima, T.; Honda, Y.; Kitao, O.; Nakai, H.; Vreven, T.; Montgomery, J. A.; Peralta, J.E.; Ogliaro, F.; Bearpark, M.; Heyd, J.J.; Brothers, E.; Kudin, K.N.; Staroverov, V.N.; Kobayashi, R.; Normand, J.; Raghavachari, K.; Rendell, A.; Burant, J.C.; Iyengar, S.S.; Tomasi, J.; Cossi, M.; Rega, N.; Millam, J.M.; Klene, M.; Knox, J.E.; Cross, J.B.; Bakken, V.; Adamo, C.; Jaramillo, J.; Gomperts, R.; Stratmann, R.E.; Yazyev, O.; Austin, A.J.; Cammi, R.; Pomelli, C.; Ochterski, J.W.; Martin, R.L.; Morokuma, K.; Zakrzewski, V.G.; Voth, G.A.; Salvador, P.; Dannenberg, J.J.; Dapprich, S.; Daniels, A. D.; Farkas.; Foresman, J.B.; Ortiz, J.V.; Cioslowski, J.; Fox, D.J. Gaussian 09, Revision B.01, 2009.

[137] Brehm, G.; Reiher, M.; Le Guennic, B.; Leibold, M.; Schindler, S.; Heinemann, F.W.; Schneider, S. Investigation of the low-spin to high-spin transition in a novel [Fe(pmea)$(NCS)_2$] complex by IR and Raman spectroscopy and DFT calculations. Journal of Raman Spectroscopy, 2006, 37, 108–122, doi:https://doi.org/10.1002/jrs.1437.

[138] McQuarrie, D.; A, M. Statistical Mechanics; Chemistry Series, Harper & Row, 1975.

[139] Hill, T. An Introduction to Statistical Thermodynamics; Addison-Wesley series in chemistry, Dover Publications, 1986.

[140] Teague, S.J. Implications of protein flexibility for drug discovery. Nature Reviews Drug Discovery 2003, 2, 527–541. doi:10.1038/nrd1129.

[141] Adamo, C.; Barone, V. Toward reliable density functional methods without adjustable parameters: The PBE0 model. The Journal of Chemical Physics 1999, 110, 6158–6170, doi:10.1063/1.478522.

[142] Weigend, F.; Ahlrichs, R. Balanced basis sets of split valence, triple zeta valence and quadruple zeta valence quality for H to Rn: Design and assessment of accuracy. Phys. Chem. Chem. Phys. 2005, 7, 3297–3305. doi: 10.1039/B508541A.

[143] Grimme, S.; Antony, J.; Ehrlich, S.; Krieg, H. A consistent and accurate ab initio parametrization of density functional dispersion correction (DFT-D) for the 94 elements H-Pu. The Journal of Chemical Physics 2010, 132, 154104, doi:10.1063/1.3382344.

[144] Pan, L.L.; Li, J.; Wang, L.S. Low-lying isomers of the B_9^- boron cluster: The planar molecular wheel versus three-dimensional structures. The Journal of Chemical Physics 2008, 129, 024302, doi:10.1063/1.2948405.

[145] Moezzi, A.; Olmstead, M.M.; Power, P.P. Boron-boron double bonding in the species $[B_2R_4]_2^-$: synthesis and structure of [(Et_{2O}) $Li_2Mes_2BB(Mes)Ph$], a diborane(4) dianion analog of a substituted ethylene. Journal of the American Chemical Society 1992, 114, 2715–2717, doi: 10.1021/ja00033a054.

[146] Zhou, M.; Tsumori, N.; Li, Z.; Fan, K.; Andrews, L.; Xu, Q. OCBBCO: A Neutral Molecule with Some Boron-Boron Triple Bond Character. Journal of the American Chemical Society 2002, 124, 12936–12937, doi:10.1021/ja026257+.

[147] de la Puente, E.; Aguado, A.; Ayuela, A.; López, J.M. Structural and electronic properties of small neutral (MgO)n clusters. Phys. Rev. B 1997, 56, 7607–7614. doi:10.1103/PhysRevB.56.7607.

[148] Pople, J.A.; Head-Gordon, M.; Raghavachari, K. Quadratic configuration interaction. A general technique for determining electron correlation energies. The Journal of Chemical Physics 1987, 87, 5968–5975, doi:10.1063/1.453520.

Section 2

New Perspectives

Chapter 5

Transformation of Drug Discovery towards Artificial Intelligence: An *in Silico* Approach

Abstract

Computational methods play a key role in the design of therapeutically important molecules for modern drug development. With these "*in silico*" approaches, machines are learning and offering solutions to some of the most complex drug related problems and has well positioned them as a next frontier for potential breakthrough in drug discovery. Machine learning (ML) methods are used to predict compounds with pharmacological activity, specific pharmacodynamic and ADMET (absorption, distribution, metabolism, excretion and toxicity) properties to evaluate the drugs and their various applications. Modern artificial intelligence (AI) has the capacity to significantly enhance the role of computational methodology in drug discovery. Use of AI in drug discovery and development, drug repurposing, improving pharmaceutical productivity, and clinical trials will certainly reduce the human workload as well as achieving targets in a short period of time. This chapter elaborates the crosstalk between the machine learning techniques, computational tools and the future of AI in the pharmaceutical industry.

Keywords: Conceptual Density functional theory, drug discovery, Artificial Intelligence, Machine learning, pharmaco-kinetic parameters

1. Introduction

Computer-aided drug design [1, 2] has the potential to lower the cost, decrease the failure rates and speed up the discovery process. Computational tools play various roles in medicinal chemistry ranging from optimization of protein-ligand interactions for drug discovery to the design of new drugs. These methods are broadly classified as structure based and ligand based methods. For structural methods the computational studies are carried out for molecular dynamics, protein–ligand docking and calculation of free binding energies. For the ligand based methods, the computational calculations help to predict the biological response about known active and inactive ligands which include quantitative structure–activity relationships, activity cliffs analysis, and similarity search. In recent years, discovery of new molecules that could be more effective with fewer unwanted side effects is a constant concern of pharmaceutical industry. So, new developed research methods are used to predict the properties and activities of molecules even before they are synthesized. The significant development of computational tools as well as theoretical studies of quantum chemistry allow researchers to obtain more precise physicochemical and quantum parameters of compounds in a shorter time. These techniques move towards the synthesis of a very large number of molecules simultaneously and to test their actions on therapeutic targets. Density Functional theory (DFT) has become a powerful tool to study the electronic and geometric characteristics of the drugs. Conceptual density functional theory (CDFT), originally developed by Parr and collaborators [3–8], with several global and local reactivity descriptors help us to understand various physicochemical processes. As Global reactivity descriptors are connected with several electronic structure principles so they play very important role in the physico-chemical information of the complexes. The understanding of the relationship between the structure and activity of the drug, the pharmacokinetic parameters responsible for bioavailability and the toxicity are evaluated by other computational tools. In this way the drugs with higher efficiency is obtained. These studies give a complete picture to design new drug molecules and its physicochemical parameters, drug-likeness and cytotoxicity evaluation in a shorter time. Recent advances in these methods increase the quantity and complexity of generated data. This massive amount of raw data needs to be stored and interpreted in order to advance the medicinal world. The correlations and patterns from large amounts of complex drug data should be performed by machine learning algorithms to extract knowledge and insights from the accumulated data. Databases are used to design new molecular descriptors and the models are validated with external test sets [9].

Modern artificial intelligence (AI) has the potential to significantly enhance the role of computational methods and machine learning in pharmaceutical industry [10]. According to the World Economic Forum, a combination of big data and AI are considered as the fourth paradigm of science and the fourth industrial revolution. Interestingly, with machine learning and AI solutions to some of the most complex drug related problems, drug discovery has created a potential breakthrough in medicinal world.

2. Method

CDFT is used to predict, analyze and interpret the reactivity properties of small drug molecules. Global reactivity descriptors, Fukui indices [11] and the Dual descriptors proposed by Morell et al. [12] are used for the analysis. These theories have been validated by a large number of studies [13–18]. A better microscopic insight to the whole interaction process can be observed by global indices and the other derived reactivity indices for the interacting complexes. In our work, we have selected design and structure-based, informatics-based, fragment-based, small-molecule microarray screening, dynamic combinatorial screening and use of phenotypic assays based 22 small drug molecules to identify RNA-binding molecules. These drug show several desirable properties as good absorption, distribution, oral bioavailability and have ability to target bulges, loops, junctions, pseudo-knots, or higher order structures. The optimized structures for these 22 small drug molecules are given in **Figure 1**. We have computed relevant electronic properties of the studied drug including global parameters such as E_{HOMO}, E_{LUMO}, Energy gap, IP, EA, electronegativity (χ), global hardness (η), global softness(S), Chemical potential (μ) Softness (S), Electrophilicity index (ω), fraction of electrons transferred (Δ), Nucleophilicity index as well as local ones (Fukui functions and dual descriptors). CDFT results predicted structural and thermodynamic stability and low reactivity for the complexes. Also few complexes were identified as fluorescent biomarkers as their emission lies in the visible region [19]. In another study, the

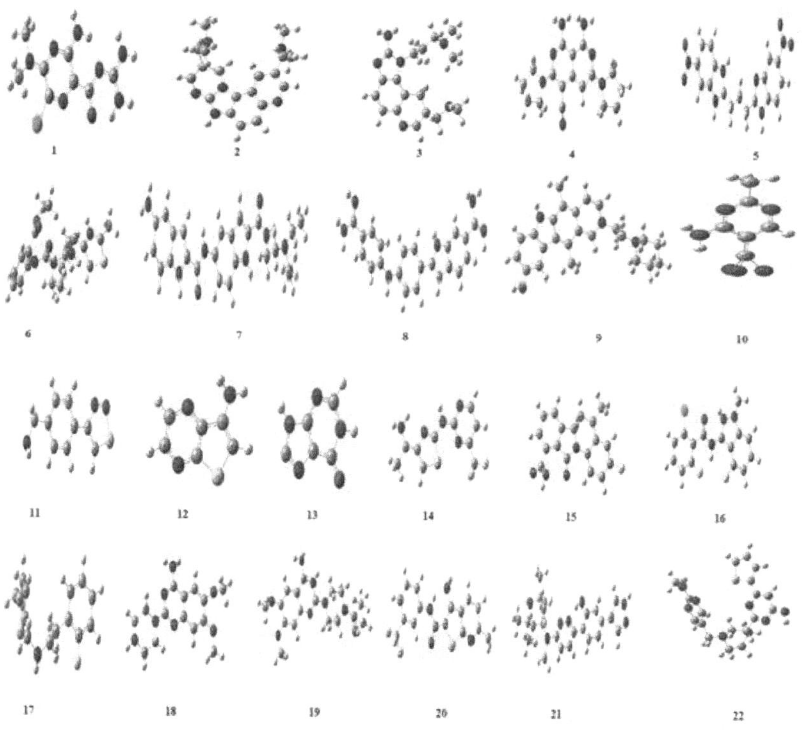

Figure 1.
Optimized structures of 22 small drug-like biologically active molecules. Adopted from Ref. [19].

global and local descriptors are calculated to study ten Anti-inflammatory steroids (AIS) to understand the structure–activity relationship. The toxicity evaluation of drug and the pharmacokinetic parameters responsible for bioavailability and bioactivity are carried by the bioinformatics Osiris/Molinspiration [20, 21] tools. The physico-chemical properties are studied by G09 software tools. As the structures of small-molecule drugs increase in complexity, the importance of synthetic and *in silico* approaches both play an important role in understanding of ligand–receptor interactions within target classes. The predictive drug discovery tools offered a high degree of specificity within molecular design. Careful arrangement of structural features about a molecule necessitated efficient and practical approaches to model these privileged structures. We have also used similar computational tools to study the 21 molecular new chemical entities (NCEs) approved for the first time by a governing body anywhere in the world during 2019 [22]. Out of 11 therapeutic areas, 10 therapeutic areas as anti-infective/antibiotic, cardiovascular and hematologic, neurological (central nervous system (CNS)), dermatologic, inflammation and immunologic, metabolic, musculoskeletal, oncologic, reproductive, and respiratory drugs were selected. See **Figures 2** and **3**. Osiris Calculations were carried out to predict the toxicity risk in the drug molecules. Results showed drug conform behavior for all studied drugs except Triclabendazole, Trifarotene, Alpilisib and Ertafinitib, which shows high risks of undesired effects like mutagenicity, tumorigenicity, irritating effects and reproductive effects [23].

The understanding of various types of interactions is also crucial for the drug-like molecule and its target. These possible interactions between a drug and target consist of covalent bonds, dipole–dipole interactions, ion-dipole interactions, ionic interactions, hydrogen bonding, hydrophobic interactions and charge transfer. The mechanism of drug action can be explained with ionic interactions as during

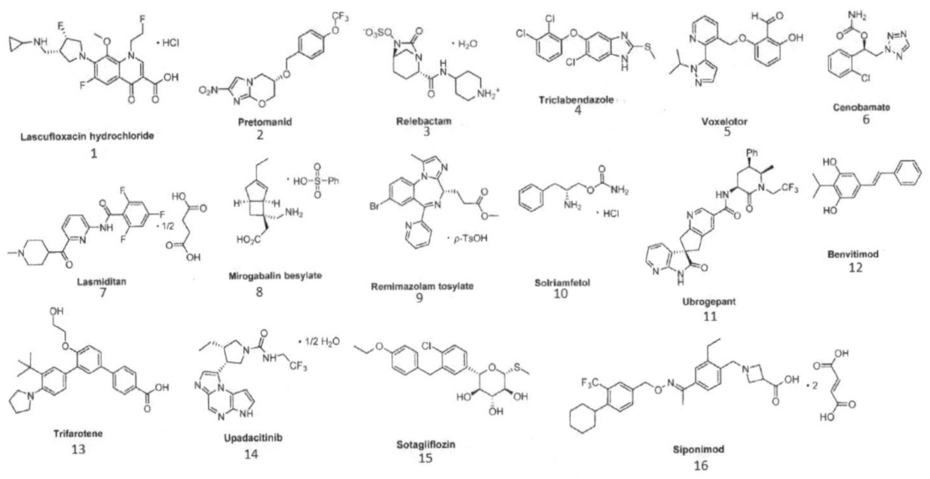

Figure 2.
Chemical structures of studied New Chemical Entities (NCEs) (1-16).

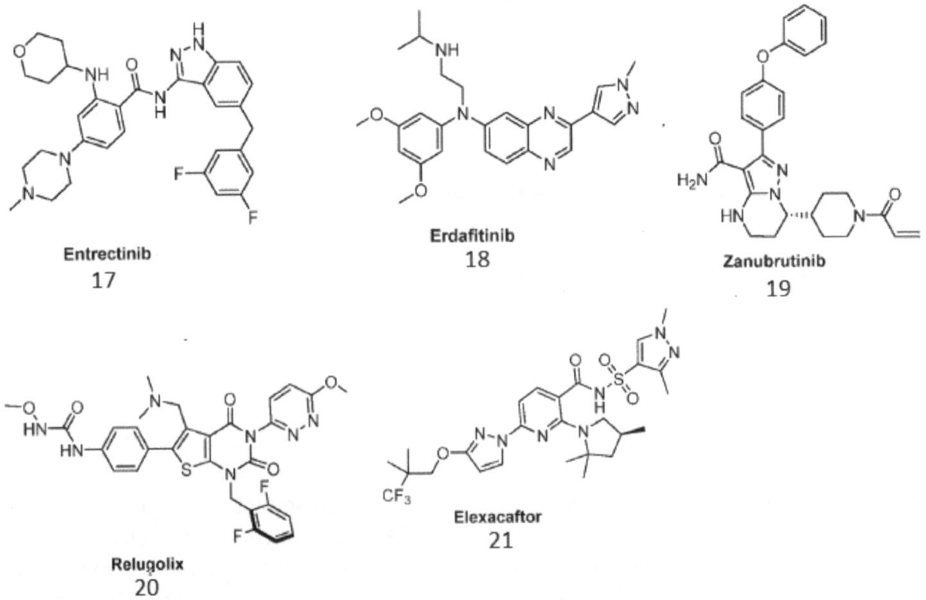

Figure 3.
Chemical structures of studied New Chemical Entities (NCEs) (17-21).

physical pH condition several functional groups undergo ionization. The weak ion—dipole and dipole–dipole interactions both plays significant role in drug-receptor binding. Weak interactions include hydrogen bonds, hydrophobic interactions and charge transfer which exist between drug and receptor to provide stability to the drug-receptor binding. DFT is utilized to understand the reaction mechanisms of the drug molecule. Various Computational tools are used to precisely calculate the transition state for drug-target complexes. Dipole moment (DM) is also an important parameter which is used to explain observable chemical and physical properties of drug molecules. DM is used to assess cell permeability and oral bioavailability of drugs as complexes with large dipole moment are more soluble in water and less likely to be absorbed through lipophilic membranes

[24, 25]. DM is included as a highly relevant descriptor in explaining the catalytic activity of enzymes in Quantitative Structure–Activity Relationships (QSAR) or Quantitative Structure–Property Relationships (QSPR) studies. For example: QSAR modeling of aromatase inhibition [26], antifungal activity [27], and HIV-1 protease/ cyclin-dependent kinases inhibition [28], QSAR modeling of aromatase inhibition [26], antifungal activity [27], and HIV-1 protease/ cyclin-dependent kinases inhibition [28], and in the estimation of micellar properties such as drug loading capacity (LC) [29] in QSPR model. Since DFT calculations are computationally too demanding for most large-scale virtual screening explorations, or for the incorporation in fast QSAR or QSPR models, so the relevant parameters are calculated by empirical or machine learning (ML) methods.

ML from data precalculated by DFT has emerged as a successful method for drugs as the results are highly accurate and has higher speed compared to the previous approaches [30, 31]. A new class of atomistic simulation techniques combining machine learning (ML) with simulation methods based on quantum mechanical (QM) calculations has emerged in the last decades. These methods can dramatically increase the computational efficiency of QM-based simulations and enable to reach the large system sizes and long timescales required to access properties with relevance for drug industry. Machine learning tools provide early stage filtering to identify promising drug molecules for further screening by computationally more intensive methods. The core quantity of atomistic simulation is the Potential energy landscape (PES), a high dimension function which is the basic ingredient for Monte Carlo (MC) simulations. Simulation methods are more or less computationally efficient depending on the degree of physical approximation. Force fields as AMBER [32], CHARMM [33], GROMOS [34], and OPLS [35] are computationally very efficient since they employ simple pairwise interaction terms and fixed atomic charges for MM based methods in drug discovery pipeline [36]. QM simulations are fully reactive and can describe the complex bonding patterns, polarization effects and charge transfer processes that govern the behavior of biological systems [37].Various machine learning tools as artificial neural networks (ANN), support vector machines (SVM) and genetic programming have been explored to predict inhibitors, blockers, agonists, antagonists, activators and substrates of proteins related to specific therapeutic targets. These methods use screening compound libraries of diverse chemical structures, “noisy” and high-dimensional data to complement QSAR methods, and in cases of unavailable receptor 3D structure to complement structure-based methods. Several open access chemical spaces as, PubChem, ChemBank, DrugBank, and ChemDB are used in virtual screening of Drug molecules. DeepVS is used for docking of 40 receptors and 2950 ligands, showed exceptional performance when 95000 decoys were tested against these receptors [38]. In another study, multiobjective automated replacement algorithm is used to optimize the potency profile of a cyclin-dependent kinase-2 inhibitor by assessing its shape similarity, biochemical activity, and physicochemical properties [39]. GLORY, an innovative tool was used to predict the metabolism of molecules, identifying chemical structures of metabolites formed by cytochrome P450 enzyme family (CYPs) [40]. In another approach, drug combination synergy was used to exploit the largest available dataset reporting synergism of anticancer drugs (NCI-ALMANAC, with over 290,000 synergy determinations) [41].

As the vast chemical space comprising $>10^{60}$ molecules, fosters the development of a large number of drug molecules [42, 43], sometimes limits the drug development process, making it a time-consuming and highly expensive. So AI is used as it can recognize hit and lead compounds and provide a quicker validation of the drug target and optimization of the drug structure design [42–44]. See **Figure 4**. AI can

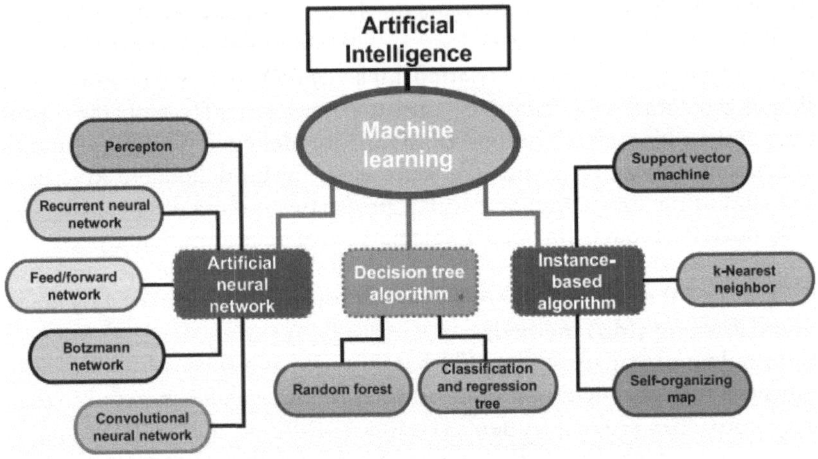

Figure 4.
Method domains of artificial intelligence (AI). Adopted from Ref. [44].

aid rational drug design [45]; assist in decision making; determine the right therapy for a patient, including personalized medicines; and manage the clinical data generated and use it for future drug development [46]. AI has done major contributions to the further incorporation of the developed drug in its correct dosage form as well as its optimization, in addition to aiding quick decision-making, leading to faster manufacturing of better-quality products. Robotic synthesis could eventually provide a fully automated drug discovery pipeline driven by AI [47, 48]. AI based approaches can also contribute to the safety and efficacy of the product in clinical trials as well as ensuring proper positioning and costing in the market through comprehensive market analysis and prediction.

3. Conclusions

The world of science has changed, and there is no question about it. The new model is for the data to be captured by instruments or generated by simulations before being processed by software and for the resulting information or knowledge to be stored in computers. The continued improvement of ML methods in chemistry, which compete with standard computational approaches and expertise are continuously developing the modern computational medicinal chemistry. Machine learning potentials are capable of carrying out high-throughput calculations in millisecond time scales with DFT accuracy and help to avoid false positives and false negatives. *De novo* molecular design are giving accurate predictions of lead compounds to target for simulation, effectively narrowing the search space for high-throughput screening applications. The advancement of AI along with its remarkable tools is continuously aims to reduce challenges faced by drug development process along with the overall lifecycle of the product as healthcare sector is facing several complex challenges, such as the increased cost of drugs and therapies, and society needs specific significant changes in this area. Though there are specific challenges remain with regards to the implementation of this technology, it is likely that AI will become an invaluable tool in the pharmaceutical industry in the near future. The vast knowledge of physics needs to be utilized to improve these advance techniques and tools while also not sacrificing speed and accuracy.

References

[1] Jorgensen WL (2004) The many roles of computation in drug discovery. Science 303(5665):1813. DOI: https://doi.org/10.1126/scien ce.1096361.

[2] Aminpour M, Montemagno C, Tuszynski JA (2019) An overview of molecular modeling for drug siscovery with specifc illustrative examples of applications. Molecules 24(9):1693. DOI:https://doi. org/10.3390/molecules24091693.

[3] Parr, R. G. and Yang, W. Density Functional Theory of Atoms and Molecules, Oxford University Press, Oxford, 1989.

[4] Kohn, W. Becke, A. D. and Parr, R. G. (1996) Density Functional Theory of Electronic Structure. J. Phys. Chem., 100, 12974-12980. DOI:10.1021/JP960669L.

[5] Ayers, P. W. and Yang, W. Density Functional Theory, in Computational Medicinal Chemistry for Drug Discovery, ed., P. Bultinck, H. de Winter, W. Langenaeker and J. P. Tollenaere, Dekker, New York, 2003, p. 571.

[6] Chattaraj, P. K. and Parr, R. G. Density Functional Theory of Chemical Hardness, in Chemical Hardness, Structure and Bonding, ed., K. D. Sen and D. M. P. Mingos, SpringerVerlag, Berlin, 1993, p. 11.

[7] Chattaraj, P. K. Poddar, A. and Maiti, B. Chemical Reactivity and Dynamics within a Density-based Quantum Mechanical Framework, in Reviews In Modern Quantum Chemistry: A Celebration of the Contributions of Robert Parr, ed., K. D. Sen, World Scientific, Singapore, 2002, p. 871.

[8] Chattaraj, P.K. Nath, S and Maiti, B. Reactivity Descriptors, in Computational Medicinal Chemistry for Drug Discovery, ed., J. Tollenaere, P. Bultinck, H. D. Winter and W. Langenaeker, Marcel Dekker, New York, 2003, p. 295.

[9] Smith JS, Roitberg, AE. and Isayev,O. (2018) Transforming Computational Drug Discovery with Machine Learning and AI. ACS Medicinal Chemistry Letters. 9 (11), 1065-1069. DOI: 10.1021/acsmedchemlett.8b00437.

[10] Butler, K. T.; Davies, D. W.; Cartwright, H.; Isayev, O.; Walsh, A. (2018) Machine Learning for Molecular and Materials Science. Nature 559 (7715), 547–555. DOI: https://doi.org/10.1038/s41586-018-0337-2.

[11] Parr, R. G., & Yang, W. (1984). Density functional approach to the Frontier electron theory of chemical reactivity. Journal of the American Chemical Society, 106 (14), 4049–4050. DOI:10.1021/ja00326a036.

[12] Morell, C., Grand, A., & Toro-Labb_e, A. (2005). New dual descriptor for chemical reactivity. The Journal of Physical Chemistry A, A109, 205–212. DOI:10.1021/jp046577a.

[13] Glossman-Mitnik, D. (2013). A comparison of the chemical reactivity of naringenin calculated with the M06 family of density functionals. Chemistry Central Journal, 7 (1), 155–161. DOI: 10.1186/1752-153X-7-155.

[14] Frau, J., & Glossman-Mitnik, D. (2017). A conceptual DFT study of the molecular properties of glycating carbonyl compounds. Chemistry Central Journal, 11 (1), 8–16. DOI: 10.1186/s13065-017-0239-7.

[15] Frau, J., & Glossman-Mitnik, D. (2018a). Conceptual DFT study of the local chemical reactivity of the dilysyldipyrrolones A and B intermediate melanoidins. Theoretical

Chemistry Accounts, 137 (5), 67. DOI: 10.1007/s00214-018-2244-x.

[16] Frau, J. & Glossman-Mitnik, D. (2018b). Conceptual DFT study of the local chemical reactivity of the colored BISARG melanoidin and its protonated derivative. Frontiers in Chemistry, 6, 136. DOI:10.3389/ fchem.2018.00136.

[17] Frau, J. Flores-Holgu_ın, N. & Glossman-Mitnik, D. (2018) Chemical Reactivity Theory (CRT) study of the melanoidin M8: Local conceptual density functional theory descriptors. Computational Molecular Bioscience, 08 (02), 80–90. DOI:10.4236/ cmb.2018.82004

[18] Frau, J., Flores-Holgu_ın, N., & Glossman-Mitnik, D. (2019). Conceptual density functional theory study of the chemical reactivity properties and bioactivity scores of the Leu-Enkephalin opioid peptide neurotransmitter. Computational Molecular Bioscience, 09 (01), 13–26. DOI: 10. 4236/cmb.2019.91002.

[19] Srivastava R. (2020) Chemical reactivity theory (CRT) study of small drug-like biologically active molecules, Journal of Biomolecular Structure and Dynamics, DOI: 10.1080/ 07391102.2020.1725642.

[20] Molinspiration Cheminformatics Nova ulica SK-900 26 Slovensky Grob Slovak Republic. https://molinspiration. com/about.html. (free online server).

[21] OSIRIS Property Explorer, https:// www.rdchemicals.com/drug-relevant-properties.html. (free online server)

[22] Andrew C. Flick, Carolyn A. Leverett, Hong X. Ding, Emma McInturff, Sarah J. Fink, Subham Mahapatra, Daniel W. Carney, Erick A. Lindsey, Jacob C. DeForest, Scott P. France, Simon Berritt, Simone V. Bigi-Botterill, Tony S. Gibson, Yiyang Liu, and Christopher J. O'Donnell (2021) Synthetic Approaches to the New Drugs Approved during 2019, JMC. DOI: https://doi.org/10.1021/acs.jmedchem. 1c00208.

[23] Srivastava R, CDFT and toxicity evaluation of 22 New Drugs Approved during 2019. (manuscript under submission).

[24] Ioakimidis L, Thoukydidis L, Mirza A, Naeem S, Reynisson J (2008) Bench- marking the reliability of QIKPROP. Correlation between experimental and predicted values. QSAR Comb Sci 27(4):445–456. DOI: https://doi.org/10.1002/qsar. 200730051.

[25] Matuszek AM, Reynisson J (2016) Defning known drug space using DFT. Mol Inf 35(2):46–53. DOI: https://doi. org/10.1002/minf.201500105.

[26] Adhikari N, Amin SA, Saha A, Jha T (2017) Combating breast cancer with non-steroidal aromatase inhibitors (NSAIS): understanding the chemicobiological interactions through comparative SAR/QSAR study. Eur J Med Chem 137:365–438. DOI: 10.1016/j. ejmech.2017.05.041.

[27] Wang D, Wu Y, Wang L, Feng J, Zhang X (2017) Design, synthesis and evaluation of 3-arylidene azetidin-2-ones as potential antifungal agents against Alternaria solani Sorauer. Bioorg Med Chem 25(24):6661–6673. DOI: 10.1016/j.bmc.2017.11.003.

[28] Fong CW (2016) The efect of desolvation on the binding of inhibitors to HIV-1 protease and cyclin-dependent kinases: causes of resistance. Bioorg Med Chem Lett 26(15):3705–3713. DOI: https://doi.org/10.1016/j.bmc l.2016.05.080.

[29] Wu W, Zhang R, Peng S, Li X, Zhang L (2016) QSPR between molecular structures of polymers and micellar properties based on block unit

auto- correlation (BUA) descriptors. Chemom Intell Lab Syst 157:7–15. DOI: https://doi.org/10.1016/j.chemolab. 2016.06.011.

[30] Jorgensen WL, Tirado-Rives J (1988) The OPLS [optimized potentials for liquid simulations] potential functions for proteins, energy minimizations for crystals of cyclic peptides and crambin. J Am Chem Soc 110(6):1657. DOI: https://doi.org/ 10.1021/ja002 14a001.

[31] Ganesan A, Coote ML, Barakat K (2017) Molecular dynamics-driven drug discovery: leaping forward with confdence. Drug Discov Today 22(2): 249. DOI: https://doi.org/10.1016/j. drudi s.2016.11.001.

[32] Cornell WD, Cieplak P, Bayly CI, Gould IR, Merz KM, Ferguson DM, Spellmeyer DC, Fox T, Caldwell JW, Kollman PA (1995) A Second Generation Force Field for the Simulation of Proteins, Nucleic Acids, and Organic Molecules. J Am Chem Soc 117(19):5179-5197. DOI: https://doi.org/ 10.1021/ja00124a002.

[33] MacKerell AD, Banavali N, Foloppe N (2000) Development and current status of the CHARMM force field for nucleic acids. Biopolymers 56(4):257-265. DOI: https://doi.org/10.1002/1097- 0282 (2000)56:43.0.CO, 2-W.

[34] Oostenbrink C, Villa A, Mark AE, Gunsteren WFV (2004) A biomolecular force field based on the free enthalpy of hydration and solvation: the GROMOS force-feld parameter sets 53A5 and 53A6. J Comput Chem 25(13): 1656-1676. DOI: https://doi.org/ 10.1002/ jcc.20090.

[35] Jones JE (1924) On the determination of molecular felds—II. From the equation of state of a gas. Proc R Soc Lond Ser A 106(738): 441-463. DOI: https://doi.org/10.1098/ rspa.1924.0082.

[36] Mortier J, Rakers C, Bermudez M, Murgueitio MS, Riniker S, Wolber G (2015) The impact of molecular dynamics on drug design: applications for the characterization of ligand-macromolecule complexes. Drug Discov Today 20(6):686-702. DOI: https://doi. org/10.1016/j.drudis.2015.01.003.

[37] Carloni P, Rothlisberger U, Parrinello M (2002) The role and perspective of Ab initio molecular dynamics in the study of biological systems. Acc Chem Res 35(6):455-64. DOI: https://doi.org/10.1021/ ar010018u.

[38] Pereira, J.C. et al. (2016) Boosting docking-based virtual screening with deep learning. J. Chem. Inf. Model. 56, 2495–2506. DOI: 10.1021/acs. jcim.6b00355.

[39] Firth, N.C. et al. (2015) MOARF, an integrated workflow for multiobjective optimization: implementation, synthesis, and biological evaluation. J. Chem. Inf. Model. 55, 1169–118. DOI: https://doi. org/10.1021/acs.jcim.5b00073.

[40] de Bruyn Kops Christina, Stork Conrad, Šícho Martin, Kochev Nikolay, Svozil Daniel, Jeliazkova Nina, Kirchmair Johannes. (2019) GLORY: Generator of the Structures of Likely Cytochrome P450 Metabolites Based on Predicted Sites of Metabolism. 7, 402. DOI: https://doi.org/10.3389/fche m.2019.00402.

[41] Sidorov P, Naulaerts S, Ariey-Bonnet J, Pasquier E, Ballester PJ. (2019) Predicting Synergism of Cancer Drug Combinations Using NCI-ALMANAC Data. Front Chem. 7:509. DOI: 10.3389/fchem.2019.00509.

[42] Mak, K.-K. and Pichika, M.R. (2019) Artificial intelligence in drug development: present status and future prospects. Drug Discovery Today 24, 773–780. DOI: 10.1016/j. drudis.2018.11.014.

[43] Sellwood, M.A. et al. (2018) Artificial intelligence in drug discovery. Fut. Sci. 10, 2025–2028. DOI: 10.4155/fmc-2018-0158.

[44] Paulz, D. Sanapz, G. Shenoyz, S. Kalyane, D. Kalia, K. and Tekade, R.K. (2021) Artificial intelligence in drug discovery and development. Drug Discovery Today, 26, 80-93. DOI: 10.1016/j.drudis.2020.10.010.

[45] Duch, W. et al. (2007) Artificial intelligence approaches for rational drug design and discovery. Curr. Pharm. Des. 13, 1497–1508. **DOI:** 10.2174/138161207780765954.

[46] Blasiak, A. et al. (2020) CURATE.AI: optimizing personalized medicine with artificial intelligence. SLAS Technol. 25, 95–105. DOI: 10.1177/2472630319890316.

[47] Dobchev DA, Pillai GG, Karelson M. (2014) *In silico* machine learning methods in drug development. Curr Top Med Chem. 14(16):1913-22. DOI: 10.2174/1568026614666140929124203.

[48] Granda, J. M.; Donina, L.; Dragone, V.; Long, D.-L.; Cronin, L. Controlling an Organic Synthesis Robot with Machine Learning to Search for New Reactivity. Nature 559 (7714), 377–381. DOI: 10.1038/s41586-018-0307-8.

Chapter 6

Ferromagnetism in Mn and Fe Doped LuN: A Potential Candidate for Spintronic Application

Abstract

Diluted magnetic semiconductor (DMS) materials have gained a lot of attention in the last decade due to their possible use in spintronics. In this chapter, the effect of transition metal (TM) i.e., Mn and Fe doping on the structural, electronic, magnetic as well as optical properties of pure and doped LuN has been presented from the first principles density functional theory (DFT) calculation with the Perdew-Burke-Ernzerhof-generalized gradient approximation (PBE-GGA) and Tran Blaha modified Becke-Johnson potential (TB-mBJ) as correlation potentials. The predicted Curie temperature is expected to be greater than room temperature in order to better understand the ferromagnetic phase stability, which has also been confirmed through the formation and cohesive energies. The calculated lattice constants for perfect LuN (rock-salt structure) are in good agreement with the experimental values. Interestingly, doping of Mn and Fe on pure LuN displays indirect band gap to a direct band gap with half metallic and metallic character. The detailed analyses combined with density of state calculations support the assignment that the Half-magnetism and magnetism are closely related to the impurity band at the origin of the hybridization of transition states in the Mn-doped LuN. Absorption spectra are blue shifted upon increase in dopant contents and absorption peaks are more pronounced in UV region. The refractive index and dielectric constant show increase in comparison to the pure LuN. According to the Penn's model, the predicted band gaps and static actual dielectric constants vary. These band gaps are in the near visible and ultraviolet ranges, as well as the $Lu_{0.75}TM_{0.25}N$ (TM = Fe, Mn) materials could be considered possible candidates for the production of optoelectronic, photonic, and spintronic devices in the future.

Keywords: density functional theory, spintronics, LuN, doped nitrides, electronic structure, magnetism, optical properties

1. Introduction

Over the past few years, half metallic ferromagnetic materials have gained a lot of interest due to their possible use in spintronic and optoelectronic applications [1, 2]. Every ferromagnetic half-metal consists of two spin versions, one spin is a semiconductor or insulator, while the other is metallic. The rare earth elements, which have atomic numbers from Z = 57 (Lanthanum, La) to Z = 71 (Lutetium, Lu), the electronic configuration of these elements is like [Xe] $6s^2 4f^n$, where n is zero for La and 14 for Lu [3, 4]. These elements have large orbitals and spin magnetic moments due to their partially filled 4f shells. The rare earth (RE) elements form wide variety of stable nitrides, monopnictide and monochalcogenide compounds such as LaN, CeN, EuN, GdN, DyN, HoN, LuN, YN, LuP, LuAs, LuBi, $La_2Fe_2S_5$, $HoAgS_2$, $GdAgS_2$, $YbAgS_2$, etc. Out of these different types of rare earth compounds, rare earth nitrides, attracted immense attention to the condensed matter physicists, material scientists in the last two decades due to their diverse structural, mechanical, magnetic, optical, electronic, and thermal properties [5–18]. The structural properties and structural phase transition at high pressure have been investigated for the rare earth nitrides by a number of groups [7–13]. The thin film of rare earth nitrides can be utilized in spintronic devices and electronic nano-devices [13–16]. The experimental research is being conducted for the realization of rare earth nitrides [7]. Being in the first and last elements of the rare earth series LaN and LuN respectively, theoretical investigation of these two nitrides gives opportunity to probe the closed shell and empty shell 4f electronic properties. Though the effect of rare earth doping in III-Nitrides has been investigated extensively, there are few reports on the investigation of the electronic, magnetic, optical properties available in the literature [19, 20].

In this chapter, we present a detailed theoretical investigation on structure stability, formation energy, electronic band structure, magnetic and optical properties for LuN using a density functional theory (DFT) approach. Most importantly, the effect of Mn and Fe doping on LuN has also been investigated vividly for their optoelectronic applications.

2. Computational method

We have performed the electronic structure calculations in the framework of full potential linearized augmented plane wave (FP-LAPW) method implemented on Wien2k [21] code employed to solve Kohn sham equation based on the density functional theory (DFT) [22, 23]. The treatment of the exchange correlation potential is done within the generalized gradient approximation (GGA) with the Perdew–Burke Ernerhof scheme [24]. Furthermore, in the process of determining the most accurate electronic band characteristics, especially the band gap E_g, which is compatible with the experiment, modified Becke and Johnson (mBJ) [25] potential is also used. The relativistic effects are taken into account. In the interstitial area, the plane wave sets were used outside the muffin tin sphere and the $R_{MT}K_{max}$ = 7.00 cutoff value was used but the spherical harmonic expansion was used within the sphere. For the incorporation of the irreducible Brillouin field, 64 k-points mesh was used. The muffin tin radii were chosen, respectively. The separation of the core and valence states in terms of the energy cut-off parameter was adjusted to −6.0 Ry to avoid charge leakage. The 1000 k-points were used for reciprocal space sampling. A series of repeated iterations were used to achieve energy convergence up to 0.01 mRy.

3. Results and discussion

3.1 Structural phase stability

To study the electronic structure and magnetic properties of TM doped LuN, a $1 \times 1 \times 1$ super cell of LuN containing 8 atoms was constructed. **Figure 1a–c** shows

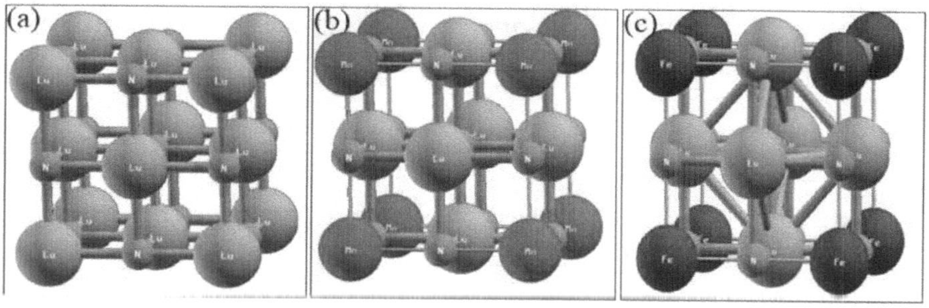

Figure 1.
The ball-and-stick model (rock-salt structure) for (a) intrinsic LuN, (b) Mn-doped LuN and (c) Mn-doped LuN unit cell.

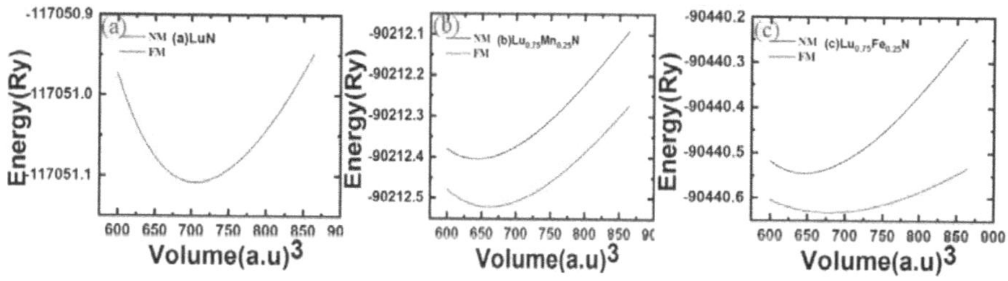

Figure 2.
Total energy vs. unit cell volume for the (a) LuN, (b) 25% Mn doped LuN (c) 25% Fe doped LuN. The plot indicates that the 25% Mn and Fe doped LuN possess lower energy in the ferromagnetic phase. The total energy calculations are obtained using PBE-GGA pure and doped LuN.

the ball-and-stickmodel for the intrinsic LuN, 25% Mn and Fe-doped LuN, respectively. We have considered the rocksalt crystal structure with space group Fm-3 m for the LuN unit cell which is the most stable structure of the bulk LuN. The corner atoms are the lutetium (Lu) atoms surrounded by six nitrogen (N) atoms as shown in **Figure 1a**. In **Figure 1b**, the Mn-doped LuN unit cell has been illustrated. The corner lattice sites have been doped with Mn atoms. In **Figure 1c**, the corner Lu atoms of the rock salt structure have been replaced by Fe atoms for achieving 25% Fe doping of LuN. The equilibrium structure of the pure LuN has been determined by the minimization of total energy vs. unit cell volume using generalized gradient approximation (GGA) method. **Figure 2a–c**, shows the total energy vs. unit cell volume plot for the intrinsic LuN, 25% Mn-doped LuN and 25% Fe-doped LuN, respectively. From the total energy minimization calculation, we have obtained the lattice parameter of the intrinsic LuN as 4.7104 Å. This lattice is very close to the experimentally measured value of 4.766 Å [26]. The calculated lattice parameter is also consistent with the theoretically calculated values by other groups [27, 28]. The equilibrium lattice constant for the 25% Mn-doped LuN in FM phase is calculated to 4.6094 Å and in the NM phase is 4.5674 Å. The equilibrium lattice constant for the 25% Fe-doped LuN in ferromagnetic (FM) phase is calculate to 4.5924 Å and in the non-magnetic (NM) phase is 4.5728 Å. In order to get the equilibrium lattice parameters of the Mn-Fe doped LuN, the total energy vs. unit cell volume size are also calculated by considering the Birch Murnaghan' equation method. The plots for the total energy vs. unit cell volume size are shown in **Figure 2b** and **(c)**, respectively. We found that the value of lattice parameters, and volumes decrease due to incorporation of 3d-TM doping in the pure LuN. It is worth mentioning that both the 25% Mn and Fe-doped LuN are found total energy minimization in two phases i.e. ferromagnetic phase and non-magnetic phase. The values of the optimized lattice parameter, volume, bulk modulus, pressure derivative of bulk modulus

Properties		**LuN**	**$Lu_{0.75}Mn_{0.25}N$**	**$Lu_{0.75}Fe_{0.25}N$**
Lattice constant, a (Å)	FM	4.7104	4.6094	4.5924
	NM		4.5674	4.5728
Other theory		4.77 [27] 4.76 [28]		
Experiment		4.76 [26]		
Volume, V (Å^3)	FM	705.2909	660.8789	653.5967
	NM		642.9726	645.2908
Bulk modulus, B (GPa)	FM	186.9883	188.7961	189.2357
	NM		206.7191	202.4745
Pressure derivative of B, B′	FM	4.2116	4.3813	4.2797
	NM		4.4348	4.4452
Total optimized energy (Ry)	FM	−117051.107870	−90212.521303	−90440.631938
	NM		−90212.403516	−90440.544004
T_c(K)		—	747	909
Bond length (Å)		Lu-N = 2.20	Mn-N = 2.3047 N-Lu = 2.3047 Mn-Lu = 3.2593	Fe-N = 2.296 N-Lu = 2.296 Fe-Lu = 3.247
Formation energy (eV/atom)		−5.36	−8.540	−8.162
Band gap, E_g (eV)		0.93	Spin up↑ = 0.43, Spin down = No gap	Spin up↑ = 1.79, Spin down = No gap
Other theory & exp.		1.173 [27]		
Exp.		1.55,1.6 [26]		
Mgnetic Moment (μ_B)		—	Int = 0.08801 Mn = 4.00700 Lu = 2.589 N = −0.1209 Tot = 4.00015	Int = −0.09335 Fe = 3.94823 Lu = −0.1032 N = 1.05447 Tot = 4.8984
Static dielectric constant $\varepsilon_1(0)$		6.40	12.69	19.99
n(0)		2.6	3.2	4.25
R(0)		0.19	0.32	0.423

Table 1.
The ground state properties, bond length (Å), formation energy (eV), band gap (eV), magnetic moment (μ_B) and optical parameters for the pure LuN, and $Lu_{0.75}TM_{0.25}N$ (TM = Fe, Mn).

and total energy of the pure LuN, Mn and Fe doped LuN in nonmagnetic and ferromagnetic phases are presented in **Table 1**.

We have used a relaxed structure in both the FM and NM phases to determine the ground-state energies of Mn and Fe-doped LuN to highlight the actual stable phase. The positive value of E ($E = E_{NM} - E_{FM}$) indicates that the FM state is more stable (see **Table 1**). The formation energies for LuN, $Lu_{0.75}Mn_{0.25}N$ and $Lu_{0.75}Fe_{0.25}N$ obtained from calculation are to −5.36 eV, −8.540 eV and −8.162 eV, respectively. The negative sign of the formation energy indicates that during creation of the compounds, energy release revealing the stability in the FM state. The cohesive energy (E_{coh}) of $Lu_{0.75}TM_{0.25}N$ (TM = Mn, Fe) compounds are responsible for holding the atoms together in the crystal structure, further confirms the stability of the compounds. The values of E_{coh} are calculated to −2.89Ry, −3.43 Ry and −3.44 Ry for pure and TM-doped LuN (TM = Mn, Fe). The computed cohesive energies exhibit higher values than their binaries, which again imply FM stability.

After geometry optimization the structure around the TM dopant is slightly suppressed with Lu atoms drawn closer to the TM the relaxed TM-Lu bond length is less than that of the N-Lu bond.

3.2 Electronic properties

The spin polarized band structures of pure LuN, Mn and Fe (25%)-doped LuN are presented in **Figure 3a–f**. An Indirect band gap to direct band gap transition in the band structure of the pure LuN is clearly observed for 25% Mn and Fe substitution at the Lu sites. In the case of pure LuN, the highest occupied state of the electrons occurs at 'Γ' point, whereas the lowest empty state of the electrons occurs at the 'X' point in the band structure. The indirect band gap of 0.93 eV is obtained from the calculation like that seen experimentally, as shown in **Table 1**. It is to be mentioned here that there is no distinguishable difference observed in the up (↑) spin and down (↓) spin bands as can be seen in **Figure 3a** and **b**. A slight overlap of the spin up (↑) band for the Mn doped LuN whereas total spin down (↓) band exhibits an direct energy gap as can be seen in the **Figure 3c** and **d**. The 25% Mn doped LuN behaves as a half-metallic ferromagnet. In case of Fe doped LuN there is a distinct energy gap in the total spin up (↑) band whereas there is a significant overlapping in the total spin down (↓) band as a result, **Figure 3e** and **f** show that the Fe doped LuN compound is magnetic compound.

The total spin polarized density of states (DOS) and partial density of states (PDOS) obtained using the mBJ-GGA potential, for the pure LuN, Mn and Fe doped LuN as well as spin density of states for the individual atoms in the unit cell are plotted in **Figure 4a–c**. From **Figure 4a**, it is clear that the total spin polarized DOS of the up and down spin states for the Lu atom (red) and N atom (blue) are identical. From the partial spin polarized DOS plots for Lu and N atom, it can be seen that the spin up (↑) band for Lu and N atoms are just symmetric to that of the spin down (↓) band. As a results total (black line) spin polarized spin up (↑) band is symmetric to that of the spin down (↓) band in total for LuN unit cell and the pure LuN exhibits the non-magnetic electronic character with an energy band gap. But from the partial spin density of state plot in **Figure 4b**, it can be seen that the spin down (↓) density of state and spin up (↑) density of state are no more symmetric upon 25% Mn doping in LuN. The spin down (↓) density of state of the Mn atom (green line) in $Lu_{0.75}Mn_{0.25}N$ is completely different in both conduction band and valance band. Due to the delocalized character of the 'd' and 's' electrons in Mn electron density of states appears from nearly 0.1 eV in the conduction band. The peak of the spin down (↓) DOS occurs at 1.2 eV. But there is no contribution to the spin up (↑) density of states of the Mn atom (green line) in $Lu_{0.75}Mn_{0.25}N$ in the conduction band. There is slightly high spin up (↑) DOS compared to that in the spin down (↓) DOS from the Mn atom (green line) in the valance band. Also the contribution is much less compared to that of the Lu atom (red line). The electronic up spin (↑) DOS from the N atom is asymmetric to that of the down spin DOS and it connects the up spin DOS of conduction band to that of the valance band. As a result, there is no gap in the total (black line) up spin DOS of the band structure of the $Lu_{0.75}Mn_{0.25}N$, but there is a gap in the total (black line) spin down DOS of the $Lu_{0.75}Mn_{0.25}N$. Thus $Lu_{0.75}Mn_{0.25}N$ exhibits a half-metallic electronic bahaviour. Due to the nature of band filling in minority spin states and majority spin states, $Lu_{0.75}Fe_{0.25}N$ is metallic nature. In metallic Lu0.75Fe0.25N, the asymmetry and occupancy of d states of Fe in majority states and p states of N in minority states is evident, led to a significant magnetic moment of Fe.

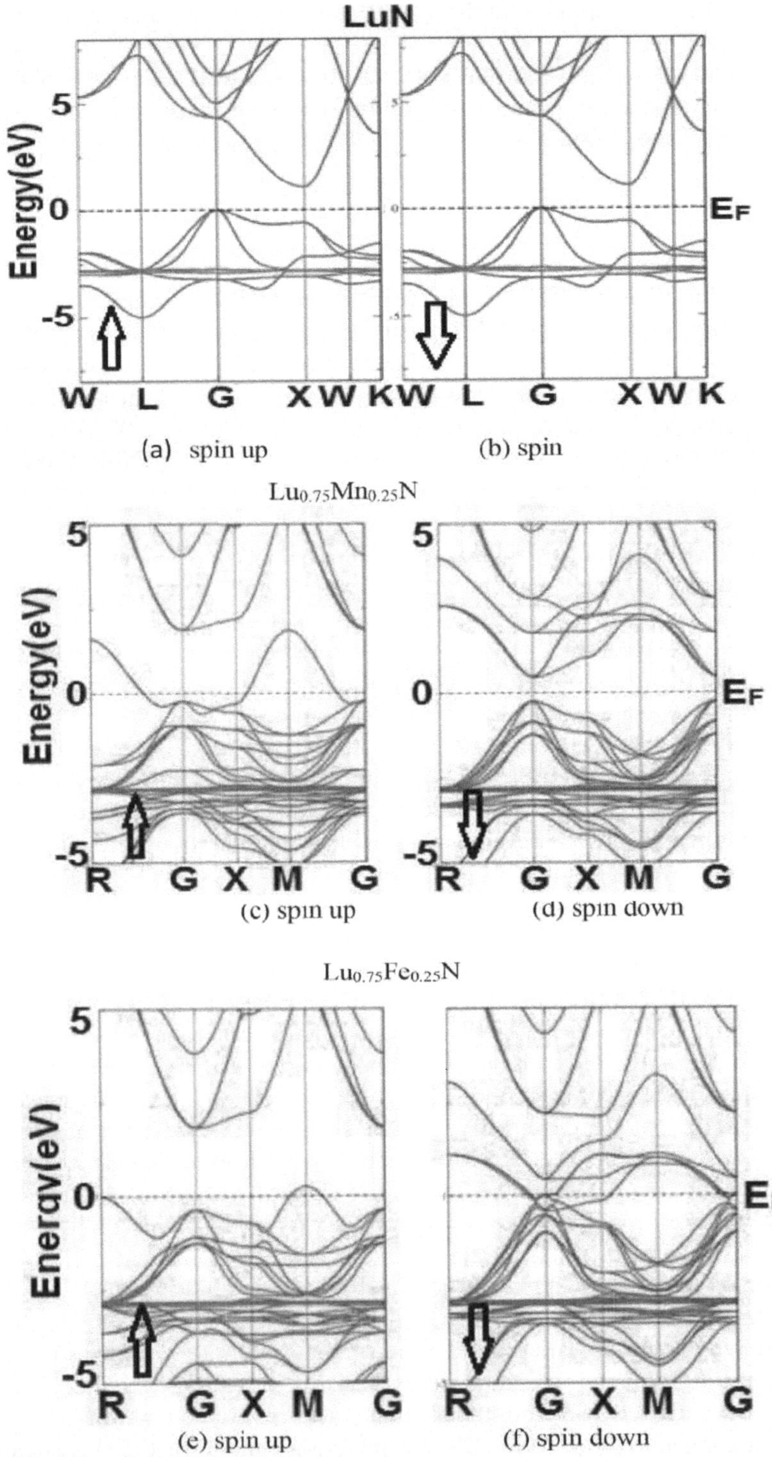

Figure 3.
Band structures of $Lu_{1-x}TM_xN$ ($x = 0.25$, TM = Mn, Fe) along the high symmetry directions at their equilibrium lattice constants, estimated using mBJ-GGA in figure (a) for $x = 0.0$, (b) for TM = Mn, $x = 0.25$ and (c) for TM = Fe, $x = 0.25$.

In Mn-Fe-doped LuN, the interaction arising between 3d-state of Mn/Fe impurities and p-states of host lattice N anions, causes the appearance of the localized states within the band gap at E_F, which induce half-metallic ferromagnetic

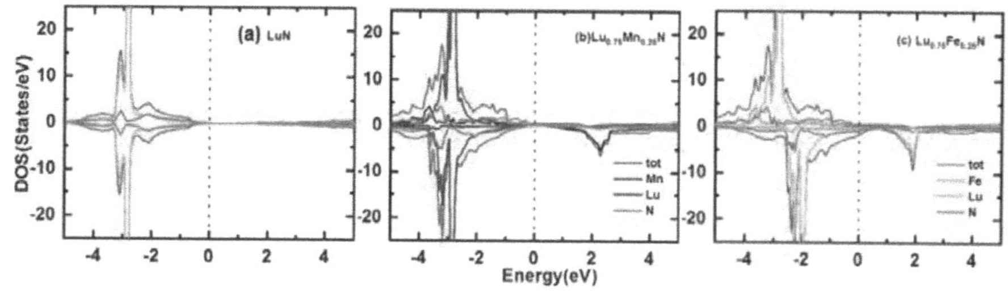

Figure 4.
Spin-polarized total and partial DOS for $Lu_{1-x}TM_xN$ (x=0.25, TM=Mn, Fe) obtained using the mBJ-GGA potential (a) for x= 0.0, (b) for TM= Mn, x= 0.25 and (c) for TM= Fe, x= 0.25.

characteristics. Therefore, the large splitting of t_{2g} states causes the double exchange mechanism that is responsible for producing ferromagnetism in Mn doped LuN half-metallic semiconductors. The investigated compounds demonstrate a strong p–d exchange interaction within the valence band of LuN, resulting in p-type carrier induced ferromagnetism.

3.3 Magnetic property

The calculated total and partial magnetic moments of individual elements of $Lu_{0.75}TM_{0.25}N$ are reported in **Table 1**. Dopant atoms Mn and Fe are the leading source of the magnetization in $Lu_{0.75}TM_{0.25}N$ (TM = Mn, Fe) as their magnetic moments are $4.00\mu_B$ and $4.89\mu_B$, respectively. In $Lu_{0.75}Mn_{0.25}N$ compound some N atoms provide negative values of magnetic moment indicating anti-parallel magnetic moment to the Mn and Lu atoms as their magnetic moment values are positive. As the interstitial site holds a positive value of magnetic moment equal to $0.08801\mu_B$, it adds its contribution to the total magnetic moment of the super cell. But in case of $Lu_{0.75}Fe_{0.25}N$ compound interstitial site holds a negative value of magnetic moment which indicates its anti-parallel effect to parallel magnetic moment containing atoms. 3 atoms of Lu contain negative value of magnetic moment which results anti-parallel magnetic moment in the cell. Therefore, it could be said that the atoms that hold negative and positive values of magnetic moment are interacting anti-ferromagnetically. In comparison, there is a difference in total magnetic moments of the two supercells due to the fact that Mn atom has greater magnetic moment value than Fe atom. Also in Mn doped supercell, only N atoms are anti-parallel but in Fe doped super cell Lu atoms, as well as interstitial site magnetic moment values are anti-parallel.

Moreover, Heisenberg model has been employed to estimate the Curie temperature (T_c) of $Lu_{1-x}TM_xN$ compounds by using the expression $Tc = 2\Delta E/3xK_B$ [29], where x represents impurity cation concentration and K_B shows Boltzmann constant. The estimated T_c values for Mn and Fe-doped LuN, as listed in **Table 1**, have been observed to exhibit highest and lowest T_c, respectively. However, all the $Lu_{0.75}TM_{0.25}N$ compounds have shown T_c above room temperature (RT). The decrease in magnetic moment per magnetic ion from $4.00\mu_B$ to $3.94\mu_B$ is caused by changing the dopant from Mn to Fe, which affects the size of the exchange interactions and thus the related T_c. As a result, all of the compounds investigated can be employed in various magnetic devices that operate above room temperature.

Therefore, the studied DMSs are most suitable for spintronic device applications, which also indicate that experimental material properties can efficiently be simulated by employing theoretical methods.

3.4 Optical properties

Motivated by the prospect of using its interesting electronic structure for optoelectronic semiconductor applications, the optical and electronic transport properties of the pure LuN, $Lu_{0.75}Mn_{0.25}N$ and $Lu_{0.75}Fe_{0.25}N$ were calculated. The frequency dependent optical properties were studied and calculated through the dielectric functions. It was realized that the optical properties of a solid can be portrayed by the complex dielectric function ε(ω), with two parts, real and imaginary as [30].

$$\varepsilon(\omega)_1(\omega)_2(\omega). \tag{1}$$

In this chapter, we focused on the optical parameters such as real part, $\varepsilon_1(\omega)$ and imaginary part, $\varepsilon_2(\omega)$ of the dielectric functions, refractive indices n(ω), extinction coefficient k(ω), absorption coefficient α(ω), reflectivity R(ω), optical conductivity σ(ω) and electron energy function L(ω) of cubic LuN and $Lu_{1-x}TM_xN$ (x = 0.25, TM = Mn, Fe) using the mBJ-GGA. **Figure 5a–h** show obtained results from the mBJ-GGA method.

The complex dielectric function $\varepsilon(\omega)$ encompasses both intraband and interband transitions. It is observed that intraband attribute mainly in the case of metal or metal

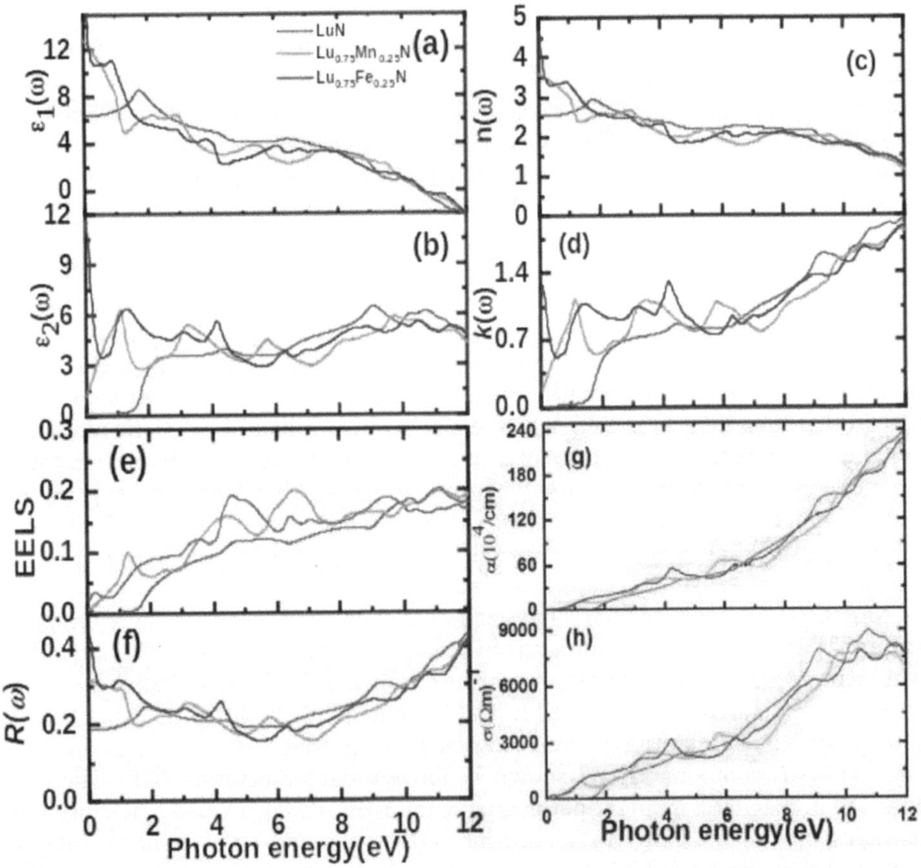

Figure 5.
Spectra of the (a) real $\varepsilon_1(\omega)$ and (b) imaginary $\varepsilon_2(\omega)$ dielectric functions, (c) refractive indices $n(\omega)$, (d) extinction coefficient $k(\omega)$, (e) Absorption coefficient $\alpha(\omega)$, (f) reflectivity $R(\omega)$, (g) optical conductivity $\sigma(\omega)$ and (h) electron energy function $L(\omega)$ of $n(\omega)$ cubic $Lu_{1-x}TM_xN$ (x = 0.25,TM = Mn, Fe) obtained using the mBJ-GGA method.

like conductors while in the case of insulating and semiconducting materials, direct as well as indirect transitions contribute to the interband transitions. In this study, we have not treated intraband excitations and indirect interband excitations, which usually contribute very little to $\varepsilon(\omega)$. The imaginary part $\varepsilon_2(\omega)$ can be obtained from the momentum matrix elements between the occupied and unoccupied wave functions by applying the selection rules, and is computed using the given expression

$$\varepsilon_2(\omega) = \frac{e^2\hbar}{\pi m^2\omega^2}\sum\int |M_{\nu,c}(k)|^2\delta[\omega_{\nu c}(k)-\omega]d^3k \tag{2}$$

whereas the real component $\varepsilon_1(\omega)$ can be derived from $\varepsilon_2(\omega)$ using Kramers-Kronig's relation with the help of following expression [31].

$$\varepsilon_1(\omega) = 1 + \frac{2}{\pi}P\int_0^\infty \frac{\omega'\varepsilon_2(\omega')d\omega'}{(\omega'^2-\omega^2)} \tag{3}$$

where 'P' represents the principal value of the given integral, 'e' is electronic charge and 'm' is the mass. Since the studied compounds crystallize in cubic phase, one dielectric tensor component is only required for the complete illustration of the optical properties. All other optical constants can be calculated using $\varepsilon_1(\omega)$ and $\varepsilon_2(\omega)$. We have predicted refractive index $n(\omega)$ from the computed dielectric function, using the following expression:

$$n(\omega) = \left[\frac{\sqrt{\varepsilon_1^2(\omega)+\varepsilon_2^2(\omega)}+\varepsilon_1(\omega)}{2}\right]^{\frac{1}{2}} \tag{4}$$

and, Kramers-Kronig equation has been used to obtain frequency dependent $k(\omega)$ using the following relation

$$k(\omega) = \frac{-2P}{\pi}\int_\infty^0 \frac{n(\omega')-1}{\omega'-\omega}d\omega' \tag{5}$$

Frequency dependent extinction function $k(\omega)$ can also be calculated from $\varepsilon_1(\omega)$ and $\varepsilon_2(\omega)$ using the expression below,

$$k(\omega) = \left[\frac{\sqrt{\varepsilon_1^2(\omega)+\varepsilon_2^2(\omega)}}{2}-\frac{\varepsilon_1(\omega)}{2}\right]^{\frac{1}{2}} \tag{6}$$

Similarly, the other frequency reliant optical functions such as absorption coefficient $\alpha(\omega)$, optical conductivity $\sigma(\omega)$ and reflectivity $R(\omega)$ can also be computed from the predicted real and imaginary parts of the complex dielectric function.

Using the above expressions these optical parameters are further plotted in **Figure 5. Figures 5(a)** and **(b)** represent the variation of real part and imaginary part of the complex dielectric function $\varepsilon(\omega)$ in the energy range 0–12 eV in the case of pure LuN and $Lu_{1-x}TM_xN$ (x = 0.25, TM = Mn, Fe). These dielectric plots depict material's capability to allow external electromagnetic field to penetrate through it when 25% concentration of Lu is subjected for doping with Mn and Fe. The dielectric trends depicted by all curves show similar behavior in the entire range except in the static values and in the high energy region, the characteristic distinctions with dielectric maxima at specific energies may correspond to various doping effects. As seen from **Figure 5(a)**, the static value of real part $\varepsilon_1(0)$ is approximately 6.40 for LuN, which drastically increased to 12.69 for Mn-doped LuN and 19.99 for Fe-doped LuN. The real part $\varepsilon_1(\omega)$ of pure LuN has a peak at 2 eV, whereas the plots correspond to Mn-doped LuN and Fe-doped LuN show red shift in their dielectric function. After the dielectric maxima, the graph of all three investigated materials starts decreasing gradually and finally falls to zero in the higher energy region. The band gap of the examined materials and the static value $\varepsilon_1(0)$ computed from the n(0) by the relation $n(\omega) = \sqrt{\varepsilon_1(0)}$ spectra are exactly connected according to Penn's model [31]. The plasma frequency and the material band gap are represented by p and E_g, respectively. The imaginary part $\varepsilon_2(\omega)$ (see **Figure 5(b)**) of pure LuN has zero value up to 1 eV, after that a rapid rise in its value and increases gradually attaining a peak around 9 eV. While, Mn doped LuN has static value of $\varepsilon_2(0)$ as 1.4 and with Fe doping, it sharply rises to 12. The peaks present in the optical spectra of $\varepsilon_2(\omega)$ usually occur due to transitions of electrons from the valence to conduction band. The electronic band structures reveal the indirect band gap of 0.93 eV for LuN so the direct optical transitions occur between valence and conduction bands, whereas in the case of Mn doped LuN, it exhibits half-metallic nature, with direct band gap in minority spin channel, but Fe doped LuN shows metallic behavior in both spin channels.

Refractive index $n(\omega)$ is an important optical entity, which shows the response of the light in any type of materials. It is observed that the light depicts different properties upon interaction with different materials, i.e., light may be reflected, refracted, transmitted and diffracted depending on the properties of the materials on which it is incident. Since the velocity of the light varies as it traverses through different materials, hence variations can be observed in the refractive index of the materials for a range of optical spectrum. Overall similar behavior can be observed for refractive index as well as real part of the complex dielectric function $\varepsilon(\omega)$ as seen in **Figures 5(a)** and **(c)**. **Figure 5c** represents the optical spectral response of the refractive index of pure LuN and that of $Lu_{1-x}TM_xN$ (x = 0.25, TM = Mn, Fe). The high refractive index is presented in the infrared region when light rays traverse through any material. In our present calculations the static refractive index of LuN is 2.5, whereas Mn doped LuN has $n(0)$ as 3.5 and Fe doped LuN has 4.6, respectively, in the low energy region and decreases in high energy region.

It is found from **Figure 5b** and **d** that the optical spectral response of extinction coefficient $k(\omega)$ and imaginary part $\varepsilon_2(\omega)$ of dielectric constant illustrate similar profile. It can be seen that $k(\omega)$ also have some threshold energy values similar to imaginary part $\varepsilon_2(\omega)$ of the dielectric constant. It is found that the threshold values of incident photon energy for extinction coefficient k(ω) are approximately 1 eV for cubic LuN and 0 eV for $Lu_{1-x}TM_xN$ (x = 0.25, TM = Mn, Fe). The pure cubic LuN shows a smooth increase in the value of extinction coefficient over the studied photon energy range, whereas both doped LuN compounds show several minor peaks in the incident photon range. **Figure 5e**, represents electron energy loss spectroscopy (EELS) function $L(\omega)$ over energy range from 0 to 12 eV, an important parameter depicting the energy loss of fast-moving electron while traversing through the material. The peak associated with the plasma resonance and the corresponding frequency is called plasma frequency. The material exhibits dielectric nature if $\varepsilon_1(\omega) > 0$ and metallic nature if $\varepsilon_1(\omega) < 0$, above and below plasma frequency. **Figure 5f** represents reflectivity spectra of cubic LuN and $Lu_{1-x}TM_xN$ (x = 0.25, TM = Mn, Fe) over an energy range up to 12 eV. The frequency or wavelength dependent reflectivity optical spectrum of the studied pure LuN has a static reflectivity $R(0)$ value as ~0.2 up to 1.5 eV, afterwards it shows slight decrease in its value at 6 eV and then gradually increases in the high energy region. With doping of Fe in LuN, it is observed from **Figure 5f** that static reflectivity $R(0)$ has increased to 0.31 and that of Mn shows that its value rose to 0.45. These results reveal that doping with Fe and Mn has increased the reflectivity of these compounds in the low energy region, which makes them potential candidates of IR reflectors.

Figure 5(g) shows the optical spectral response of the absorption coefficient of LuN and $Lu_{1-x}TM_xN$ (x = 0.25, TM = Mn, Fe) over an energy range up to 12 eV. The absorption coefficient provides information about how much incident light energy can be absorbed by the material, when it is exposed to the electromagnetic radiations. All three compounds show similar behavior and it displays the maximum absorption in the high energy region, which indicates that the absorption quality is good for these compounds in ultra violet region of the electromagnetic spectrum, predicting the usage of these materials as good UV absorbers.

When photons are incident on the material, the number of free charge carriers is increased due to absorption of the incident photons, which results in the rise of electrical conductivity, and also results in enhanced photoconductivity. The incident photons should possess adequate energy to excite electrons from the valence band, helping them to cross the band gap and reach the conduction band in the material. **Figure 5(h)** shows the photoconductivity spectrum of the cubic LuN and $Lu_{1-x}TM_xN$ (x = 0.25, TM = Mn, Fe). It is seen from **Figure 5(h)** that the photo conductivity of LuN starts 1.8 eV, clearly depicting its semiconducting nature whereas with doping, the photoconductivity starts with zero photon energy due to their metallic nature. The optical conductivity of pure as well as doped compounds increases due to the absorption of photons. The optical conductivity spectra of $Lu_{1-x}TM_xN$ (x = 0.25, TM = Mn, Fe) have few maxima and minima peaks in the calculated energy range as also shown in **Figure 5(h)**. The optical spectra studies predicts that the doping of Fe and Mn in LuN makes them potential candidates of UV absorbers as well as IR reflectors, which may be used for fabrication of optical filters in the optoelectronic industry.

4. Conclusions

The structural, electronic, magnetic and optical properties of pure LuN, Mn/Fe doped LuN were investigated under equilibrium conditions by DFT using the PBE-GGA and mBJ-GGA potentials. The positive energy difference between NM and FM states has verified the stability of FM state. The stability of FM state has also been validated by TM-doped DMSs having a lower enthalpy of formation and a higher cohesive energy than the binary un-doped LuN host semiconductor. The Heisenberg classical model, which predicts above-room temperature ferromagnetism, has also predicted the Curie temperature. The calculated electronic properties showed pure LuN to be an indirect band gap semiconductor, and it transforms to a half-metal upon 25% Mn doping, while Fe doping its reveals magnetic nature. The findings presented in this work encourage further experimental research of the electronic structures of RE nitrides. Additionally, the optical characteristics of TM = Mn, Fe-doped LuN DMSs have been predicted in order to investigate future optical applications. The static value of dielectric constants and optical band gaps are observed to vary according to the Penn's model indicating the accuracy of the presented calculations. It has been noted from the imaginary part of the dielectric constant that the studied materials are red-shifted with maximum absorption in the visible as well as in the ultraviolet energy. Therefore, the studied compounds are the best candidates for optoelectronic and spintronic devices.

References

[1] Furdyna JK. Dilute magnetic semiconductors. J. Appl. Phys. 1988;**64**: R29. DOI: 10.1063/1.341700

[2] Ando K. Seeking room temperature ferromagnetic semiconductors. Science. 2006;**312**:183. DOI: 10.1126/science.1125461

[3] Mir SH, Jha PC, Islam MS, Banerjee A, Luo W, Dabhi SD, Jha PK, and Ahuja R. Static and dynamical properties of heavy actinide monopnictides of lutetium. Scientific Reports. 2016;**6**:29309. DOI: 10.1038/srep29309

[4] Natali F, Ruck BJ, Plank NOV, Trodhal HJ, Granville S, Meyer S, et al. Rare-earth mononitrides. Prog. Mater. Sci. 2013;**58**:1316-1360

[5] Klemm W, Winkelmann G. To know about the nitrides of the rare earth metals. Z. Anorg. Allg. Chem. 1956;**288**: 87. DOI: 10.1002/zaac.19562880112

[6] Larson P, Lambrecht WRL. Electronic structure of rare-earth nitrides using the LSDA+U approach: Importance of allowing 4f orbitals to break the cubic crystal symmetry. Phys. Rev. B. 2007;75:045114. DOI:10.1103/PhysRevB.75.045114

[7] Painter GS, Averill FW, Becher PF, Shibata N, Benthem KV, Pennycook SJ. First-principles study of rare earth adsorption at β-Si_3N_4 interfaces. Phys. Rev. B. 2008;78:21420. DOI:10.1103/PhysRevB.78.214206

[8] Sreeparvathy PC. Gudelli Vijay Kumar, Kanchana V, Vaitheeswaran G, Svane a, Christensen NE. Thermoelectric properties of binary LnN (ln = La and Lu). First principles study. AIP Conf. Proc. 2015;**1665** (110008). DOI: 10.1063/1.4918064

[9] Oualdine A, Bentouaf A, Chebli A, Nouamane B, Bouyakoub AZ, Aissa B. Structural, elastic, and electronic properties of CeN and LuN using: Ab initio study. Journal of Superconductivity and Novel Magnetism. 2018;**31**:3323-3330. DOI: 10.1007/s10948-018-4604-0

[10] Singh SK, Verma UP. Investigation of high pressure phase transition and electronic properties of lutetium nitride. Journal of Physics: Conference Series. 2015;**640**. DOI: 012029, 10.1088/1742-6596/640/1/012029

[11] Winiarski MJ, Kowalska D. Electronic structure of REN (RE = Sc, Y, La, and Lu) semiconductors by MBJLDA calculations. Mater. Res. Express 2019;6:095910. DOI:10.1088/2053-1591/ab31c2

[12] Vaitheeswaran G, Kanchana V, Rajagopalan M. Structural phase stability and superconductivity of LaN. Solid State Communications. 2002;**97–102**(2002):124. DOI: 10.1016/S0038-1098(02)00481-7

[13] Ludbrook BM, Farrell IL, Kuebel M, Ruck J, Preston ARH, Trodahl HJ, et al. Growth and properties of epitaxial GdN. J. Appl. Phys. 2009;**106**:063910-063914. DOI: 10.1063/1.3211290

[14] Griebel M, Smet JH, Driscoll DC, Kuhl J, Deiz CA, Freytag N, Kadow, Gossard AC, Klitzing KW. Tunable subpicosecond optoelectronic transduction in superlattices of self-assembled ErAs nanoislands. Nat. Mater. 2013;2:122–126. DOI:10.1038/nmat819

[15] Jha PK, Sanyal SP. Lattice vibrations in Yb-pnictide compounds. Phys. Rev. B. 1995;**52**:15898-15902. DOI: 10.1103/PhysRevB.52.15898

[16] Larson P, Lambrecht WRL, Chantis A, Van SM. Electronic structure of rare-earth nitrides using the LSDA+U

approach: Importance of allowing 4f orbitals to break the cubic crystal symmetry. Phys. Rev. B. 2007;7:1-14. DOI: 10.1103/PhysRevB.75.045114

[17] Pagare G, Chouhan SS, Soni P, Sanyal SP, Rajagopalan M. First principles study of structural, electronic and elastic properties of lutatium mono-pnictides. Comput. Mater. Sci. 2010;**50**: 538-544. DOI: 10.1016/j.commatsci.2010.09.016

[18] Legar JM. Chalcogenides and pnictides of cerium and uranium under high pressure. Physica. B. 1993;**84–91** (1993):190. DOI: 10.1016/0921-4526(93)90447-E

[19] O'Donnell K, Dierolf V. Rare earth doped III-nitrides for optoelectronic and spintronic applications. Springer. . DOI: 10.1007/978-90-481-2877-8

[20] Maity T, Trodahl HJ, Natali F, Ruck BJ, Vézian S. Electron transport in heavily doped GdN. Phys. Rev. Materials. 2018;**2**. DOI: 014405, 10.1103/PhysRevMaterials.2.014405

[21] P. Blaha, K. Schwarz, G.K.H. Madsen, D. Kvasnicka, J. Luitz, WIEN2k, an Augmented Plane Wave plus Local Orbitals Program for Calculating Crystal Properties Vienna University of Technology, Vienna (2001).

[22] Kohn W, Sham LJ. Self-consistent equations including exchange and correlation effects. Phys. Rev. 1965;**140**: A1133. DOI: 10.1103/PhysRev.140.A1133

[23] Hohenberg P, Kohn W. Inhomogeneous Electron gas. Phys. Rev. 1964;**136**:B864. DOI: 10.1103/PhysRev.136.B864

[24] Perdew JP, Burke K, Ernzerhof M. Generalized gradient approximation made simple. Phys. Rev. Lett. 1996;**77**: 3865. DOI: 10.1103/PhysRevLett.77.3865

[25] Tran F, Blaha P. Accurate band gaps of semiconductors and insulators with a semilocal exchange-correlation potential. Phys. Rev. Lett. 2009;**102**: 226401. DOI: 10.1103/PhysRevLett.102.226401

[26] Duan CG, Sabirianov RF, Mei WN, Dowben PA, Jaswal SS, Tsymbal EY. Electronic, magnetic and transport properties of rare-earth monopnictides: J. Phys: Condensed. Matter. 2007;**19**: 315220. DOI: 10.1088/0953-8984/19/31/315220

[27] Suehiro T, Hirosaki N, Yamamoto Y, Nishimura T, Mitomo M. Preparation of lutetium nitride by direct nitridation. J. Mater. Res. 2004;**19**:959. DOI: 10.1557/jmr.2004.19.3.959

[28] Gupta DC, Bhat IH. Electronic, ductile, phase transition and mechanical properties of Lu-monopnictides under high pressures. J. Mol. Model. 2013;**19**: 5343. DOI: 10.1007/s00894-013-2021-7

[29] Wood DW, Dalton NW. Ferromagnetic curie temperatures of the Heisenberg model with next-nearest-neighbor interactions. Phys. Rev. 1967; **159**:384. DOI: 10.1103/PhysRev.159.384

[30] Ambrosch-Draxl C, Sofo JO. Linear optical properties of solids within the full-potential linearized augmented planewave method. Comp. Phys. Commun. 2006;**175**:1-14. DOI: 10.1016/j.cpc.2006.03.005

[31] Penn D. Wave-number-dependent dielectric function of semiconductors. Phys. Rev. B. 1962;**128**:2093. DOI: 10.1103/PhysRev.128.2093

Chapter 7

Pancake Bonding Seen through the Eyes of Spectroscopy

Abstract

From local mode stretching force constants and topological electron density analysis, computed at either the UM06/6-311G(d,p), UM06/SDD, or UM05-2X/6-31++G(d,p) level of theory, we elucidate on the nature/strength of the parallel π-stacking interactions (i.e. pancake bonding) of the 1,2-dithia-3,5-diazolyl dimer, 1,2-diselena-3,5-diazolyl dimer, 1,2-tellura-3,5-diazolyl dimer, phenalenyl dimer, 2,5,8-tri-methylphenalenyl dimer, and the 2,5,8-tri-t-butylphenalenyl dimer. We use local mode stretching force constants to derive an aromaticity delocalization index (AI) for the phenalenyl-based dimers and their monomers as to determine the effect of substitution and dimerization on aromaticity, as well as determining what bond property governs alterations in aromaticity. Our results reveal the strength of the C···C contacts and of the rings of the di-chalcodiazoyl dimers investigated decrease in parallel with decreasing chalcogen···chalcogen bond strength. Energy density values H_b suggest the S···S and Se···Se pancake bonds of 1,2-dithia-3,5-diazolyl dimer and the 1,2-diselena-3,5-diazolyl dimer are covalent in nature. We observe the pancake bonds, of all phenalenyl-based dimers investigated, to be electrostatic in nature. In contrast to their monomer counterparts, phenalenyl-based dimers increase in aromaticity primarily due to CC bond strengthening. For phenalenyl-based dimers we observed that the addition of bulky substituents steadily decreased the system aromaticity predominately due to CC bond weakening.

Keywords: local stretching force constant, dimerization, pancake bonding, aromaticity, 2e/mc bonding

1. Introduction

The initial concept of "pancake bonding" was constructed by Mulliken and Person as to characterize the overall shape and bonding mechanisms of donor-acceptor π systems [1]. More recently the term "pancake bonding" has primarily been used to describe the formation of stabilizing parallel π–π interactions between two or more open-shell free radicals, those of which are typically planar and/or consist of light-atoms [2–4]. Such interactions have received a considerable amount of interest as they allow one to synthesize novel radical-based materials, via electron or hole through-space delocalization, that exhibit unique magnetic [5], optical [6], and electronic properties (i.e. conductive polymers, organic conductors) [7].

Generally, free radical species are short lived and exist in low concentration as two radicals will typically react to form a single covalently bonded dimer, or σ-dimer. However, when radicals are sterically hindered against approaching within a covalent bonding distance, they can exist as a stable, spin-paired, open shell species. Unlike general non-covalent interactions between closed-shell species (i.e. van der Waals), the open-shell radicals have been said to undergo stabilization with each other via through-space π-stacking 2e/mc distributed interactions (i.e. pancake bonding). This 2e/mc bonding (i.e. pancake bonding) is a result of overlapping antibonding (π^*) singly occupied molecular orbitals (SOMO) of the two monomer radicals with highly delocalized π-electrons [8]. It is noted that magnetic experimental analysis has found the spin pairing of pancake bonded dimers to be diamagnetic with an overall spin density of zero (i.e. singlet electronic state) [9]. The overlapping of antibonding (π^*) SOMOs is the basis of pancake bonds as this interaction leads to the following distinctive features [4]: i) contact bond distances that are beyond the usual C(sp^3)–C(sp^3) bond length (1.54 Å) but are also much shorter than the bonds of closed shell dimers that are held together by vdW forces (sum of vdW radii = 3.40 Å) (ii) due to direct atom-to-atom overlap, SOMO-SOMO overlapping strongly favors configurations that yield maximum overlap orientations which lower the energy of the two radical SOMOs iii) low lying singlet (-singlet-singlet) and triplet (singlet-triplet) electronic excited states, iv) negative singlet-triplet splitting energies (i.e., ΔE_{ST} = E(singlet) – E(triplet)) for stable open shell singlet pancake bonded complexes [10] and v) interaction energies larger than those of vdW interactions. Bond dissociation energies (BDE) of pancake bonded system have been estimated to be smaller than those of a normal covalent system but larger than dimers subject to typical π-stacking where this type of π-stacking is observed for DNA base pairs [11] (vdW π stacking interactions and pancake bonds are different). Several works analyzed the related binding energies (BE), splitted into two contributions, a destabilizing stabilizing vdW part, E_{vdW}, and a stabilizing energy, E_{SOMO}, associated with the bonding overlap of the singly occupied SOMO [12]. E_{SOMO} yields a reasonable description of the SOMO-SOMO overlap contribution to BE and it has been suggested that E_{SOMO} can be estimated from the difference between E(singlet) – E*(triplet), where E*(triplet) is the triplet energy evaluated for the singlet geometry [12].

BE, E_{SOMO} and SOMO-SOMO overlap have been utilized as to further explain the nature of these systems [8, 13]. It was argued that the dimerization of such radicals exhibit covalent bonding character as the spin-pairing of the electrons in the SOMO leads to a filled highest occupied molecular orbital (HOMO) and a corresponding empty antibonding LUMO [14]. In this situation, the interaction occurs at rigid rotational geometries, due to SOMO-SOMO overlapping, which is different from π-stacking in which various rotational orientations are possible [15]. On the other hand, dispersion and/or van der Waals interactions have been suggested to play important roles in the overall stabilization of these dimers [14]. Thus, the nature of pancake bonds between 1,2-chalcogen-3,5-diazol radicals and phenalenyl-based radicals remains in debate to the present day.

A CSD database survey based upon 35 cis-cofacial dimers composed of HCNSSN radicals, with C–C contact distances ranging between 2.75 to 3.50 Å, showed that S···S contact bond distances ranged from 2.93 to 3.30 Å [8]. These S···S contact bond are much shorter than the vdW distance between two sulfur atoms (4.06 Å) [16], in the case of two spherical sulfur atoms the vdW distance has been computed to be 3.60 Å. A CSD database survey based on 12 cis-cofacial 1,2-diselena-3,5-diazolyl dimers, with C···C contact distances between 2.80 and 3.50 Å, found the average Se···Se contact distance to be 3.26 (s = 0.05) [8]. This average Se···Se contact distance is slightly smaller than the vdW distance between spherical Se atoms (3.32 Å). Previously computed dissociation energies have suggested that dimers of R-CNSeSeN radicals dimers are more binding than dimers of R-CNSSN radicals; relative binding energy values were also observed to be analogous to vdW interactions [8].

1,2-chalcogen-3,5-diazole dimers: Within the past two decades di-chalcogen-diazole radicals, such as 1,2-dithia-3,5-diazolyl (i.e. HCNSSN) and 1,2-diselena-3,5-diazolyl (i.e. HCNSeSeN) radicals, and their derivatives have been a subject of many investigations [17]. The rings of HCNSSN and HCNSeSeN are rich in π-electrons and have π^* singly occupied molecular orbitals (SOMO). The 1,2-dithia-3,5-diazolyl and 1,2-diselena-3,5-diazolyl radicals have been experimentally observed to result in stable dimerizations in the solid state where, in most cases, the neutral radicals prefer to be oriented with their faces parallel to one another (cis-cofacial) in order to achieve a configuration that supports maximum π^*-π^* (SOMO-SOMO) overlapping observed as two electron/eight-center (2e/8c) π-stacking (i.e. pancake bonding) interactions. A notable feature of HCNSSN and HCNSeSeN dimers are their four long chalcogen-chalcogen bonds (i.e. contacts) ranging between 2.2 and 4.0 Å. HCNSSN and HCNSeSeN dimers have been suggested to stabilize via a combination of π and σ aromaticity [13].

Phenalenyl-based dimers: In solution, phenalenyl radicals maintain chemical equilibrium via the formation of a σ-bonded dimer [18]. Due to the very high symmetry of the radical phenalenyl monomer, a unpaired electron is delocalized across all α-positions of the phenalenyl framework excluding the central carbon atom of the monomers [19]. As noted in the work of Kubo [19], the thermodynamic stability of such carbon-centered radical species increases as the delocalization of unpaired electrons across a system increases [19]. Another interesting feature of phenalenyl dimers and their derivatives (i.e. carbon-centered hydrocarbon radicals) is due to the formation of unique two-electron/twelve-center (2e/12c) π-stacking interactions between these spin-delocalized hydrocarbon radicals [20] as verified by NMR [21]. The hexagonal arrangement of the SOMO of the phenalenyl radicals enables perfect π-π overlap in both eclipsed and staggered stacking configurations, the staggered stacking configuration is favored over the eclipsed configuration due to shorter π-π contacts as a result of less atom-atom repulsion [19]. It is mentioned, that various phenalenyl derivatives, which demonstrate π-π stacking (i.e. pancake bonding), have been experimentally identified via single crystal X-ray diffraction (XRD) [22]. The formation of σ-bonded phenalenyl radical dimer can be inhibited by substituting the carbon atoms of the phenalenyl rings, at the 2,5,8-positions, with *tert*-butyl groups as a π-bonded dimer results from the sterically hindered phenalenyl radicals [19]. Moreover, X-ray studies have revealed that the application of sterically hindered substituents (i.e. *tert*-butyl groups) on phenalenyl radicals prevent σ-dimerization and results in a π-bonded dimer with a face-to-face stacking distance, twice that of the σ-bonded dimer, at a length of of 3.2 Å [23]. This π-π contact (face-to-face) stacking distance is characteristic to pancake bonding as this length is shorter than that of a vdW complex and is beyond the length of a coventional covalent bond. Bond dissociation energy (BDE) for systems containing carbon radicals such as phenalenyl have been estimated to be around 10 kcal/mol [11]. Because σ-bonded and π-bonded phenalenyl-based dimers are close in energy the existence of the pancake bonded dimer as a fluxional molecule has been reviewed [12].

Although many experimental and computational have been conducted for the dimerizations of 1,2-chalcogen-3,5-diazol and phenalenyl-based radicals, the intrinsic strength of these interactions remains unclear. While popular BDE and its decomposition [24] provides valuable information about the stabilizing forces involved in bond formation (in the case of pancake bond in particular in the formation of 2e/mc interactions), BDE does not adequately describe the intrinsic strength of a bond [25–27]. Because BDE measures the overall effect of bond breakage it contains the electronic reorganization and geometrical relaxation of the fragments upon dissociation. Therefore, we introduced in this work an intrinsic bond strength measure based on vibrational spectroscopy. Unlike BDE, the local stretching force constant (k^a), derived from local vibrational modes [25], conserves the geometry and electronic structure of all bonds/interactions. k^a provides a direct description of intrinsic bond strength and has been applied successfully applied to assess the intrinsic bond strengths for a variety of covalent interactions including ultra long C–C bonds, carbon-halogen bonds and non-covalent interactions such as hydrogen, tetrel, pnicogen, chalcogen and halogen bonds; see Ref. [25] and citations therein.

In this study, we applied the local mode analysis [25] complemented with the RING puckering analysis of Cremer and Pople [28] and Bader's quantum theory of atoms in molecules (QTAIM) analysis of the electron density [29] to quantify the strength of the pancake bonds in six spin-paired, open-shell singlet state dimers **1–6** (shown in **Figure 1**) and and to learn more about their nature. Species **1–3** are 1,2-chalcogen-3,5-diazole dimers which contain sulfur (**1**), selenium (**2**), and tellurium atoms (**3**); it is noted that **3** is a prototypal (i.e. theoretical) species. Species **4–6** are phenalenyl-based dimers in which the bulkiness of substituents increase as follows: phenalenyl dimer (**4**) < 2,5,8-trimethylphenalenyl dimer (**5**) < 2,5,8–*tert*-butylphenalenyl (**6**). The aromatic character of the dimer species (**4–6**) was also explored, in particular the role of the aromaticity for the stabilization of phenalenyl-based dimers. In summary, special focus was on: i) to assess the intrinsic bond strengths of the 2e/mc interactions for selected species, ii) to quantify the ring strengths of the selected species, iii) to determine if the pancake bonds of these species are covalent in nature, iv) to elucidate on the effect of substituents on the aromaticity of phenalenyl-based species, v) to determine, for phenalenyl-based dimers, the effect of dimerization on the aromaticity for phenalenyl-based species,

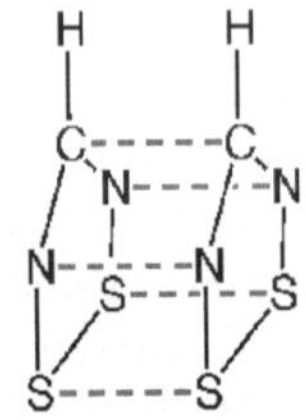

1

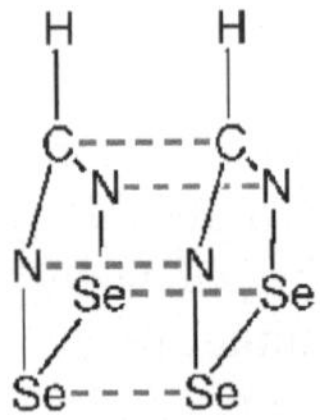

2

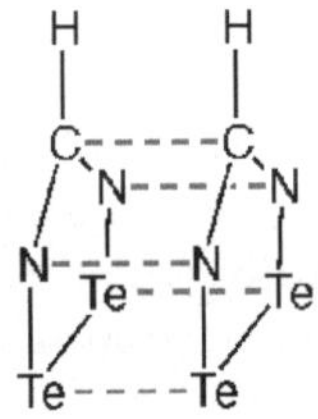

3

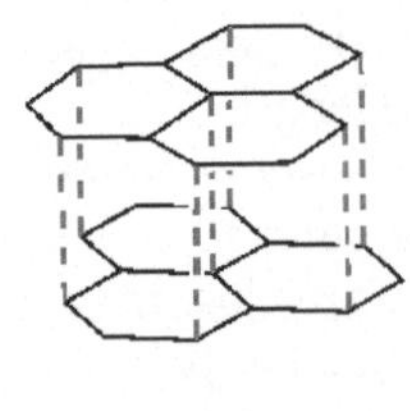

4

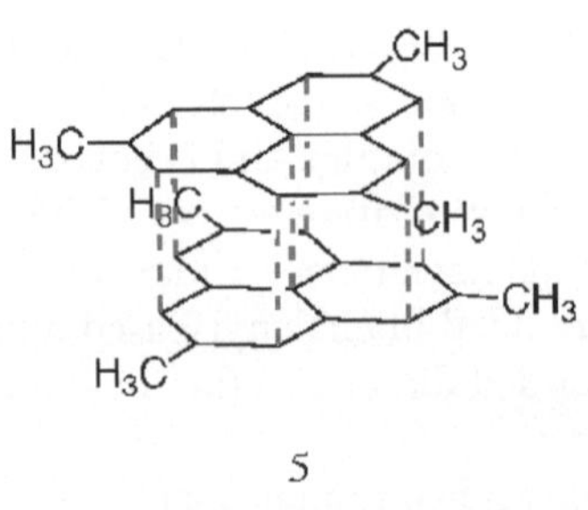

5

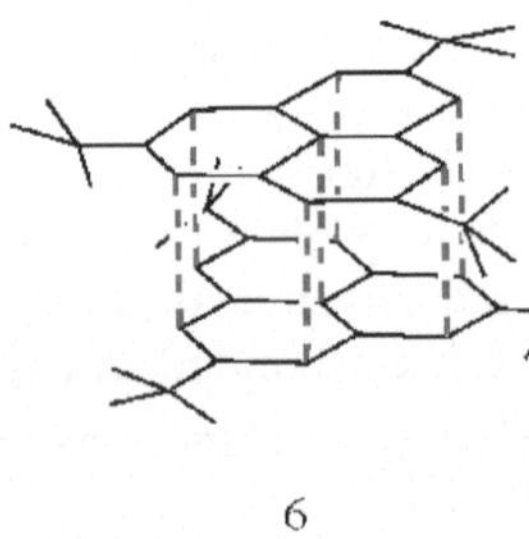

6

Figure 1.
Species investigated in this work. 1) 1,2-dithia-3,5-diazolyl (HCNSSN) dimer 2) 1,2-diselena-3,5-diazolyl (HCNSeSeN) dimer. 3) 1,2-tellura-3,5-diazolyl (HCNTeTeN) dimer 4) phenalenyl dimer. 5) 2,5,8-tri-methylphenalenyl dimer. 6) 2,5,8-tri-t-butylphenalenyl dimer. Detected pancake bonds (2e/mc) (i.e. targeted contact bonds and interdimer CC bonds) are denoted in red.

and vi) to determine what bond property, of the phenalenyl-based species investigated, predominately governs changes in aromaticity.

2. Computational methods

Local mode theory: Since the underlying theory behind the derivation of local vibrational modes is elaborated on in Ref. [25] the following text briefly covers the fundamental aspects. Every vibrational mode, being associated with potential and kinetic energy contributions, is subject to two mode-to-mode coupling mechanisms, electronic coupling and kinematic (mass) coupling [30]. As a result the normal modes remain perpetually delocalized over a molecule and cannot be directly used to assess chemical bond strength [25]. Solution of the vibrational secular equation (i.e. the Wilson equation) eliminates the electronic coupling as a result of force constant matrix diagonalization. The kinematic coupling which remains is eliminated in the local mode theory via a modified version of the Wilson equation that uses mass-decoupled Euler–Lagrange Equations [25]. This leads to local vibrational modes, associated with local mode frequencies ω^a and local mode force constants k^a that can serve as a quantitative bond strength measure [25] which we applied to assess the strength all 2e/mc interactions (i.e. pancake bonds) of species **1–6** (see **Figure 1**). Stretching force constants k^a can be transformed into relative bond strength orders (BSO) n which are more convenient for comparison, via a generalized Badger rule [31], leading to the following power relationship between these two quantities: BSO $n = A\ (k^a)^B$. Constants A and B can be determined from two reference molecules with known k^a and BSO n values and the requirement that for a zero k^a value the BSO n is also zero.

In our study we used the CC single bond of ethane with BSO $n = 1$ and the CC double bond of ethene with BSO $n = 2$ as references [32]. In addition to BSO n values for the C···C contacts, BSO n values for N···N, S···S, Se···Se, and Te···Te bonds of the dichalcodiazolyl species **1–3** were derived using the same power relationship. For dimers **4–6**, aside from deriving the BSO n values for the central C–C bonds, we also computed the BSO n values for the outer C···C contacts which are established between six carbon atoms of each monomer (see **Figure 1**).

Aromaticity index based on local modes: π delocalization in species **4–6** was determined via an aromatic delocalization index (AI) derived from local force constants following the procedure of Kraka, Cremer and co-workers [33, 34]. In contrast to the HOMA index [35] which is based upon optimal bond lengths, which sometimes tend to fail for this purpose [33], the AI is based on local stretching force constants and bond strength orders (BSO n). As a reference, we used benzene with an AI value of 1.00 and assigned BSO n value of 1.451 [33].

BDEs for **1–6** were derived via potential energy curves by varying the interdimer distance from 2.5 to 8.0 Å, using increments of 0.1 Å around and 1.0 Å further away from the equilibrium geometry, followed by a constrained optimization. By calculating BDEs via potential energy curves any basis set superposition errors can be avoided, such errors have been reported to as large as 16 kcal/mol in these complexes when the BDE is calculated from the differences between dimer and monomer energies [14]. The covalent character of the pancake bonds was assessed with the Cremer-Kraka criterion [36, 37] of covalent bonding within the framework of Bader's QTAIM [29]. The Cremer-Kraka criterion is composed of two conditions; (i) existence of a bond path and bond critical point $r_b = b$ between the two atoms under consideration; (ii) sufficient condition: the energy density $H(r_b) = H_b$ is smaller than zero. $H(r)$ is defined as $H(r) = G(r) + V(r)$, where $G(r)$ is the kinetic energy density and $V(r)$ is the potential energy density. A negative $V(r)$ corresponds to a stabilizing accumulation of density whereas the positive $G(r)$ corresponds to depletion of electron density [36]. As a result, the sign of H_b indicates which term is dominant [37]. If $H_b < 0$, the interaction is considered covalent in nature, whereas $H_b > 0$ is indicative of electrostatic interactions.

Model chemistry used: To describe the spin-paired open shell singlet states, we applied a single determinant broken-symmetry (BS) unrestricted ansatz, which works well for systems with small singlet-triplet gaps [38, 39], combined with a density functional theory (DFT) approach. We refrained from a multi-reference description, such as CASSCF, which has been mostly applied to unsubstituted species **4** with a relatively small active space and basis sets [40]. We also refrained from post-SCF methods, such as Møller-Plesset perturbation theory of second order, which has shown to over-bind in the case of dimer complexes with pancake bonds and may results in an unrealistic C···C contact distance of 2.8 Å [14].

A reliable description of pancake bonding requires a careful choice of DFT functional. The popular B3LYP functional [41, 42] does not describe dispersion well whereas the dispersion corrected ωB97X-D [43] functional sometimes leads to inconsistent results [44]. It was reported that the M06-2X functional [45] yields generally shorter C···C contact distances [46] whereas the C···C contact distances based off the M05-2X functional [47] agree well for complexes for **4–6** with experimental values [48]. On the other hand, the M06 functional has shown to be well parameterized for describing chalcogens (i.e. sulfur, selenium and tellurium atoms) [45]. Another important part of the model chemistry is the basis set. We tested both, Pople's augmented 6–31++G(d,p) double zeta [49, 50] and 6-311G(d,p) triple zeta basis sets [51]. For the Te atom we applied the SDD basis set [52] which uses the Stuttgart-Dresden pseudopotentials [53] to account for relativistic effects. Guided by our test calculations, we decided to use for our study the BS-UM06/6-311G(d,p) model chemistry for **1–2**, BS-UM06/SDD for **3**, and BS-UM05-2X/6–31++G(d,p) for **4–6**.

Software used: All DFT geometry optimizations and frequency calculations were carried out using the Gaussian program package [54]. The following local mode analysis and the aromaticity delocalization index (AI) study was carried out with the LModeA software [55]. The QTAIM analysis was performed with the AIMALL program [56] For the rings of the di-chalcodiazoyl dimers (**1–3**), which do not contain a central atom, we used the ring puckering program [57] followed by LMA, as to obtain the local mode properties of the rings.

3. Results and discussion

It is noted that in regard to the text which follows, the terms contact bonds, π-π stacking interactions, and face-to-face interactions loosely refer to pancake bonds while interdimer/central C-C bonds refer to the C-C bond established in the center of two monomers. **Table 1** summarizes the calculated bond distances (R_{calc}), experimental bond distances (R_{exp}), calculated bond dissociation energies (BDE_{calc}), experimental bond dissociation energies (BDE_{exp}), local stretching force constants (k^a), local mode vibrational frequencies (ω^a), bond strength orders (BSO n), electron densities (ρ_b), and energy densities (H_b) for the targeted CC bonds of of targeted contacts bonds of dimers **1–6** and rings of **1–3**. **Table 2** summarizes symmetry, singlet and triplet C···C contact distance (R(CC)), energy values of SOMOs (E_{SOMO}), and triplet/singlet (ΔE_{ST}) for all species investigated in this work (**1–6**). **Figure 2** shows the equilibrium geometries for the HCNTeTeN **3** dimer (C_2) in

No.	Species	R_{calc}	R_{exp}	BDE_{calc}	BDE_{exp}	k^a	ω^a	BSO n	ρ_b	H_b
1	HCNSSN									
	Ring	3.071		−5.8	−5.3 [8]	0.657	147	0.214	0.016	0.005
	C–C	3.036	3.18			0.208	243	0.083	0.041	0.007
	N-N	3.034				0.128	176	0.056	0.052	0.004
	S-S	3.125				0.192	143	0.078	0.104	−0.000
2	HCNSeSeN									
	Ring	3.210		−4.7	N/R	0.302	72	0.113	0.015	0.004
	C–C	3.119	3.31			0.080	151	0.038	0.034	0.006
	N-N	3.152				0.074	134	0.036	0.042	0.003
	Se-Se	3.313				0.151	80	0.064	0.098	−0.000
3	HCNTeTeN, C_{2v}									
	Ring	3.514		−6.0	N/A	0.049	23	0.021	0.013	0.001
	C–C	3.219	N/A			0.029	83	0.014	0.036	0.006
	N-N	3.333				0.032	29	0.016	0.039	0.006
	Te-Te	3.840				0.045	123	0.022	0.073	0.002
4	HCNTeTeN, C_2									
	Ring	3.413		−8.4	N/A	0.162	43	0.062	0.018	0.002
	N-N	3.342				0.112	165	0.045	0.046	0.009
	Te-Te	3.820				0.038	65	0.018	0.086	0.003
	N-Te	3.510				0.045	78	0.021	0.069	0.007
5	Phenalenyl									
	Peripheral C–C	3.110	N/A	−11.0	N/A	0.366	123	0.136	0.072	0.005
	Central C–C	3.152	N/A			0.293	288	0.113	0.063	0.006
6	tMP									
	Peripheral C–C	2.997	3.053	−14.8	N/R	0.172	64	0.074	0.090	0.006
	Central C–C	3.093	3.145			0.167	217	0.072	0.070	0.007
7	tTBP									
	Peripheral C–C	3.391	3.306	−12.4	−9.5 [59]	0.194	68	0.081	0.047	0.003
	Central C–C	3.287	3.201			0.147	204	0.065	0.050	0.005

The UM06/6-311G(d,p) methodology used for ***1*** *and* ***2****, UM06/SDD for* ***3****, and UM05-2X/6–31++G(d,p) for* ***4****,* ***5*** *and* ***6****. N/A, not applicable; N/R, not reported.*

Table 1.
Summary of calculated bond distances (R_{calc}) in Å, experimental bond distances (R_{exp}) in Å, bond dissociation energies (BDE_{calc}) in kcal/Mol, experimental bond dissociation energies (BDE_{exp}) in kcal/Mol, vibrational spectroscopy data, electron densities (ρ_b) in e/Å³, and energy densities (H_b) in h/Å³ of the targeted contacts bonds and rings of dimers ***1–6*** *(see* ***Figure 1****).*

singlet and triplet states. **Figure 3** shows the various conformations of the phenalenyl dimer in the triplet state where the red lines indicate detected C···C contacts. **Figure 4** show the generated Morse potential curves of dimers **1–6**. **Figure 5** shows the correlation between BSO n values and the local stretching force constants k^a of **1–6**. **Figure 6** showcases the BSO n(CC) values, corresponding CC bond lengths, AI values, bond weakening/strengthening parameters (WS), and bond alteration parameters (ALT) for the carbon ring structures and the outer ring

No.	Species	Dimer	Monomer	Singlet	Triplet	E_{SOMO}	ΔE_{ST}
		Symmetry	Symmetry	R(CC)	R(CC)		
1	HCNSSN	C_{2v}	C_{2v}	3.036	3.452	−15.61	−2.17
2	HCNSeSeN	C_{2v}	C_{2v}	3.119	3.208	−13.90	−2.09
3	HCNTeTeN	C_{2v}	C_{2v}	3.165	3.362	−13.26	0.96
3	HCNTeTeN	C_2	C_2	3.563	3.104	−8.46	−1.35
4	Phenalenyl	D_{3d}	C_{3H}	3.152	3.622	−12.97	−5.98
5	tMP	D_{3d}	C_{3H}	3.093	3.744	−19.26	−5.44
6	tTBP	S_6	C_{3H}	3.281	3.855	−6.11	−3.13

Table 2.
*Symmetry of dimer and monomer, singlet and triplet face-to-face distances (R(CC)) in Å, energy values of SOMOs (E_{SOMO}) in kcal/Mol and triplet/singlet splitting (ΔE_{ST}) in kcal/Mol for comlexes **1–6** (see **Figure 1**) calculated at corresponding levels of theory.*

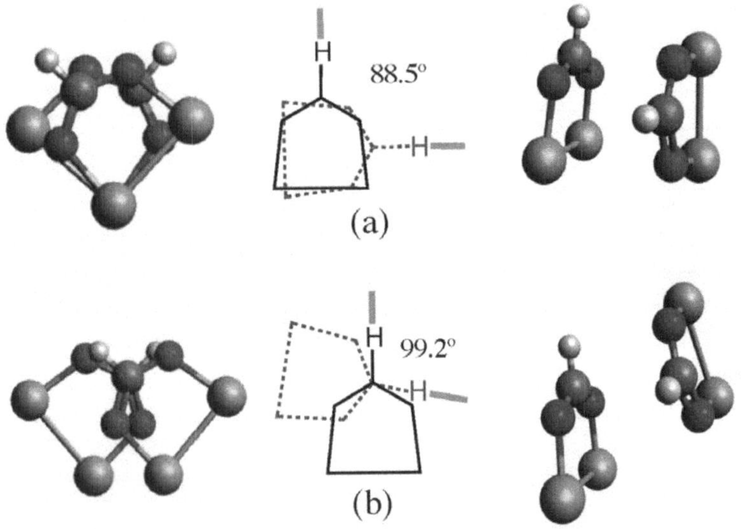

Figure 2.
Equilibrium geometries for HCNTeTeN (3) dimers in C_2 symmetry. a) Singlet. b) Triplet.

structures of phenalenyl, 2,5,8-trimethylphenalenyl, and 2,5,8-tri-t-butylphenalenyl monomer radicals and dimers.

3.1 Energetics

Identifying pancake bond interactions: As shown in **Table 2**, the E_{SOMO} values for dimers **1–6** range between −6.11 and −19.26 kcal/mol where **5** acquires the largest E_{SOMO} value. We note that the E_{SOMO} value of **6** is in good agreement with the ST-splitting of −6.64 kcal/mol derived from ESR experiments [21]. As shown in **Table 2** the $\Delta E(_{ST})$ is small and negative for dimers **1–6** with **3** in C_2 symmetry. These results are in line with the notion that the formation of pancake bonded dimers requires the spin-paired singlet state to be energetically favored over the triplet state.

From singlet to the triplet state, the central C–C bond distances in dimers **1** and **2** increase from 3.04 Å and 3.12 Å to 3.45 Å and 3.21 Å, respectively. No alterations in

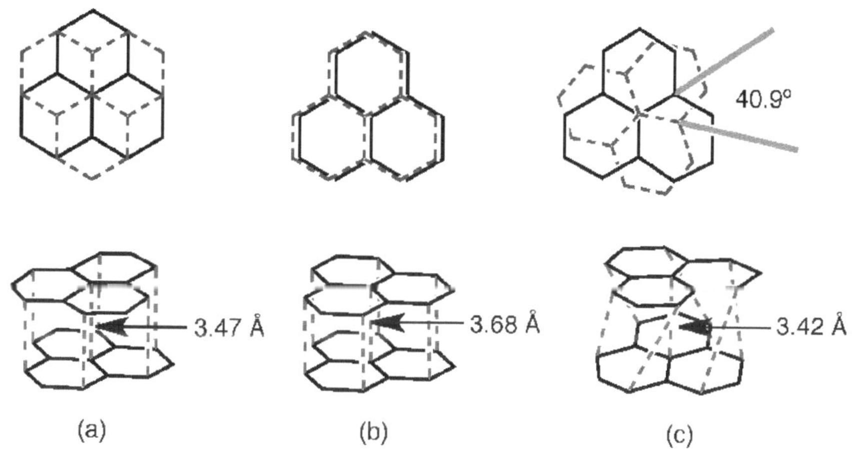

Figure 3.
Conformations of the phenalenyl dimer in the triplet state. a) Staggered. b) Eclipsed. c) Intermediate geometry. The red lines indicate detected π-π contacts. Bond distances for the central CC bond between the two monomers are given.

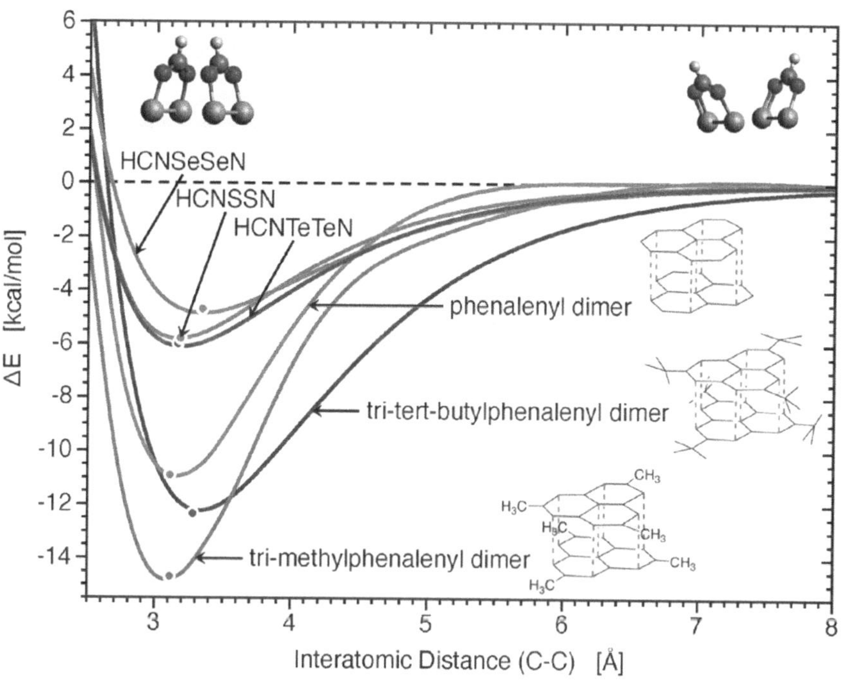

Figure 4.
Dissociation curves for dimers 1 and 2 (UM06/6-311G(d,p)), 3 (C_{2v}) (UM06/SDD), and 4–6 (UM05-2X/6–31++G(d,p)).

the rotational alignments amongst these two species were observed. Unlike for dimers of **1** and **2**, we observe that, in the singlet state of the HCNTeTeN dimer (**3**) one monomer rotates about the CC central axis by 88.5°, resulting in a C_{2v} symmetry for the dimer. Moreover, the triplet state of the HCNTeTeN dimer (**3**) involves the rotation of a monomer, about the central C-C axis, by 99.2° and results in a C_2 symmetry for the dimer (see **Figure 2**).

The ΔE_{ST} values of **1–3**, where **3** is in C_2 symmetry, indicate stable arrangements (see **Table 2**). In the case of **3**, which is common in symmetry to dimers **1** and **2** (C_{2v}),

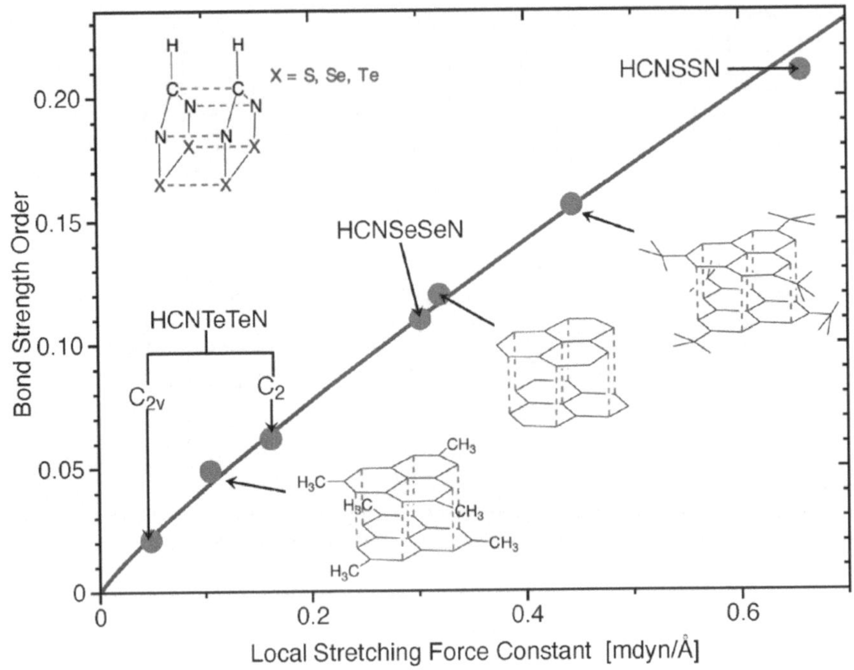

Figure 5.
*The relationship between BSO n and force constants of dimers **1–6** calculated with UM06/6-311G(d,p) (**1** and **2**), UM06/SDD (**3**), and UM05-2X/6–31++G(d,p) (**4–6**). BSO n(ring) values for **1–6** were computed via in accordance to the level of theory used.*

the triplet state is lower than the singlet state ($\Delta E_{(ST)}$ = 0.96 kcal/mol) reflecting an unstable dimer structure as no pancake bonding is formed. We note that for the lower energy structure of **3** (C_2) BCP's were detected for Te···Te, Te···N, and N···N contacts, being consistent with the observations of Gleiter and Haberhauer [58], in which the reorientations of dithiatriazine molecules favored the formation of S···N and S···C interactions over the the formation of a C···C contacts. Notably, unlike the other dichalcodiazoyl dimers, the central C–C distance of the **3** (C_2) dimer, from the singlet to the triplet state, decreases from 3.56 Å to 3.10 Å. Going from a C_{2v} symmetry to C_2 symmetry the E_{SOMO} value of **3** changes from −13.26 kcal/mol to −8.46 kcal/mol. These results indicate that there are attractive interactions between the monomers of **3** (C_2) that are unrelated to SOMO-SOMO overlap. Overall, the results based on **3** in C_{2v} and symmetry C_2, suggest that chalcogen···chalcogen bonds and the electrostatic attraction between a chalcogen and a less electronegative atom play significant roles in the stabilization of such dimers.

For the phenalenyl dimer (**4**), the triplet geometry exhibits two local minima and one global minima (see **Figure 3**). The staggered configuration of **4** is −1.7 kcal/mol lower in energy than the eclipsed conformer. The central C···C distance of both the staggered and eclipsed conformer of **4** are longer than the sum of the van der Waals radii where the central C···C bond of the staggered configuration is shorter than that of the eclipsed configuration by 0.27 Å (see **Figure 4**). The most stable arrangement of **4** is represented by an intermediate structure with a rotational dihedral of 40.9° which, in contrast to the staggered and eclipsed geometries, has a shorter central C···C distance (3.42 Å) and is −0.4 kcal/mol lower in energy than the staggered configuration. These results suggest that the triplet state of **4** is a π-complex.

Though the interatomic distances of **5** and **6**, when going from a singlet to triplet state, increase from values of 3.09 Å and 3.28 Å to values of 3.74 Å and 3.86 Å, we

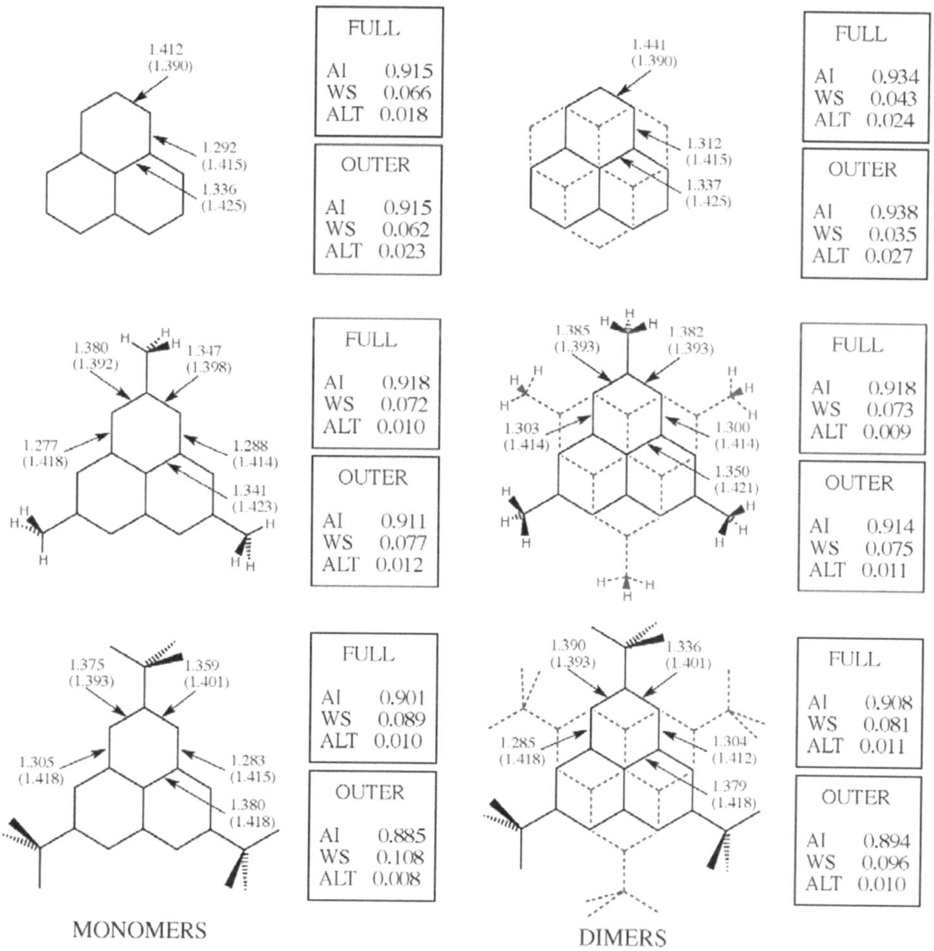

Figure 6.

Bond strength orders (BSO) and bond lengths (in parentheses, Å) for the phenalenyl, 2,5,8-trimethylphenalenyl and 2,5,8-tri-t-butylphenalenyl radical monomers and dimers (4 through 6). The aromaticity delocalization index (AI), bond weakening (strengthening) parameters (WS) and alteration parameters (ALT) for the FULL carbon ring structures (FULL) and the OUTER ring structure (OUTER) are indicated in boxes. The term FULL accounts for all CC bonds while the term OUTER accounts only for outer CC bonds and does not account for the inner most CC bonds.

observe no change in the rotational alignment between the monomers of the two species. These results suggest that any change in the orientations of **5** and **6** monomers are hindered by their substituent groups. We also note that the ΔE_{ST} values of **4** to **6** steadily decline as substituent size increases (see **Table 2**).

Dissociation energies: From the Morse potential curves of the dichalcodiazoyl dimers **1–3** (C_2) bond dissociation energy (BDE_{calc}) values of −5.8, −4.7 and −6.0 kcal/mol are obtained, these values being more analogous to the BDE values of electrostatic interactions. The calculated BDE of **1** is in good proximity to the experimental value reported by Beneberu et al. (see **Table 1**). The bond dissociation energy of species **3**, in C_2 symmetry, in comparison to **3** in C_{2v} symmetry, is more negative by −2.4 kcal/mol.

The computed bond dissociation energy values for species **4** through **6** are −11.0, −14.8 and −12.4 kcal/mol, respectively. The computed dissociation energy value of **6** is in good agreement with the previously reported experimental enthalpy change (ΔH_D) of −9.5 kcal/mol [59]. We observe the BDE_{calc} of the

2,5,8-trimethylphenalenyl dimer (**5**) to be larger than that of both **4** and **6** by values of 3.8 and 2.4 and kcal/mol suggesting that the addition of three methyl groups to each monomer of the phenalenyl dimer (**5**) yields a more stable dimer as dispersion contributions are enhanced (see **Tables 1** and **2**). In contrast to **5**, the addition of three tert-butyl groups (C4H9) to each monomer of the phenalenyl dimer (**6**) results in a decreased stabilization due to increased steric repulsion between the bulky C4H9 substituents. Moreover, we observe 2,5,8-tri-t-butylphenalenyl dimer (**6**) to be more stable than the phenalenyl dimer (**4**) by 1.4 kcal/mol, indicating that, within **6**, there is a trade-off amongst the steric repulsion of the tert-butyl groups and stabilizing dispersion (see **Table 1** and **Figure 4**).

3.2 Bond parameters and derived bond strength orders *n*

Di-chalcodiazoyl dimers: As the chalcogen atoms (S, Se, and Te) of the di-chalcodiazoyl dimers (**1–3**) increase in atomic radius (see **Figure 5**), the BSO *n* values of the C···C contacts of **1–3** decrease (see **Table 1**). It is noted that C···C contact distances of **1** and **2** are in excellent agreement with experiment (see **Table 1**). The chalcogen···chalcogen contacts within the 1,2,3,5-ditelluradiazolyl dimer (**3**), in C_{2v} symmetry, acquire a k^a value that is smaller than that of the chalcogen···chalcogen interactions of dimers **1** and **2** by 0.147 and 0.109 mdyn/Å, respectively. In the case of the **3**, in C_2 symmetry, N···Te, rather than C···C contacts as observed in **3** (C_{2v}), are seen to coexist alongside Te···Te contacts. Moreover, we find that the BSO *n* value for the hetero-chalcogen (N···Te) bond of **3** (C_2) is larger than that of the Te···Te contact (see **Table 1**).

In regard to individual aromatic rings of **1–3** (i.e. HCNSSN, HCNSeSeN, and HCNTeTeN) we observe the overall bond strength order of each ring (i.e. BSO *n* (ring)) to decrease as the strength of the chalcogen···chalcogen interactions between corresponding rings decrease in the following order: S···S > Se···Se > Te···Te [BSO *n*(ring) = 0.214 (**1**), 0.113 (**2**), 0.021 (**3**, C_{2v}), 0.062 (**3**, C_2)]. Moreover, as depicted in **Figure 5**, the dimer 1,2,3,5-dithiadiazolyl (**1**) is more stable than the 1,2,3,5-diselenadiazolyl dimer (**2**) by 0.355 mdyn/Å (see **Figure 1**); this result indicates that a greater extent of π-stacking is present within **1** which results in the C···C, N···N, and chalcogen···chalcogen contacts of **1** being shorter than those of **2** (see **Table 1**). Furthermore, the k^a values for the chalcogen···chalcogen contacts (i.e. S···S, Se···Se, and Te···Te) reveal that S···S and Se···Se interactions contribute large amounts of π-delocalization primarily towards the rings, where the overall rings strength of **2** is stronger than that of **3** due to a greater amount of π-delocalization from the corresponding chalcogen···chalcogen interactions (Se···Se) (see **Table 1**). From **Figure 5**, in addition to the individual k^a values of the NN, TeTe, NTe, and CC contacts of **3** in C_2 and C_{2v} symmetry, we can see that the C_2 configuration of **3** results in a greater amount of stabilizing π-delocalization, dominantly due to the N···N contacts, towards the rings (see **Table 1**). Alongside a decrease in ring strength from **1** to **3** the overall bond length of the aromatic rings, which, in the case of **1** is equivalent to the summation of all R(C-N), R(S-S), and (N-S) bond lengths of a HCNSSN ring, decreases from **1** to **3** (see **Table 1**).

The energy density (H_b) values at the chalcogen···chalcogen (i.e. S···S, Se···Se, and Te···Te) bond critical points r_b of **1–3** are negative for **1** and **2** and positive for **3** (see **Table 1**). The negative energy density H_b values at the bond critical points r_b of the chalcogen···chalcogen contacts within **1** and **2** (i.e. S···S, Se···Se) indicate the presence of chalcogen···chalcogen covalent bonding [60]. Positive H_b values of the Te···Te interactions for **3**, in both C_{2v} and C_2 symmetries, indicate that the Te···Te contacts are much weaker than the S···S and Se···Se contacts of **1** and **2** which are of an electrostatic nature. We note that in all cases (**1–3**), the H_b values of C···C and

N···N contacts are positive. The non-detection of a bond critical point for the C···C contacts of **3**, in C_2 symmetry, reveal that such interactions disappear when the C···C bond distance stretches slightly beyond that for the equilibrium geometry of **3** (C_{2v}) (see **Table 1**).

From our results we observe that the stabilization of molecules **1** and **2** is primarily due to the large magnitude of π-delocalization from their corresponding chalcogen interactions (i.e. S···S and Se···Se) where the extent of π-delocalization is seen to correlate in parallel with the strength of the C···C contacts and the overall strength of an aromatic rings (see **Table 1**). In contrast to dimers **1** and **2**, **3** (C_{2v}) acquires a much weaker C···C contact strengths and an overall weaker aromatic ring strength due to a lesser extent of π-delocalization from the Te···Te interactions as revealed from the much smaller k^a(chalcogen···chalcogen) values (see **Table 1**). Our results show that the chalcogen bonding does play a stabilizing role in the dimers such as **1** and **2** as suggested by Gleiter and Haberhauer [13, 58, 61], which observe that as pancake bonded species (dimer) are drawn apart the monomers tip outward in such a way that the chalcogen atoms, on each monomer, undergo separation at a slower rate in contrast to their carbon and nitrogen atoms.

Phenalenyl-based dimers: Unlike dimers **1–3**, the phenalenyl dimers (**4–6**) contain central (interdimer) C–C bonds (see **Figure 1**). As mentioned earlier, in addition to the central C–C bonds of **4–6**, we also analyze all peripheral C···C bonds which are established between six carbon atoms of each monomer that comprise the corresponding phenalenyl-based dimers (see **Figure 1**). We observe that the central C–C bonds of **4–6** decrease in strength from **4** to **6** due to a lesser extent of π-delocalization from peripheral C···C as observed from corresponding k^a(C···C) values (see **Table 1**). The relative BSO n values of the peripheral C···C interactions for all phenalenyl-based dimers (**4–6**) are stronger than the corresponding central C–C bonds (see **Table 1**). The k^a values of the central C–C bonds within **4–6** are within a range 0.16 and 0.70 mdyn/Å; these bonds are weaker than the C–C single bond in ethane, a classical C–C bond prototype (k^a(C–C) = 4.3 mdyn/Å).

Moreover, the peripheral C···C bonds of the phenalenyl dimer (**4**) and of the trimethylphenalenyl dimer (**5**) are shorter than their central C–C bonds (see **Table 1**). For the tri-*tert*-butylphenalenyl dimer (**6**), the interdimer C–C bond is distance is smaller than that of the peripheral C···C bonds (see **Table 1**) due to the steric repulsion between the bulky *tert*-butyl groups of the monomers as this repulsion "locks" the dimer into a staggered configuration. The steric repulsion between the *tert*-butyl groups groups of **6** results in a concave pyramidalization of the central CC bonds of the monomers [40], causing the central interatomic C–C bond to be shorter than the outer CC interactions (see **Table 1**). Moreover, the electron density values (ρ_b) of the peripheral C···C bonds of **4** and **5** are less than those for the corresponding central C–C bonds and an opposite trend is observed for that **6** (see **Table 1**). We observe both the C···C contacts and interdimer C–C interactions of **4–6** to have positive energy density values H_b indicating that both interactions acquire an electrostatic nature, rather than a covalent character (see **Table 1**).

3.3 Aromaticity and ring strength of phenalenyl-based monomers and dimers

In order to assess the effect of substitution and dimerization on the monomers and dimers of **4–6** we conduct aromaticity delocalization index (AI) analysis. Two AI were determined for each monomer and dimer of **4–6**, one AI value considers all CC bonds while the second AI value considers only the outer most CC bonds which trace the species (the inner/central most CC bonds are not considered). In addition to AI values, **Figure 6** lists corresponding WS and ALT parameters, WS gives the weakening/strengthening parameter of the bonds in and *ALT* reflects the magnitude of bond strength alteration. Overall, the WS and ALT parameters reflect the loss of aromaticity which is attributed to increased structure irregularity. Therefore, the more symmetrical an aromatic perimeter, the greater the aromaticity (i.e. AI) of the system. For example, in the case of benzene, which is planar and very symmetrical as all CC sides (bonds) are identical, the parameters are as follows: WS = 0, ALT = 0, and AI = 1. In general, the smaller the AI the weaker the aromatic character of a species.

Phenalenyl-based monomers: We observe the six outer most CC bonds of the phenalenyl monomer (BSO n(CC) = 1.412) to be identical in strength to those of benzene (BSO n(CC) = 1.451). The addition of methyl substituents to the phenalenyl monomer, in the form of 2,5,8-trimethylphenalenyl, favors a skewed arrangement which places one H atom of every CH_3 group in plane with the phenalenyl rings and the other two H atoms of every CH_3 group above and below the plane of the rings (see **Figure 6**). From the BSO n values and bond distances of the six outermost CC bonds of the 2,5,8-trimethylphenalenyl monomer we observe the outer bonds to be dissimilar (see **Figure 6**). For the CC outer bonds, that are on the same side of the coplaner hydrogen atom of the CH_3 group, CC bond distances and BSO n values increase by 0.002 Å and decrease by 0.032 while that for the CC outer bonds, that are on the same side of the two CH_3 hydrogen atoms above and below the ring, increase by 0.008 Å and decrease by 0.065 in contrast to that of the phenalenyl monomer. A similar trend is observed for the substitution of phenalenyl with t-butyl substituents in the form of 2,5,8-tri-t-butylphenalenyl, where the six outer CC bonds become slightly longer and weaker in contrast to 2,5,8-trimethylphenalenyl (see **Figure 6**). In comparison to the phenalenyl monomer the CC outer bonds of 2,5,8-tri-t-butylphenalenyl, which are on the same side of the coplaner methyl group, become longer by 0.003 Å and weaker by 0.037 BSO n units while that for the outer CC bonds, that are on the same side of the methyl groups above and below the ring, stretch by 0.011 Å and decrease in strength by 0.053 units. For the outer CC bonds, not affiliated with the point of substituent attachment (periphery CC bonds), the effect of substitution is too a lesser extent with bond lengths ranging between 1.412 to 1.415 Å and the BSO n(CC) values ranging from 1.283 to 1.312. We note that **6** acquires the weakest outer and periphery CC bonds. Conversely, the three bonds which radiate from the central C (i.e. inner CC bonds) increase in strength from **4** to **5** and from **5** to **6** (see **Figure 6**). This indicates that electron density lost by the deformation of the outer CC bonds, occurring from monomer of **4** to **6**, redistributes to the inner bonds.

The AI (full/outer) values of the phenalenyl monomer are both 0.915. From monomers **4** to **6** we observe the AI, based upon the outer CC bonds, to decrease steadily while the AI, based upon all CC bonds, fluctuates. From the AI outer/full values of the phenalenyl (AI (full, outer) = 0.915), 2,5,8-trimethylphenalenyl (AI (full, outer) = 0.918, 0.911) and 2,5,8-tri-t-butylphenalenyl monomers (AI (full, outer) = 0.901, 0.885) we observe that the outer rings have a larger degree of π-delocalization than the full ring. From WS and ALT parameters we can see that the decrease in the aromatic character of the outer CC bonds from monomer **4** (WS, ALT = 0.062, 0.023), to **5** (WS, ALT = 0.077, 0.012), to **6** (WS, ALT = 0.108, 0.008) is predominantly due to bond weakening (as indicated by WS). Overall we observe that as the **4** monomer is substituted with CH_3 (**5**) and *tert*-butyl groups (**6**) the outer aromaticity decreases steadily and is predominately governed by bond weakening effects which are attributed to smaller magnitudes of π-delocalization as additional π-delocalization (i.e. electron density) is pushed away from the points of substitution and adjacent (periphery) CC bonds towards the inner most CC bonds as reflected from the increasing inner CC bond strength from **4** to **6**.

Phenalenyl-based dimers: We note that the trend in BSO *n* values observed amongst the CC bonds of the monomers discussed in the previous section is similarly observed for the CC bonds of their dimers (**4–6**). It is worth mentioning that the AI (outer/full) values for the dimers are greater than that of their monomer components (see **Figure 6**). The phenalenyl dimer **4**, in contrast to its monomer counterpart, has larger outer, peripheral, and central CC bond strength orders (BSO *n*(CC)) of 1.441, in very close proximity to that of benzene (1.451). We observe that the the *bigger* aromaticity of dimer **4** is predominately attributed to bond strengthening as revealed from a comparison between the WS parameters of the phenalenyl monomer (WS (full/outer) = 0.066, 0.062) and dimer (**4**) (WS (full/outer) = 0.043, 0.035).

From **Figure 6** it is shown that dimers **5** and **6** favor configurations which position the six methyl or *tert*-butyl groups amongst the dimers in an alternating manner yielding a symmetrical arrangement and in turn a stable species. We note that the methyl groups within the lowest energy rotational isomer of dimer **5** do not have the same orientation as those within its monomer as six hydrogen atoms of the CH_3 groups are rotated inward, towards the center of the molecule (see **Figure 6**). From WS and ALT parameters we see that the dimer of 2,5,8-trimethylphenalenyl (5) has a greater outer CC aromaticity (AI (outer) = 0.911 (monomer), 0.914 (dimer)) than its monomer due to bond strengthening (WS (full, outer) = 0.077 (monomer), 0.075 (dimer)). We note that this result is consistent with the BSO *n* values of the peripheral and central CC bonds of dimer **5**, which are greater than those of the monomer by 0.012 to 0.026 units (see **Figure 6**). In contrast to the phenalenyl dimer (**4**), **5** has much larger WS (full/outer) and smaller ALT (full/outer) parameters, where the WS parameters are more altered than the ALT parameters (see **Figure 6**). These results reveal that the aromaticity of the 2,5,8-trimethylphenalenyl dimer (5) (AI (full/outer) = 0.918, 0.914) is *less* than that of the phenanlenyl dimer (AI (full/outer) = 0.934, 0.938) primarily due to bond weakening (indicated by WS, see **Figure 6**). The outer/full AI values of the 2,5,8-tri-t-butylphenalenyl dimer (6) are both *bigger* than its monomer counterpart being primarily due to bond strengthening as observed from the smaller WS (outer/ full) parameters of the dimer in contrast to that of it monomer (see **Figure 6**). It is also notable that changes in AI (outer/full), when comparing monomer to monomer, monomer to dimer, or dimer to dimer, do not correspond directly to changes in CC bond lengths, in some instances these lengths stay the same or do not drastically change unlike BSO *n* (CC) orders (see **Figure 6**).

From our results, it is clear that substituents not only prevent σ-dimer formation but reduce the overall aromaticity of both phenalenyl-based monomers and the dimers. As noted, the dimeric systems display a higher AI than the monomeric systems indicating that the dimerization of phenalenyl-based species enhances the aromaticity of the species. Our observation is in line with the nucleus-independent chemical shift (NICS) NMR analysis of Suzuki et al. [21], which suggets that SOMO-SOMO overlap in the dimerized system, overall, supports and stabilizes the aromaticity of the molecules. Furthermore, our work supports the suggestions of Gleiter and Haberhauer who propose that dimers which are pancake bonded undergo stabilization via electron combination as to create a Hückel-allowed [62] (4n + 2 electron) 3-dimensional aromatic system as we observe that, despite the fact that the dimers, unlike their monomers, are not planar (which reduces orbital overlap), the dimers exhibit higher aromaticity. Ultimately, from AI, WS, ALT, and BSO *n* parameters, we discover that the dimerization of phenalenyl-based monomers increases the aromaticity of the phenalenyl rings predominantly through CC bond strengthening while the substitution of the phenalenyl dimer, alongside inhibiting σ-dimerization, reduces the overall aromaticity of the system predominantly through CC bond weakening.

4. Conclusions

In this work, we conducted local mode analysis, electron density analysis, and aromaticity delocalization index (AI) calculations (based upon vibrational frequencies) for a set of six neutral pancake-bonded systems, di-chalcodiazoyl dimers (**1–3**) and phenalenyl-based dimers (**4–6**), as to elucidate on the strength of pancake bond interactions within dimers, the ring strength of their monomers, the nature of the pancake bond interactions, the effect of substituents on the aromaticity of phenalenyl-based species, and the effect of dimerization on the aromaticity for phenalenyl-based species. The local stretching force constants, being suitable descriptors of bond strength and π-delocalization, are used to describe the pancake bond interactions of **1–6** and the degree of π-delocalization amongst these bonds and their corresponding dimer species. Directly from computed local stretching force constants we derived bond strength orders. We use measures of AI, and corresponding WS and ALT parameters, to determine what bond property, of the phenalenyl-based species investigated, predominately governs changes in aromaticity. From the results of our work we draw the following: [**1**] We find that dimer species **1** (1,2,3,5-dithiadiazolyl) and **2** (1,2,3,5-diselenadiazolyl) are significantly stabilized by their chalcogen···chalcogen contacts. Unlike **1** and **2**, which have C_{2v} symmtery, the 1,2,3,5-ditelluradiazolyl (**3**) dimer is found to be stable in C_2 symmetry as the singlet state is energetically favored over the triplet state, revealed from a negative ΔE_{ST}. [**2**] In regard to the phenalenyl-based dimers, as the substituent size increased from **4** to **6** the stability of the system steadily declined as the steric repulsion between the substituent groups hindered the monomers of these dimers from changing into a orientation of lower energy. [**3**] As the radius of the chalcogen atoms di-chalcodiazoyl dimers **1–3** increase (Te $<$ Se $<$ S) the strength of the C···C contacts decreases. As the strength of the chalcogen···chalcogen interactions (i.e. contacts) decrease from **1** to **3** the overall ring strength decreases and the strength of the central (i.e. interdimer) C–C bond decreases [**4**]. For all phenalenyl-based dimers (**4–6**) we observed that the BSO n values of peripheral C···C are stronger that of their corresponding central C–C bonds. Revealing that pancake bonding interactions contribute largely to the stability of these species [**5**]. From energy density analysis H_b, following the Cremer-Kraka criteria, we observe the chalcogen···chalcogen pancake bonding interactions of the 1,2-dithia-3,5-diazolyl dimer (**1**) and 1,2-diselena-3,5-diazolyl dimer (**2**) are covalent in nature as they have negative (stabilizing) H_b values at their bond critical point r_b. [**6**] Unlike the other 1,2-chalcogen-3,5-diazole dimers (**1** and **2**) the chalcogen···chalcogen contacts (i.e. Te···Te) of **3** are much weaker in strength and have a positive (destabilizing) energy density value H_b at the Te···Te bond critical point r_b revealing that the Te···Te do not have a typical pancake bond nature as we observed **1** and **2**. [**7**] All pancake bonding interactions within the phenalenyl dimer (**4**), 2,5,8-trimethylphenalenyldimer (**5**), and the 2,5,8-tri-t-butylphenalenyl dimer (**6**) were observed to have postive (destabilizing) H_b values revealing that their pancake interactions are electrostatic in nature. [**8**] From BSO n(CC) values, the calculated AI, and related WS and ALT parameters we found that the dimerization of phenalenyl-based monomers leads to an increased aromaticity primarily due to CC bond strengthening. [**9**] From the same parameters mentioned above we observed that the substitution of the phenalenyl dimer, which is necessary for inhibiting

σ-dimerization, results in an overall reduction of system aromaticity predominantly through CC bond weakening.

References

[1] Mulliken RS, Person WB. Molecular Complexes: Chapter 16 – Inner and Outer Complexes with π-Acceptors. Hoboken, NJ: Wiley-Interscience; 1969.

[2] Gleiter R, Haberhauer G. Chapter 3: Aromaticity and Other Conjugation Effects. VCH, Weinheim: Wiley; 2012.

[3] Boeré RT. Experimental and computational evidence for "double pancake bonds": The role of dispersion-corrected DFT methods in strongly dimerized 5-aryl-1λ^2,3λ^2-dithia-2,4,6-triazines. ACS Omega. 2018;3(12): 18170–18180.

[4] Kertesz M. Pancake bonding: An unusual pi-stacking interaction. Chem Eur J. 2018;25(2):400–416.

[5] Lemes MA, Mavragani N, Richardson P, Zhang Y, Gabidullin B, Brusso JL, et al. Unprecedented intramolecular pancake bonding in a Dy_2 single-molecule magnet. Inorg. 2020;7(14):2592–2601.

[6] Tonami T, Nagami T, Okada K, Yoshida W, Nakano M. Singlet-fission-induced enhancement of third-order nonlinear optical properties of pentacene dimers. ACS Omega. 2019;4 (14):16181–16190.

[7] Yuan D, Liu W, Zhu X. Design and applications of single-component radical conductors. Chem. 2021 feb;7(2):333–357.

[8] Beneberu H, Tianza Y, Kertesz M. Bonds or not bonds? Pancake bonding in 1,2,3,5-dithiadiazolyl and 1,2,3,5-diselenadiazolyl radical dimers and their derivatives. Phys Chem Chem Phys. 2012;14:10713–10725.

[9] Takano Y, Taniguchi T, Isobe H, Kubo T, Morita Y, Yamamoto K, K Nakasuji, et al. Hybrid density functional theory studies on the magnetic interactions and the weak covalent bonding for the phenalenyl radical dimeric pair. J Amer Chem Soc. 2002;124:11122–11130.

[10] Mouesca JM. Density Functional Theory–Broken Symmetry (DFT–BS) Methodology Applied to Electronic and Magnetic Properties of Bioinorganic Prosthetic Groups. In: Methods in Molecular Biology. Humana Press; 2014. p. 269–296.

[11] Hobza P. The calculation of intermolecular interaction energies. Annu Rep Prog Chem, Sect C: Phys Chem. 2011;107:148.

[12] Mou Z, Uchida K, Kubo T, Kertesz M. Evidence of σ- and π-dimerization in a series of phenalenyls. J Chem Soc. 2014;136:18009–18022.

[13] Gleiter R, Haberhauer G. Electron-rich two-, three- and four-center bonds between chalcogens – New prospects for old molecules. Coord Chem Rev. 2017; 344:263–298.

[14] Mota F, Miller JS, Novoa JJ. Comparative analysis of the multicenter, long bond in $[TCNE]^{\cdot -}$ and phenalenyl radical dimers: A unified description of multicenter, long bonds. J Am Chem Soc. 2009;131:7699–7707.

[15] Takamuku S, Nakano M, Kertesz M. Intramolecular pancake bonding in helical structures. Chem Eur J. 2017;23: 7474–7482.

[16] Nyburg SC, Faerman CH. A revision of van der Waals atomic radii for molecular crystals: N, O, F, S, Cl, Se, Br and I bonded to carbon. Acta Crystallogr B Struct Sci. 1985;41(4):274–279.

[17] Brown JT, Zeller M, Rosokha SV. Effects of structural variations on π-dimer formation: Long-distance multicenter bonding of cation-radicals

of tetrathiafulvalene analogues. Phys Chem Chem Phys. 2020;22(43):25054–25065.

[18] Reid DH. Stable π-electron systems and new aromatic structures. Tetrahedron. 1958;3(3):339–352.

[19] Kubo T. Synthesis, physical properties, and reactivity of stable, π-conjugated, carbon-centered radicals. Molecules. 2019 feb;24(4):665.

[20] Zhong RL, Gao FW, Xu HL, Su ZM. Strong pancake 2e/12c bond in π-stacking phenalenyl derivatives avoiding bond conversion. ChemPhysChem. 2019;20(14):1879–1884.

[21] Suzuki S, Morita Y, Fukui K, Sato K, Shiomi D, Takui T, et al. Aromaticity on the pancake-bonded dimer of neutral phenalenyl radical as studied by MS and NMR spectroscopies and NICS analysis. J Am Chem Soc. 2006;128:2530–2531.

[22] Groom CR, Bruno IJ, Lightfoot MP, Ward SC. The Cambridge structural database. Acta Crystallogr B Struct Sci Cryst Eng Mater. 2016;72(2):171–179.

[23] Small D, Rosokha SV, Kochi JK, Head-Gordon M. Characterizating the dimerizations of phenalenyl radicals by ab initio calculations and spectroscopy: σ-bond formation versus resonance π-stabilization. J Phys Chem A. 2005;109(49):11261–11267.

[24] Andrés J, Ayers PW, Boto RA, Carbó-Dorca R, Chermette H, Cioslowski J, et al. Nine questions on energy decomposition analysis. J Comput Chem. 2019:doi.org/10.1002/jcc.26003.

[25] Kraka E, Zou W, Tao Y. Decoding chemical information from vibrational spectroscopy data: Local vibrational mode theory. WIREs: Comput Mol Sci. 2020;10:1480.

[26] Krapp A, Bickelhaupt FM, Frenking G. Orbital overlap and chemical bonding. Chem Eur J. 2006;12(36):9196–9216.

[27] Zhao L, Hermann M, Schwarz WHE, Frenking G. The Lewis electron-pair bonding model: Modern energy decomposition analysis. Nat Rev Chem. 2019;3(1):48–63.

[28] Cremer D, Pople JA. General definition of ring puckering coordinates. J Am Chem Soc. 1975;97:1354–1358.

[29] Bader RFW. Atoms in Molecules: A Quantum Theory. Oxford: Clarendron Press; 1995.

[30] Wilson EB, Decius JC, Cross PC. Molecular Vibrations. The Theory of Infrared and Raman Vibrational Spectra. New York: McGraw-Hill; 1955.

[31] Kraka E, Larsson JA, Cremer D. Generalization of the badger rule based on the use of adiabatic vibrational modes. In: Grunenberg J, editor. Computational Spectroscopy. New York: Wiley; 2010. p. 105–149.

[32] Delgado AAA, Humason A, Kalescky R, Freindorf M, Kraka E. Exceptionally long covalent CC Bonds – A local vibrational mode study. Molecules. 2021;26:950-1–950-25.

[33] Kalescky R, Kraka E, Cremer D. Description of aromaticity with the help of vibrational spectroscopy: Anthracene and phenanthrene. J Phys Chem A. 2013;118:223–237.

[34] Setiawan D, Kraka E, Cremer D. Quantitative assessment of aromaticity and antiaromaticity utilizing vibrational spectroscopy. J Org Chem. 2016;81:9669–9686.

[35] Krygowski TM, Cyrański MK. Structural aspects of aromaticity. Chemical Reviews. 2001;101(5):1385–1420.

[36] Cremer D, Kraka E. A description of the chemical bond in terms of local properties of electron density and energy. Croatica Chem Acta. 1984;57: 1259–1281.

[37] Kraka E, Cremer D. Chemical implication of local features of the electron density distribution. In: Theoretical Models of Chemical Bonding. The Concept of the Chemical Bond. vol. 2. Z.B. Maksic, ed., Springer Verlag, Heidelberg; 1990. p. 453–542.

[38] Gräfenstein J, Kraka E, Filatov M, Cremer D. Can unrestricted density-functional theory describe open shell singlet biradicals? Int J Mol Sci. 2002;3: 360–394.

[39] Crawford TD, Kraka E, Stanton JF, Cremer D. Problematic *p*-benzyne: Orbital instabilities, biradical character, and broken symmetry. J Chem Phys. 2001;114:10638–10650.

[40] Cui Zh, Gupta A, Lischka H, Kertesz M. Concave or convex π-dimers: The role of the pancake bond in substituted phenalenyl radical dimers. Phys Chem Chem Phys. 2015;17:23963–23969.

[41] Becke AD. Density-functional exchange-energy approximation with correct asymptotic behavior. Phys Rev A. 1988;38(6):3098–3101.

[42] Lee C, Yang W, Parr RG. Development of the Colle-Salvetti correlation-energy formula into a functional of the electron density. Phys Rev B. 1988;37(2):785.

[43] Chai JD, Head-Gordon M. Long-range corrected hybrid density functionals with damped atom–atom dispersion corrections. Phys Chem Chem Phys. 2008;10(44):6615–6620.

[44] Mardirossian N, Head-Gordon M. Thirty years of density functional theory in computational chemistry: An overview and extensive assessment of 200 density functionals. Mol Phys. 2017; 115(19):2315–2372.

[45] Zhao Y, Truhlar DG. The M06 suite of density functionals for main group thermochemistry, thermochemical kinetics, noncovalent interactions, excited states, and transition elements: Two new functionals and systematic testing of four M06-class functionals and 12 other functionals. Theor Chem Acc. 2008;120(1–3):215–241.

[46] Mou Z, Tian YH, Kertesz M. Validation of density functionals for pancake-bonded π-dimers: Dispersion is not enough. Phys Chem Chem Phys. 2017;19(36):24761–24768.

[47] Zhao Y, Schultz NE, Truhlar DG. Design of density functionals by combining the method of constraint satisfaction with parametrization for thermochemistry, thermochemical kinetics, and noncovalent interactions. J Chem Theory Comput. 2006;2(2):364–382.

[48] Slepetz B, Kertesz M. Volume change during termal [4 + 4] cycloaddition of [2.2](9,10) anthracenophane. J Am Chem Soc. 2013; 135:13720–13727.

[49] Hehre WJ, Ditchfield R, Pople JA. Self—Consistent molecular orbital methods. XII. Further extensions of Gaussian—Type basis sets for use in molecular orbital studies of organic molecules. J Chem Phys. 1972;56(5): 2257–2261.

[50] Clark T, Chandrasekhar J, Spitznagel GW, Schleyer PVR. Efficient diffuse function-augmented basis sets for anion calculations. III. The 3-21+ G basis set for first-row elements, Li–F. J Comput Chem. 1983;4(3):294–301.

[51] Krishnan R, Binkley JS, Seeger R, Pople JA. Self-consistent molecular orbital methods. XX. A basis set for

correlated wave functions. J Chem Phys. 1980;72(1):650–654.

[52] Andrae D, Häußermann U, Dolg M, Stoll H, Preuß H. Energy-adjusted ab initio pseudopotentials for the second and third row transition elements. Theoret Chim Acta. 1990;77: 123–141.

[53] Stoll H, Metz B, Dolg M. Relativistic energy-consistent pseudopotentials-recent developments. Int J Quant Chem. 2002;23:767–778.

[54] Frisch MJ, Trucks GW, Schlegel HB, Scuseria GE, Robb MA, Cheeseman JR, et al. Gaussian 09, Revision C.01. Gaussian, Inc.; 2009.

[55] Zou W, Tao Y, Freindorf M, Makoś MZ, Verma N, Cremer D, et al. Local Vibrational Mode Analysis (LModeA); 2020. Computational and Theoretical Chemistry Group (CATCO), Southern Methodist University: Dallas, TX, USA.

[56] Keith TA. AIMAll (Version 17.01.25). TK Gristmill Software, Overland Park, KS, USA; 2017.

[57] Dimitry Izotov WZ, Cremer D, Kraka E. Ring Puckering Analysis (RING); 2021. Computational and Theoretical Chemistry Group (CATCO), Southern Methodist University: Dallas, TX, USA.

[58] Haberhauer G, Gleiter R. Double pancake versus chalogen-chalogen bonds in six-membered C,N,S-heterocycles. Chem Eur J. 2016;22: 8646–8653.

[59] Small D, Zaitsev V, Jung Y, Rosokha SV, Head-Gordon M, Kochi JK. Intermolecular π-to-π bonding between stacked aromatic dyads. Experimental and theoretical binding energies and near-IR optical transitions for phenalenyl radical/radical versus radical/cation dimerizations. J Am Chem Soc. 2004;126(42):13850–13858.

[60] Oliveira V, Cremer D, Kraka E. The many facets of chalcogen bonding: Described by vibrational spectroscopy. J Phys Chem A. 2017;121:6845–6862.

[61] Gleiter R, Haberhauer G. Long chalogen-chalogen bonds is electron-rich two and four-center bonds: Combination of π- and σ-aromaticity to a three-dimensional σ/π-aromaticity. J Org Chem. 2014;79:7543–7552.

[62] Hückel E. Quantentheoretische Beiträge zum Benzolproblem. I. Die Elektronenkonfiguration des Benzols und Werwandter Verbindungen. Z Phys Chem. 1931;70:204–286.

Chapter 8

The Formation Mechanism and Structure of Organic Liquids in the DFT Challenges

Abstract

In the paper the experimental and theoretical approaches to problem of organic liquids formation mechanism and its structure are reviewed. It was shown that all presented models have the advantages and disadvantages at interpretation of molecular interaction and arrangement in liquid phase. The DFT calculation in different variant of models including paired interaction hydrogen atom transfer, model of transformation and the general conclusion following from this consideration are presented.

Keywords: DFT calculation, organic liquids, model of formation

1. Introduction

Chemistry have stepped in the last 40–50 years to the incredible mountain. Nowadays the scientists can synthesize large composite organic molecules including important biological active species and apply physicochemical instruments with digital operating and rapid registration to the insight of pure and technological processes.

However, the current formation mechanism theory and organic liquid structure models can be questioned. Even in the late 50s–early 60s, attempts were made, basically by spectral methods, to study the nature of interactions in liquid phase of organics that lead to its high resistance to external actions and at the same time preserve its mobile properties. After numerous studies on this topic, at one point there was a pause. However, no intermolecular forces, except for classical hydrogen bonds, have been discovered.

The hope arose that with the discovery of new diffraction wave scattering methods with Fourier transformation for studying liquid substances, this problem would be clarified. Nevertheless, back in 2005, S. Ballint, studying liquid dichloromethane, concluded that the internal parameters of individual molecules that form a liquid-phase system are well defined, but their intermolecular arrangement is ambiguous.

This is strange, but we have not found a single comprehensive review that would include a comparative analysis of different approaches, including *ab initio* calculations, in a unified vision. The theory of dipole–dipole coupling, in other words Coulomb interactions, is generally accepted, although in this theory there are many contradictions with the modern theoretical and experimental findings. At the same time, the concept of specific intermolecular interactions existence in organic liquids is verified by structural, thermodynamic and spectroscopic studies. These studies are presented and discussed in Section 2.

Ten years ago, when conducting the IR investigation of simplest organic solvents to account their role in the reactivity of some organic substrates, we have paid attention to the IR bands, which could not be assigned to any known internal mode of substance. In terms of classical vibrational spectra theory these bands can be interpreted as overtones or combination bands, if any, or Fermi resonance. However, we have suggested another assignment that was combined with DFT calculations and allowed us to make the assumption that these bands are a manifestation of specific interactions in organic liquids. In first regard, our arguments were based on the principle that mutually exclusive approaches of classical and quantum mechanics cannot be mixed in the same approach. The results of our investigations are presented and discussed in detail across Sections 2–3.

The analysis of current theoretical and experimental methods exhibits the value of DFT in the knowledge progress in the field of organic liquid phase molecular arrangements. We have presented the traditional DFT calculations and our approach to this problem as well in Section 3. Unlike quoted literature that use the paired interactions model for description of condensed phase, which arises, as a rule, under hydrogen bond, in our approach, the mechanism of liquid phase formation is considered in terms of molecular transformations. This insight applies not only to the substances with hydrogen bond, but also to the systems of identical molecules without hydrogen bonds. The conclusions of our research are formulated in Section 4.

2. Theoretical and experimental approach to the molecular arrangement in organic liquids

In this section the fundamentals of molecular structure and its formation mechanism in liquid organic phase are presented, including the experimental data and theoretical study (dynamic simulation theory) in combination with the authors analysis of this approach.

Although there are numerous papers and books devoted to this problem, none of them fully resolved the problem of organic liquids molecular arrangement, and the character of long-range molecular interaction in liquid bulk is still ambiguous. Therefore, in the first part of this chapter we tried to present the current state of considered problem in terms of theoretical and experimental approach to evaluate the role of DFT calculation in the solution of the presented problem.

However, at the beginning of this consideration we would like to review the dipole–dipole interaction concept briefly because it is the oldest and most common theory of liquid formation. In fact, this theory is based on the model of Coulomb interactions between polar systems or temporary dipoles caused by fluctuation in electronic distribution.

2.1 Classical dipoles interaction model

Cooled and pressed real gases become liquid due to the intermolecular interactions called van der Waals. This intermolecular interaction energy can be defined as the heat of a liquid evaporation or rather the difference between the vaporization temperature and the work of expanding one mole of gas at the atmospheric pressure. Value of this interaction at the boiling point of gases is 1–3 kJ/mol. As it was shown by the quantum mechanical calculations, the energy of the van der Waals interactions consists of electrostatic, inductive, and dispersive components [1].

The so-called orientational interaction of polar molecules is the most important part of the electrostatic interaction. Its essence, known as the Keesome effect, consists in the orientation of two interacting polar molecules with identical dipole moments that leads to the minimization of the system energy. In this case, the head-to-tail orientation becomes the most advantageous configuration. The energy of the orientational binding can be calculated as the sum of the Coulomb attractions and repulsion of the pole charges in the dipoles, expressed in terms of the dipoles [2].

Inductive interaction, or the Debye effect [3], is the interaction of the constant dipole moment and the induced dipole moment, arising due to an additional charge separation. Inductive interactions occur with the formation of noble gases hydrates with the dissolution of polar substances in the non-polar liquids and they are valuable only for the molecules with significant polarizability (ex. molecules with conjugated bonds) [3].

The dispersion interaction, or the London effect [4], arises between the electrons in the interacting molecules. There are molecules that have no dipole moment – for example, homoatomic molecules of noble gases. The additivity of dispersion forces is manifested in adsorption, in the processes involving gas condensation, etc. Dispersion forces play the important role not only in the individual molecules, but also in the macroscopic particles (ex. colloids). However, these forces are relatively weak.

The intermolecular interaction is the summarizing action of attraction and repulsion forces. At large distances, the attraction prevails, and at very short distances the repulsion is the main contributor. The attractive forces of Van der Waals are long ranged [5, 6], and the attraction energy rather decreases with distance. Van der Waals interaction forces at the equilibrium distance is small: ~ 1–5 kJ/mol, which is considerably less than the chemical bond energy.

The main conclusion is that Coulomb and Van der Waals interactions can relate to the formation of liquid from the gaseous substances at the cooling or pressing. However, this theory is very limited for organic liquids. There are at least three fundamental reasons for that: i) the intermolecular forces in organic liquids are sufficiently stronger, accounting their vaporization enthalpy [7]; ii) there is no direct dependence of liquid stability on dipole and molecular volume of consisting molecules in many cases [7]; iii) the interaction of dipoles in organic fluids should have a certain orientation, while the molecules in liquid do not have any anisotropy [5, 6]; iv) this conception ignores the specific non-valence interaction in liquid systems [7].

2.2 Thermodynamic data source

The concept of the specific interaction in liquid systems based on the thermodynamic data is developed in [7]. The regularities in the homological series exhibit no correlation between the molecular mass and the vaporization enthalpy in many cases. The numerous thermodynamic data indicate the existence of specific non-valence interaction in organic liquids. In case of substances with the different variant of hydrogen bond (ex. with OH–, NH_i–, C=O– or COOH–group) the specific interaction providing molecular structure can be explained by classical H–donor–H–acceptor interaction [7].

However, for the liquid saturated hydrocarbons, alkenes, and alkynes this approach is not applicable. The author of [7] has suggested the scheme of hydrocarbon interactions. The main idea is that the carbon atoms in chains have a different negative charge owing the shift of 2 *s*–electronic pair between the carbon atoms. The electron density of carbon with a larger negative charge transfers partially to the carbon atom with the reduced negative charge.

In other words, the alkane molecules can be bound in the liquid state so that the carbon atoms with an enhanced negative charge would be closer to the carbon atom with a reduced charge. The carbon atoms of the methyl groups in hydrocarbons that have the enhanced negative charge can participate in the specific interaction between molecules under the donor–acceptor mechanism resembling the hydrogen bond.

In author's opinion, the non-equivalence charges distribution in hydrocarbon chains leads to the appearance of charge shift not only for the carbon atom, but for the hydrogen atom as well. It means that the hydrogen atom of the neighboring molecules can form the donor–acceptor C···H bond and even the dihydrogen bond. The comparison of thermodynamic data indicates the relatively high contribution of similar pair interactions in vaporization energy of organic liquids. These specific interactions are the nature of liquid saturated hydrocarbons formation mechanism and their stability without involving of the specific hydrogen bon. The evaluation of specific interaction energy is given in [7] and relates to the range 3–10 kJ/mol. These values correspond to the weak hydrogen bond [1, 2, 6].

The thermodynamic analysis of the vaporization enthalpy of unsaturated hydrocarbons leads to the conclusion that the presence of a double or a triple bond in the molecule results in a stronger electron density shift from one or two hydrogen atoms to the carbon atoms of neighboring molecule [7]. For example, the hydrogen atoms of the CH_2–group in propene and CH–group in propyne possess enhanced positive charges, and the carbon atoms acquire higher negative charges. Therefore, the acceptor and donor properties of these substances are more pronounced than in the propane molecule (this statement are cited fully in accordance with [7]).

2.3 Molecular light scattering (MLS) and X-ray data source

In the 70s – 80s the professor at Moscow University M.I. Shachparonov has published a work, which was, unfortunately, limitedly known in the scholar community [8, 9]. In this paper he has predicted the existence of the specific interaction in the non-polar liquids, in which the hydrogen bond lacks. For benzene, the existence

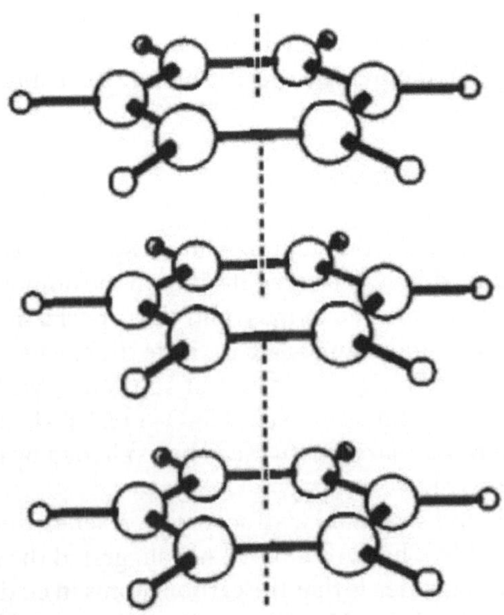

Figure 1.
The stack of aromatic rings in liquid benzene.

of a new specific interaction in liquids – the formation of molecular π-complexes (molecules stacks) between aromatic rings – was suggested (**Figure 1**).

At present, this concept was confirmed and developed in the works of I.A. Abramovich and the coauthors [10–17], devoted to the structure of the liquid system in benzene and its substituted species. The specific interaction with different spatial geometry can form in a liquid bulk of molecules without hydrogen bonds. For example, in solution of *di*-chlorobenzene, there are strong interactions in neighboring molecules between chlorine atom and carbon in aromatic ring and between carbon atoms.

Since the arrangements of molecules in a crystal always correspond to the potential energy minima of the system, it can be used for the verification of the obtained data in LS experiments combined with the modeling procedure. This approach allowed to reveal the structure and characteristics of the mutual arrangement in molecules as well as the types of intermolecular contacts in the liquid phase [18, 19].

For the determination of intermolecular arrangement in liquid *di*-chlorobenzene and other compounds with chlorine atoms, the Cambridge crystallographic database for these organics in solid state were used. The intermolecular distances between molecules, obtained by MLS study, are 4–5 Å, which is considerably longer than the internal distances in molecules. However, the liquid system has a quite certain spatial geometry realized by this long-ranged molecular binding.

X-ray studies allow to define the internal parameters of molecules, but the intermolecular structure of organic liquids stays in an ambiguous conclusion: on one hand, in the liquid phase the distances between nearest molecules are longer than 4 Å, and on the other hand, the peak of distribution function was observed in dichloromethane at distance about 2 Å. However, in these experiments it was confirmed that the organic liquids have certain intermolecular composition and labile binding between identical molecules. However, this does not exclude the spatial transformation in liquid bulk [20–22].

2.4 IR-data source

The IR spectra of organic liquids in the middle- and high-frequency regions have "additional" bands [23] that cannot be expected by the normal coordinates' analysis [24–26]. The observed spectral phenomenon was interpreted in literature as a manifestation of the vibration anharmonicity or Fermi resonance (ex. [27, 28]). Such a version has a few contradictories; firstly, it is non-logic to mix the approach using the vibrational theory based on the classical mechanics of vibration (normal coordinates' analysis) and the quantum mechanical interpretation; secondly, many systems that are considered in this conception include heavy molecular weight atoms and their vibration anharmonicity is negligible; thirdly, the overtones and combination bands should have considerably lower intensities than the basic bands. At last, these bands remain in the solid state (in a low-temperature spectra), and they are observed for homological analogs [20]. We have suggested another variant of assignment based on the conception observing spectral features relating to the manifestation of specific interaction in liquids.

In the spectra of liquid CCl_4, in accordance with the selection rules for the T_d symmetry in IR spectra, only one stretching band should be active, but for the C_{3V} symmetry, two bands of E– and A_1–symmetry species are permitted. In the IR spectrum of liquid carbon tetrachloride two overlapping bands at 786 and 761 cm^{-1} having an approximately equal intensity were observed. These bands are well-resolved in IR spectrum of the low-temperature film recorded at 20 K [29].

In the gas phase two bands were also observed (at 795 and 776 cm^{-1}), but they have a different counter structure: the first of them consists of overlapped components and another one is a single band. Therefore, we can conclude that carbon tetrachloride exists in gas phase not only in the single molecular shape that has T_d symmetry, but also in the transformed shape having C_{3V} symmetry. The first one is manifested by the band at 776 cm^{-1} and another one – by the band at 795 cm^{-1}. The spectra recorded for different temperatures of gaseous CCl_4 confirm this assumption [29].

It is reasonable to resume that the pyramidal structure relates to the cluster shape, in which the chlorine atom provides the binding between molecules. The shift of the chlorine atom in the cluster can occur in the condensed phase owing to the association of the molecules [29]. It leads to transformation of the molecular geometry to the almost planar D_{3h} symmetry, in which A_1 stretching band is forbidden in IR spectra.

In the high-frequency region of chloroform in the liquid phase, the bands at 3020 cm^{-1} (stretching vibration of the CH bond) and at 2401 cm^{-1} with a shoulder at 2435 cm^{-1} are detected (**Figure 2**, left side, spectrum 2) [30]. A similar spectrum is observed for bromoform: the bands at 3021 и 2256 cm^{-1} are shown (**Figure 2**, left side, spectrum 1). The bands in 2400–2200 cm^{-1} range cannot be assigned to overtones 2δ (bending of angle CHalH, where Hal-halogen), because the band have the isotopic D/H shift (**Figure 2**, right side), corresponding to the theoretical values (1.32–1.34) and their intensities relative to CH stretching band at 3020–3021 cm^{-1} are considerably stronger than it could be expected for anharmonic components. Besides, in bromoform molecule, the contribution of anharmonic components should decrease due to a significant increase of molecule mass. However, the opposite picture is observed in the spectra: the relative intensity of the band at 2256 cm^{-1} for bromoform is stronger than the band at 2401 cm^{-1} of chloroform [30].

The similar spectral picture is detected for water associates in liquid phase: the band of OH stretching vibration in 3400–3600 cm^{-1} region is combined with the band in 2200–2400 cm^{-1} range, assigned to stretching vibration of hydronium ion. This band manifests the hydrogen atom transfer in hydrogen bonded structures [31, 32].

The presented results indicate that the specific interaction between molecules exists in liquid haloforms due to the proton binding and its intermolecular shift, leading to the transformation of initial geometry.

The geometry of single benzene molecule is taken as a planar ring due to the conjugation of p_z-orbitals and π-aromatic configuration appearance. This state corresponds to D_{6h} symmetry point group. According to selection rules, only one stretching CH band (E-specie) should be active in IR spectra. However, in the real spectra of liquid benzene there are three bands (3092, 3071, 3036 cm^{-1}) and in the

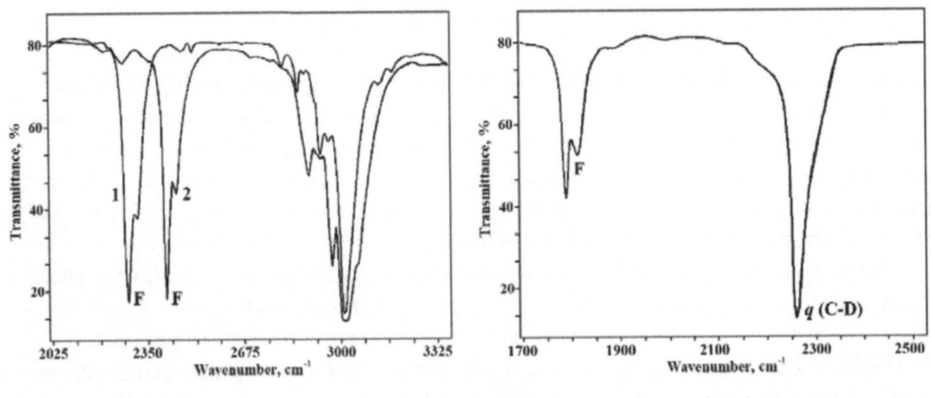

Figure 2.
Fragments of FTIR spectra of chloroform and bromoform (left side) and chloroform-d in liquid phase (right side).

solid benzene (spectra were recorded at 20 K) there are four bands (3090, 3071, 3034, 3005 cm^{-1}) (**Figure 3**, right side) in CH stretching region [33].

Three bands that have isotopic H/D shift (**Figure 3**, left side) close to the theoretical prediction also were detected in benzene-d_6 spectrum. This spectral picture can be assigned to the existence of two molecular shapes existing in the liquid phase: planar, in which one IR band is active (D_{6h} symmetry), and shape with two IR bands (A_1- and E-species), corresponding to the C_{3V} symmetry.

For the interpretation of the observed effect, we can assume that the benzene molecule exists in the liquid state as a cluster system. In this shape CH bonds deviate from the ring plane to the neighboring molecule (**Figure 4**). This leads to a distortion of the symmetry of the stretching vibrations of the CH bond.

In the middle IR range two bands at 1952, 1814 cm^{-1} were revealed (**Figure 5**, bands A), which cannot be assigned to the internal vibration modes. They also cannot be assigned to combination modes, because these bands have an isotopic shift close to the CH bond vibrations of aromatic ring, and the same bands are observed in substituted homologs of benzene as well (**Figure 5**). Besides, their amount in C_6H_5X spectra corresponds to the non-equivalent *o*-, *m*- and *p*-position in molecules [23]. These data confirm the idea that the benzene can form the π-stacks structure in the liquid phase (see section 2.3). The hydrogen atom in this system, as it was mentioned above, can interact with the carbon atom of the neighboring molecule in the stacks. Therefore, CH stretching band can shift to the middle IR interval due to the mixing of ν(C–C) and ν(C–H) stretching vibrations in the intermolecular bond (C–C–H···C–C).

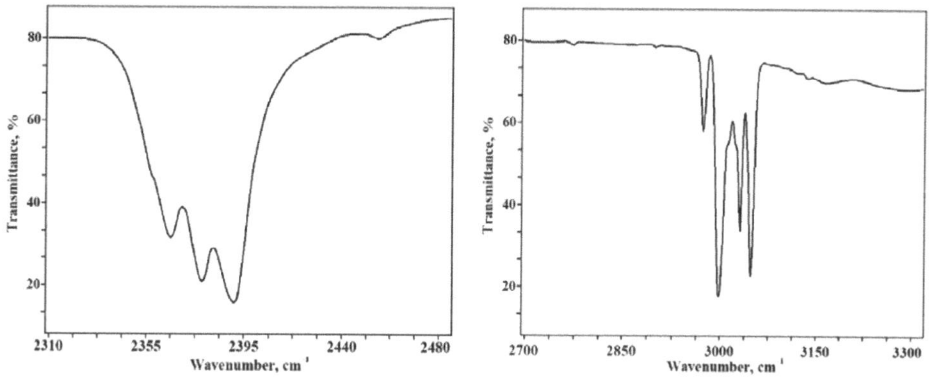

Figure 3.
Fragment of liquid benzene-d_6 FTIR spectrum at 295 K (left side) and solid benzene film at 20 K (right side) in CH/CD stretching region.

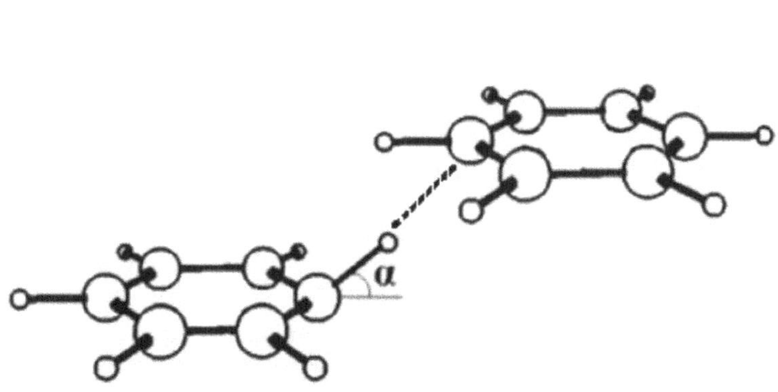

Figure 4.
Hydrogen atom shift in molecular stacks of liquid benzene.

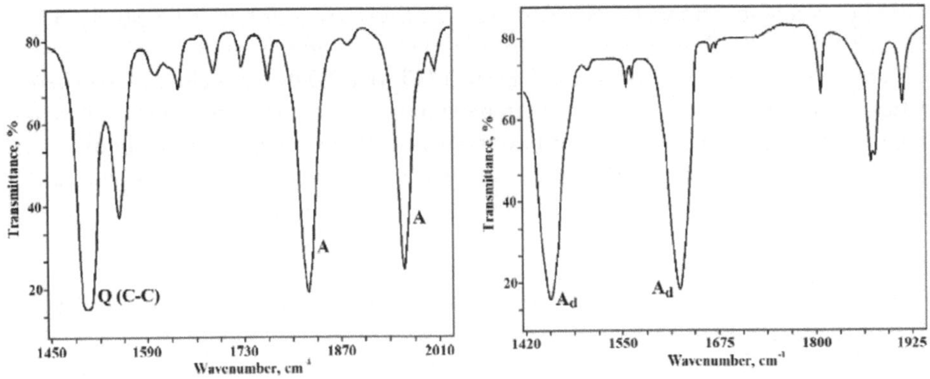

Figure 5.
Fragment of liquid benzene (left side) and benzene-d_6 (right side) FTIR spectrum in middle IR region.

2.5 Dynamic simulation data source

This approach was developed in numerous works and their complete citation was beyond the scope of this study. Therefore, we would like to highlight the studies that present the molecular arrangement of organic systems as a supramolecular structure, forming not only under hydrogen bond, but also existing in the shape of identical molecular associates [34–37].

The obtained data allow to discuss the thermodynamic and kinetic aspects of the processes in liquid phase, in terms of the supramolecular organization. These investigations give the knowledge about the composition of the aggregates as well as the general approach to their transformation due to the intermolecular binding. In author's opinion, the dynamic simulation model combined with the experimental findings allows to explain the supramolecular formation mechanism under the long-ranged interactions. The model of the homogeneous molecules' association is a key to manifold conclusions of the liquid properties' nature [37].

This approach gives the complementary information about the structure of aggregates, which is not possible by other methods, especially when long-ranged molecular binding is present. The application of this method is useful to interpret some regularities of organic liquids, although it does not detail the specific interactions' appearance [34, 37].

2.6 Section conclusion

As we have shown in the short review of the modern approaches to the problem of molecular arrangement in organic liquids, all data sources agree that the structure of liquid phase can be described as a supramolecular system with non-polar or weak polar long-ranged interactions. This binding arises at the distances close to 4–5 Å, but its energy contribution is comparable with vaporization enthalpy of liquids [7, 17, 18, 20]. The thermochemical, wave scattering, X-Ray and dynamic simulation data interpret well the lability of the system, but do not explain its high stability.

Unlike the mentioned methods, IR data predict that the initial molecular geometry transformation in comparison to a single molecule occurs in the liquid system. These changes are caused by the intermolecular forces between neighbors in bulk, even if they do not correspond to the hydrogen bond [23, 29, 30].

Applied to this problem, the DFT calculation could be useful for explaining the reasons of the transformation under weak interactions in terms of general electronic distribution. Therefore, in the following section we consider the essential results of the DFT study referring to the problems discussed in this paper.

3. DFT approach to the molecular binding in organic liquids

In this section, we have presented the traditional methods of DFT calculations together with our approach. In most works, the application of DFT to the problem of formation mechanism and arrangement of organic liquids is reduced to the non-covalent interactions' model of dimers with hydrogen bond. Although this approach is available for the gas phase binding only, such interactions arise quite rarely in the gaseous systems. Nevertheless, this concept is conventionally extrapolated to the condensed phase structure. However, the geometry and molecular interactions in this system sufficiently differ from the gaseous one. Therefore, we have suggested the DFT approach based on our IR spectral observations that indicates the certain changes of the molecular structure in the organic liquids in comparison to a single molecular state (see Section 3.4).

3.1 Geometry optimization (GO) procedure in DFT calculations of molecular system

The basic idea of system geometry optimization (GO) is that in *ab initio* calculations the SCF energy optimization procedure often leads not to a global minimum but to the local extremum. Therefore, besides of the SCF procedure, the additional mathematical method named GO is practically used.

For large molecules, which to a certain extent can simulate a supramolecular system, a method of delocalized internal coordinates within the framework of the DFT has been proposed. It was shown that the combination of the trust radius and line search method gives good accuracy in the geometry calculation of a single molecule. A performance analysis of the new geometry optimizer using different start Hessian matrices, basis sets and grid accuracies is given in [38]. This scheme has been successfully used for the study of enzyme reactions, treating the active site by a high-level method, particularly DFT, and the protein environment by molecular mechanics.

The minimum of energy is reached when the geometry characteristics of molecule is close to the experimental values. A few variants of the GO procedure for the DFT calculation can be found in the other papers, in which the findings are in a good agreement with experimental data [39, 40].

Considering these results, one can ask, firstly, why the calculation referring to single molecule can reproduce the geometry of condensed phase so well while the thermodynamic, kinetic, and spectral properties of molecular bulk differ from the individual species; secondly, the used regularity of energy change from geometric parameters, for example, from bond length, is strictly suitable only for diatomic molecules. For the polyatomic ones, this procedure can be attributed only with a sufficient approximation, because the bonds and angles in a molecule cannot vary independently. This situation is well known for the small vibrations of point masses near the interatomic equilibrium [24–26]. The solution of vibrational problem in, so called, internal natural coordinates (bonds and angles between neighboring bonds), requires involving the coefficients of the interactions between them. Besides, in organic liquids, the potential energy surface consists of the infinite closely located energy extremums of the single molecules. Therefore, in the real system the minimum of energy is broadly smoothed. Since the electron density in these systems is strongly delocalized, the geometric optimization procedure loses its real physical meaning. The arguments above show that the GO is an additional mathematical fit in the standard iterations shape, and it can be applied for the real systems only with the reasonable restriction. For this reason, the DFT calculations with the GO for the simplest organic molecules forming the stable liquid bulk (for example, tetrachloromethane, dichloroethane and non-substituted hydrocarbons etc.), does not lead to the energy minima even for two-molecule interaction (see Section 3.4). Thus, the DFT calculation without the GO procedure seems to be applicable in this field as well.

3.2 Paired interaction model in DFT calculation for liquids

The study of the formation mechanism and structure of organic liquids needs to involve the models and approximations to simulate molecular interactions. One of them can be defined as a model of paired bonds, which is based on a few reasonable assumptions. Firstly, the specific interaction in a liquid is a non-covalent interaction between molecules [41–44]. Secondly, for the definition of the pair, we can choose the strongest molecular interaction in the system. Thirdly, the liquid structure is formed by the paired molecules. Fourth, the nature of this interaction is a set of the different hydrogen bond types.

Sometime, halogen bond [45] and dihydrogen bond [46] provide a non-covalent binding, but it appears quite seldom, basically, between oppositely charged hydrogen or halogen atoms and in a solid state.

Hydrogen bond can exist in many organic and inorganic species and plays a crucial role in fields of chemistry and biology. It has been considered in many reviews; in this section, we analyze only a few recent ones [47–51]. Since the hydrogen bond existence is considered as a main aspect of the condensed phase formation theory, we would like to make a few comments regarding our conception presented in this section. Arising of the hydrogen bond presumes the occurring of an interacted pair – a hydrogen donor and a hydrogen acceptor. This type of binding is known not only in the condensed phase, but also for gases. The energy of hydrogen bond changes in the interval of 8–60 kJ/mol and more, in some cases, while its length is 2–3,5 Å.

Many authors that studied the organic liquids with a hydrogen bond in their structure, draw the statement that this bond provides the stability of liquids and the high mobility as well. This concept is quite satisfactory, because the hydrogen bond theory can be considered as a universal approach to the structure of liquid organics: amines, acids, alcohols, and related compounds; many bioactive systems can also assign to this system. Besides, it is known that some species of organic liquids can turn into a shape with a mobile proton, ex. nitro-substituted hydrocarbons, aldehydes, and ketones – they can form the hydrogen-bonded liquid system as well.

At the same time, the hydrogen-bonded pair has a linear binding, where the mobile proton is bound simultaneously with its own molecule and with a neighboring one, although to a greater extent with its own. This feature of binding allows to imagine the system, in which the initial structure of molecules remains in the combination with a thermodynamic stable molecular bulk. Therefore, the hydrogen bond is the suitable instrument to simulate the formation mechanism in a liquid phase.

Besides, it is well known that the organic matter contains of water impurities that cannot be completely removed [23, 52]. Water forms the clusters with the organic molecules due to the strong hydrogen bonds. The presence of water molecules can be considered as a variant of the mechanism for the formation of a stable organic liquid (see Section 3.3). This mechanism is valid for bioactive systems as well.

The prediction of benzene dimer formation is essential for this work because the high stable liquid in this case exists without the intermolecular hydrogen bond. The DFT calculations of the binding mechanism in the benzene dimer predicts four most stable shapes of the self-association: two variants of π-shaped and two variants of T-shaped structures [53–55].

The pair interactions' model is the most common viewpoint on the mechanism formation of organic liquids. However, this concept has a few disadvantages. Firstly, this model cannot explain, why there are many organics that exist without hydrogen bond while their stability (with the vaporization enthalpy as a criterium) is higher than in a system with hydrogen bond [7]. Secondly, this model cannot reproduce the formation mechanism of the spatial structure in organic liquids. Thirdly, the benzene dimer is one of the few examples where the DFT calculations can explain the formation of associates without involving a hydrogen bond, but they cannot explain the high stability of benzene, since the energy of the intermolecular interactions in it does not exceed 0,301 kJ/mol. The mechanism of water complex formation in organic liquids combined with the hydrogen atom transfer theory (HATT) is a positive step to clarify this problem, as it is presented in the next section.

3.3 Hydrogen atom transfer model for organic liquids

The hydrogen atom transfer theory (HATT) and the results of calculations obtained in terms of this model for gases and the condensed phase exhibit a wide area of investigations. The main data and the approaches referring to this theory are presented and reviewed in [56–60]. For our consideration, we have chosen the essential aspects of this concepts, applied for the formation mechanism and the structure of organic liquids.

The hydrogen atom transfer term is used to define the proton, hydrogen and hydride ion transfer which occur in many organic liquids. Two energy minima with a maximum between them arise in the system, determining the process of transfer.

The barrier of transfer changes considerably for different organics and can vary from 10 to 12 up to 100–200 kJ/mol [57]. The nature of these interactions is either activation, such as heating and irradiation of a substance, or a tunneling process. The last one is caused by the decrease of X–H bond energy under the hydrogen atom coupling with the Y-atom of the proton acceptor in the linear chain (X–H···Y) and the hydrogen atom vibrations with a large amplitude.

The main value of this theory for the considered problem is related to the description of the spatial arrangement formation. Presuming that the hydrogen atom transfer between the donor and the acceptor of the protons (in other words between the Bronsted acid A and the several molecules of basic B_1, B_2) occurs in a liquid phase like: ($AH + B_1 \rightarrow A^- + B_1H^+ + B_2 \rightarrow A^- + B_2H^+ + B_1$).

It means that in contrast with the dimer approach, the hydrogen atom transfer model predicts the multidirectional movement of hydrogen atom in a liquid space. These transformations can provide the spatial arrangement in the system. Besides, the HAT model gives a perspective idea that the molecular tautomerism involving different hydrogen atom transfers can be taken as a basic element of the liquid formation mechanism (see the next Section).

Another approach to hydrogen atom transfer mechanism can be formulated if organic liquid contains of the water molecules. As it was mentioned above, water often inserts in a liquid matter and then the hydrogen atom transfer can occur between the water molecules with hydronium ion formation [23, 32]. The water complexes with organics, where the interaction can be assisted with the hydrogen atom transfer, was confirmed by IR experiments [52].

The DFT calculations in the different parametrizations show that in CH_3NO_2 and CH_3CN complexes with water molecules several binding variants can be realized: first of them is the interaction of the hydrogen atom of methyl group with the oxygen atom of water, and the second one is the coupling of the hydrogen atom of water with the nitrogen atom of CN group or the oxygen atom of NO_2 group (**Figure 6**, the bonds' distances are given in Å).

The calculations predict the hydrogen atom transfer between the water molecules with the barrier about 10–12 kJ/mol and stabilizing of the hydronium cluster structure

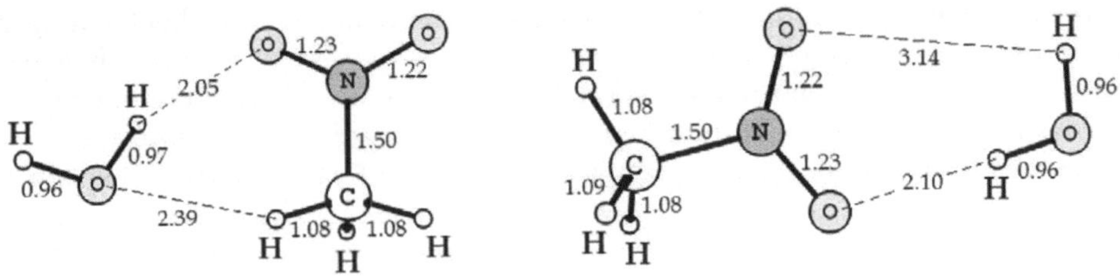

Figure 6.
The interaction model of water complex with CH_3NO_2 and CH_3CN molecules.

(**Figure 7**). In the hydronium ion, all hydrogen atoms are equal in their interaction with the organic molecules, and within the associates they can shift in all three-dimensional space, realizing the 'relay race' mechanism and forming the equilibrium spatial structure of the liquid. Thus, we can resume that in the liquid organics the water molecule can be a bonding agent in the formation mechanism of phase states.

The similar model was predicted by the DFT calculations for the aromatic substance – hexafluoride of benzene (**Figure 8**). In this complex, the barrier of the hydrogen atom transfer is even less than the one for CH_3X (X = CN, NO_2) systems (about 6 kJ/mol).

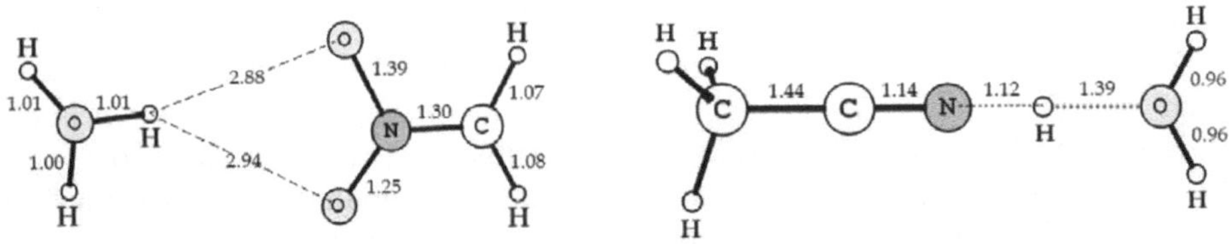

Figure 7.
The interaction model of hydronium ion interaction with CH_3NO_2 and CH_3CN molecules.

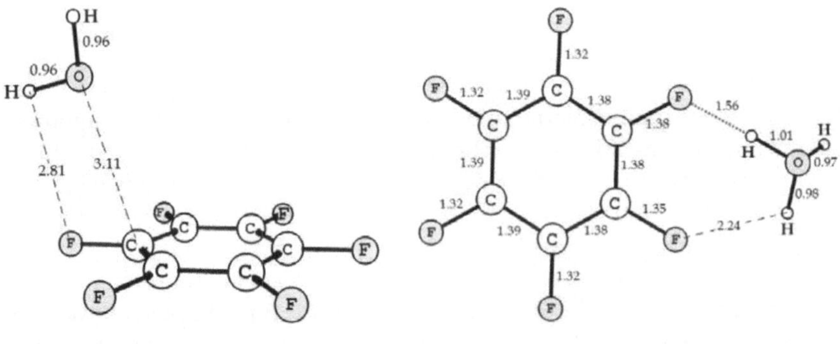

Figure 8.
The interaction model of water molecule and hydronium ion interaction with C_6F_6 molecule.

3.4 Transformations' model in DFT calculation

None of the described experimental and theoretical models can explain the formation mechanism and the structure of organic liquids completely, although they present some useful data to insight this problem. All of them proceed from the fact that the geometry of the gas phase, or, in other words, of a single molecule, is preserved in the liquid phase as well. Although it is assumed that the values of bond lengths and angles shift to some extent under the condensation, these changes are considered insignificant, without profound transformations of the molecular structure. At the same time, as it was shown above, the IR data source indicates the possibility of the symmetry vibrations violation caused by the structural factors. In this section we consider as an example the interpretation of IR findings for two classes of organic liquids, which are often-used solvents and initial products for many organic syntheses – methane halides and benzene – in terms of the DFT calculations in the B3LYP variant with the 6–311++G(2d, 2p) basis set.

3.4.1 Methane halides

For the interpretation of IR spectral effects, we have applied the symmetry point groups theory. In tetrachloromethane IR spectrum there should be one active C–Cl stretching band corresponding to the T_d symmetry. However, two approximately equal band intensities in 750–850 cm^{-1} region are observed (see Section 2.4). The appearance of two bands in the IR spectra can be assigned to the pyramidal C_{3V} and the biplanar D_{2h} or D_{2V} symmetry groups.

The DFT calculation predicts the transformation (**Figure 9**) of the isomer (1) into biplanar (2) and pyramidal (3) isomers. The energy barriers ΔE_1 and ΔE_2 are close to each other and relatively low: ΔE varies in 4–12 kJ/mol range. The calculated frequencies of C–Cl stretching vibrations (776, 713 cm^{-1}) are agreed with the experimental ones (786, 761 cm^{-1}) as far as it can be expected in such a calculation. Formally, the structures (1)–(3) do not differ so much, however, the distribution of charges in them changes in such a way that not only the interaction between a positively charged carbon and a negatively charged chlorine in the neighboring molecules is allowed, but also the interaction in the other direction, between two oppositely charged chlorine atoms. The calculations have shown that the atoms' charges depend on the value of (CClC) angles and not on the bond lengths. The suggested scheme sufficiently describes both spectral observation and formation mechanism of the spatial structure in liquid.

For trichloromethane, the DFT calculations were carried out for the chloroform and bromoform molecules. The obtained data predict the small barrier of transformation from a pyramidal C_{3V} isomer to a biplanar isomer not exceeding 10 kJ/mol (**Figure 10**). As is known, such energies are not sufficient obstacles for the transformation even in the gas phase and a fortiori for the liquid state.

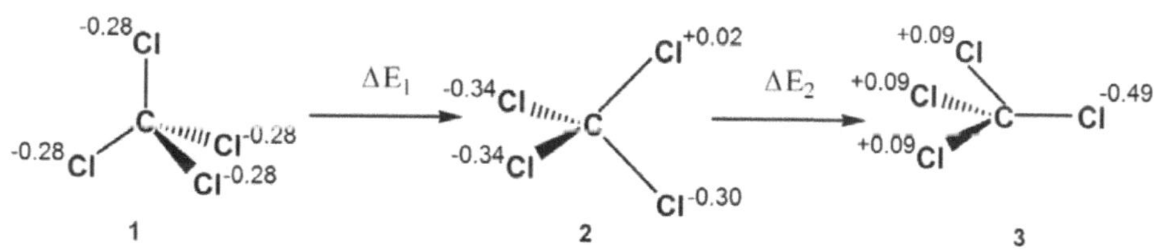

Figure 9.
The tautomeric isomers of CCl_4, predicted in the DFT calculations.

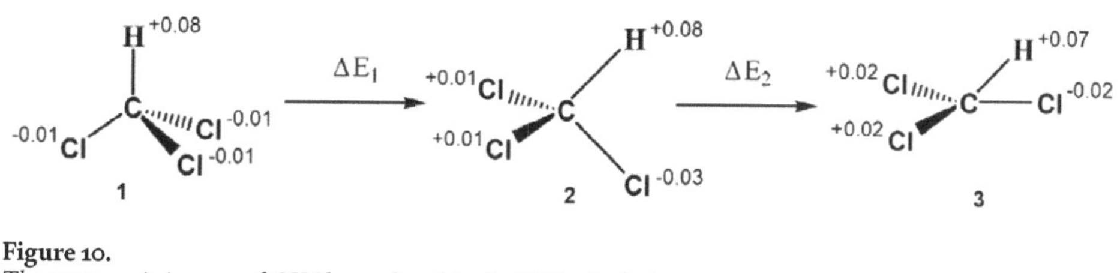

Figure 10.
The tautomeric isomers of $CHCl_3$, predicted in the DFT calculations.

Atom	Atomic charges in isomer 1	Atomic charges in isomer 2	Atom	Atomic charges in isomer 1	Atomic charges in isomer 2
Cl_1	−0,01	+0,01	Br_1	+0,26	+0,06
Cl_2	−0,01	+0,01	Br_2	+0,26	+0,06
Cl_3	−0,01	−0,03	Br_3	+0,26	−0,02
C	−0,05	−0,08	C	−1,03	−0,24
H	+0,08	+0,09	H	+0,25	+0,14

Table 1.
The charge distribution in chloroform and bromoform molecules.

The calculations reveal a few directions of the intermolecular binding: one of them is the hydrogen bond between a positive charged hydrogen atom and a negative charged carbon or chlorine atom. At the same time, the obtained results allow the intermolecular binding between the oppositely charged chlorine atoms reminding the interaction in tetrachloromethane molecule. The different types of binding can provide the spatial structure of the liquid phase.

For chloroform and bromoform, the detected charge distribution is similar (**Table 1**), but the atomic charges for the bromoform molecule in the C_{3V} isomer are considerably larger than in the chloroform molecule. In the biplanar configuration of bromoform, the atomic charges are significantly less, closer to the values of chloroform. This effect means that the basic interaction in bromoform is the hydrogen bond (C–H···C) with the hydrogen atom transfer in the C_{3V} isomer, in accordance with the IR spectra interpretation (see Section 2.4). Besides, the charge growing (along with the heavier molecular mass) can lead to a stronger intermolecular binding in bromoform and, as a result, to a higher vaporization enthalpy [61].

DFT calculation predict four directions of the tautomeric transformations for dihalogenomethane. In this case, the total energies of (1)–initial biplanar isomer {(CClCl) and (CHH)}, (2)–biplanar {two CHCl planes}, (3)–non-symmetrical pyramid with (HClCl) base and (4)–non-symmetrical pyramid with (HHCl) base (**Figure 11**) are close; ΔE_1, ΔE_2, and ΔE_3 are about 20 kJ/mol. Therefore, we conclude that the intermolecular binding can be realized in all these directions, including the hydrogen bond (C–H···Cl) and the interaction of two chlorine atoms.

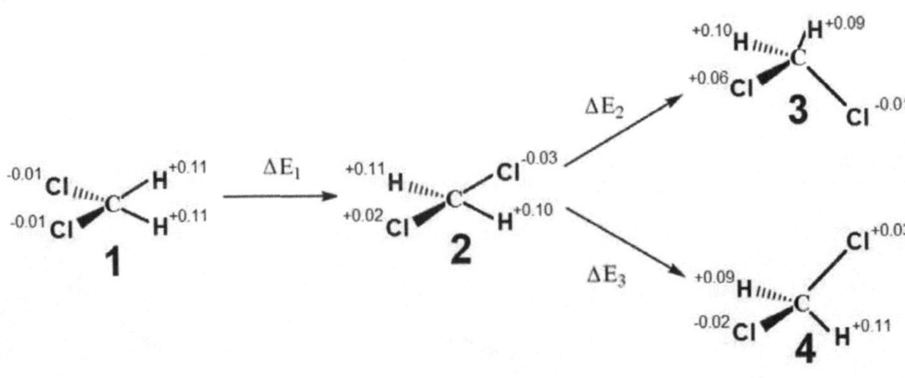

Figure 11.
The tautomeric isomers of CH_2Cl_2, predicted in the DFT calculations.

CH_2Cl_2			CH_2Br_2			CH_2I_2		
Atoms	**Atomic charges in initial isomer**	**Atomic charges in biplanar isomer (2)**	**Atoms**	**Atomic charges in initial isomer**	**Atomic charges in biplanar isomer (2)**	**Atoms**	**Atomic charges in initial isomer**	**Atomic charges in biplanar isomer (2)**
C	−0,20	−0,20	C	−0,22	−0,21	C	−0,26	−0,25
H_1	+0,11	+0,11	H_1	+0,13	+0,10	H_1	+0,09	+0,09
Cl_1	−0,01	+0,02	Br_1	−0,02	−0,02	I_1	+0,04	+0,04
H_2	+0,11	+0,10	H_2	+0,13	+0,12	H_2	+0,09	+0,09
Cl_2	−0,01	−0,03	Br_2	−0,02	+0,01	I_2	+0,04	+0,03

Table 2.
The charge distribution in CH_2X_2 molecules.

Atom	Atomic charges in C_{3V} pyramidal isomer	Atomic charges in C_{3V} planar isomer
C	−0,24	−0,19
H_1	+0,08	+0,11
H_2	+0,08	+0,11
H_3	+0,08	+0,11
I	+0,01	−0,14

Table 3.
The charge distribution in CH_3I molecule.

For CH_2Br_2 and CH_2I_2 molecules, ΔE_2 and ΔE_3 energies are considerably higher than ΔE_1. In this molecules, the transformation into isomers (3) and (4) is less expected than the C–H···X hydrogen bond in isomer (2). Since the charge distributions for all these molecules are similar (**Table 2**), the direction of the transformation is determined by the energy factor.

For CH_3I molecule, the calculated charge distribution in the initial C_{3V} specie pyramidal isomer allows the interaction between the iodide atom and the carbon atom in the chains of neighboring molecules, while the charge distribution in molecule can provide the hydrogen bond between the iodide and the hydrogen atom of the methyl group in the neighboring molecules as well (**Table 3**). Thus, the different binding variants in liquid CH_3I can provide the formation of the stable spatial arrangement.

3.4.2 Benzene

The structural element that determinates the molecular arrangement in a benzene liquid phase, was presented in the quoted literature as a set of dimers that have several geometry configurations (see Section 3.2). Unlike this approach, we have considered the trimers as a formed element in the 'stack model' of the benzene liquid phase. The conducted DFT calculations with different transformations of initial molecular geometry predict that the most optimal configuration is the stack with a 'chair' shape of the central ring and two planar rings (**Figure 12**). The trimers are bonded in a spatial structure by the hydrogen bridges. This concept can explain the IR data outside the scope of traditional assignment (see Section 2.4). In the trimer spectrum, one CH stretching band assigning to both planar rings should be observed, while the CH stretching in the central ring have to exhibit two bands: the first of them assigns to a pair of equivalent (C_1–H_1) and (C_4–H_4) bonds, and the second one – to a quartet – (C_i–H_i), i = 2.3, 5.6 (**Figure 12**). The stretching bands of the hydrogen bridges

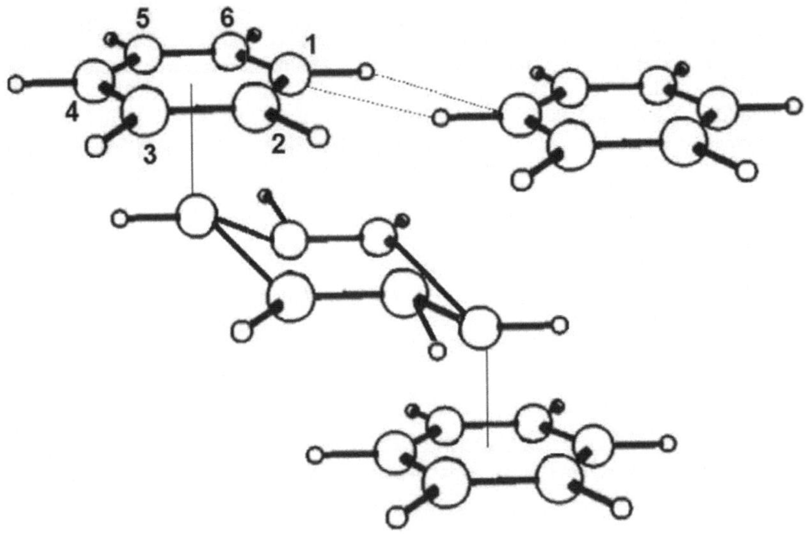

Figure 12.
The fragment of the benzene spatial arrangement in a liquid phase.

should shift in the middle IR region due to the mixing of the CH and CC stretching. Therefore, the pair of bands at 1800–2000 cm^{-1} corresponds to the stretching of two bridging C–H bonds in the stacks bound in two mutually perpendicular plains.

3.5 Section conclusion

The DFT calculations with the GO procedure that were used for the dimers with the hydrogen bond, had described the enthalpy of the molecular interaction and the bond lengths quite satisfactory. For most systems, the intermolecular bond energy is even overestimated. However, this model cannot explain the spatial arrangement in the condensed phase because the electronic density should be localized within a dimeric shape in this case.

Unlike the dimer model, the transformation model suggests the variants of mutual molecular disposition under the bonding between the chains or the stacks. However, it requires the use of suitable experimental data. In this case, as a criterium of the results' validation, the absence of the negative vibrational frequencies can be taken. Although these calculations are somewhat intuitive, they allow to predict the reasonable structure arrangement of the supramolecular system.

4. General conclusion

The presented experimental and theoretical findings confirm the conclusion that none of the approaches can explain completely the structure and the formation mechanism of organic liquids. However, these approaches complement each other and help to solve the problem to some extent. In **Table 4** we have listed the advantages and disadvantages of each method and model.

A promising direction in the development of the DFT approach for investigating the problem of the formation mechanism and structure of organic liquids is the elaboration of the methods for calculating molecular systems with delocalized and smoothed electron density. Studying these systems can help to clarify the nature and structure of organic liquids.

Approach / Model	Advantages	Disadvantages
Dipoles' interaction	universal; predicts the nature of formation mechanism and spatial arrangement for organic liquids	cannot explain the high stability of weak polar and non-polar liquids
Thermochemistry	reveals the specific long-ranged interaction in the non-polar organic liquids (with the energy values comparable to those of the vaporization enthalpy)	cannot explain the high stability of organic liquids with no hydrogen bonds; the formation mechanism interpretation is based on the intuitive model
MLS and X-ray study	reveal the specific long-ranged interactions and (approximately) the distances between molecules	cannot describe the formation mechanism and explain the high stability of liquid organics
Dynamic simulation	explains the spatial arrangement and the structural stability of the liquids	cannot describe the nature of organic liquids' phase arrangement
DFT paired interactions model	establishes the nature of liquid phase formation as a paired non- covalent interaction	does not predict the stable spatial arrangement in the organic liquids
HAT model	predicts the mechanism of stability and lability in the liquids	considers only the structure of hydrogen-bonded organic liquids
DFT transformations' model	predicts the mechanism of arrangement not only for hydrogen-bonded liquid organics	a set of certain experimental data and a concept for their interpretation is required

Table 4.
Advantages and disadvantages of approaches to organic liquids structure.

References

[1] Kaplan IG, Fraga S, Klobukowski M, Kryachko ES. Theory of Molecular Interactions. Ser. Studies in Physical and Theoretical Chemistry. Amsterdam: Elsevier; 1986. 416 p.

[2] Gavezzotti A. Molecular Aggregation: Structure Analysis and Molecular Simulation of Crystals and Liquids. Oxford: Oxford University Press; 2010. 326 p. DOI:10.1093/acprof:oso/9780198570806.001.0001

[3] Solomons TWG, Fryhle CB, Snyder SA. Organic Chemistry, 12th ed. New York: John Wiley & Sons; 2016. 1208 p.

[4] Szalewicz K. Hydrogen Bond. In: Meyers RA, editor. Encyclopedia of Physical Science and Technology. 3rd ed. New York: Academic Press; 2003. P 505-538. DOI:10.1016/B0-12-227410-5/00322-7

[5] Fisher LZ, Switz TM, Rice SA, Gray P, Andrews FC. Statistical Theory of Liquids, Chicago: Chicago University Press; 1966. 478 p. DOI:10.1115/1.3625104

[6] Wormer PES, Avoird Avd. Intermolecular Potentials, Internal Motions, and Spectra of van der Waals and Hydrogen-Bonded Complexes, Chem. Rev. 2000;100:4109-4144. DOI:10.1021/cr990046e

[7] Baev AK. Specific Intermolecular Interactions of Organic Compounds, Springer-Verlag Berlin-Heidelberg, 2012, 434 p. DOI:10.1007/978-3-642-21622-0

[8] Shakhparonov MI. Intermolecular binding. Moscow: Znanie; 1983, 64 p.

[9] Shakhparonov MI. Contemporary Problems of Physical Chemistry. Moscow: Izd. Mosk. Gos. Univ; 1970, 112 p.

[10] Byakov VM, Stepanov SV, Zorkii PM, Lanshina LV, Stepanova OP. Microheterogeneity in mixtures of benzene with alcohols and hydrocarbons: Positron annihilation and molecular light scattering Russ. J. Phys. Chem. A. 2007;81:638-643. DOI:10.1134/s0036024407040231

[11] Lanshina LV, Abramovich AI. Microheterogeneity of benzene solutions of cyclohexane, 1,4-dioxane, and morpholine. Russ. J. Phys. Chem. A. 2007;81:87-192. DOI:10.1134/s0036024407020082

[12] Lanshina LV, Abramovich AI. The Dependence of intermolecular interaction parameters in the chlorobenzene-o-chlorotoluene system on the composition of solutions. Russ. J. Phys. Chem. A. 2008;82:1851-1856. DOI:10.1134/s0036024408110113

[13] Abramovich AI, Lanshina LV. The microheterogeneous structure of solutions of cyclohexane, 1,4-dioxane, and morpholine in chlorobenzene. Russ. J. Phys. Chem. A. 2010;84:1147-1153. DOI:10.1134/s0036024410070125

[14] Abramovich AI, Alekseev ES, Bogdan TV, Lanshina LV. Chlorine aggregation in the supramolecular structure of chlorobenzene-o-chlorotoluene solutions. J. Struct. Chem. 2014;55:651-659. DOI:10.1134/s002247661404009x

[15] Kargin ID, Lanshina LV, Abramovich AI. Studying the local structure of liquid in chloro- and alkyl-substituted benzene derivatives via the molecular scattering of light. Russ. J. Phys. Chem. A. 2017;91:1737-1742. DOI:10.1134/s003602441709014x

[16] Abramovich AI, Lanshina LV, Kargin ID. Peculiarities of molecular light scattering in a chlorobenzene-o-dichlorobenzene system and their

relationship to the solution structure. Russ. Chem. Bull. 2017;66:828-832. DOI:10.1007/s11172-017-1814-8

[17] Abramovich A. Features of molecular light scattering and structure of the chlorobenzene o-dichlorobenzene solutions, Struct. Chem. 2019;30:545-549. DOI: 10.1007/s11224-019-01293-1

[18] Abramovich AI, Alekseeva ES, Bogdan TV. Structure and Physicochemical Properties of Acetonitrile–o-Dichlorobenzene Solutions. Russ. J. Phys. Chem. A. 2019;93:2108-2116. DOI:10.1134/s0036024419110025

[19] Alekseev ES, Bogdan TV. Halogen aggregation in chlorobenzene-o-dichlorobenzene solutions J. Struct. Chem. 2016;57:1568-1576. DOI:10.1134/S0022476616080114

[20] Bálint S, Bakó I, Grósz T, Megyes T. Structure of liquid methylene chloride: Molecular dynamics simulation compared to diffraction experiments. J. Mol. Liq. 2007;136:257-266. DOI: 10.1016/j.molliq.2007.08.020

[21] Megyes T, Bálint S, Grósz T, Radnai T, Bakó I. Structure of liquid nitromethane: comparison of simulation and diffraction studies. J. Chem. Phys. 2007;126:164507. DOI:10.1063/1.2721559

[22] Vincent J, Andersson M, Eklund M, Wöhri AB, Odelius M, Malmerberg E, Kong Q, Wulff M, Neutze R, Davidsson J. Solvent dependent structural perturbations of chemical reaction intermediates visualized by time-resolved X-ray diffraction. J. Chem. Phys. 2009;130:154502. DOI:10.1063/1.3111401

[23] Kalagaev IYu, Grinvald II. Formation of supramolecular structures in organic solvents. Pure Appl. Chem. 2013;85:135-148. DOI:10./PAC-CON-12-03-06

[24] L. M. Sverdlov, M. A. Kovner, E. P. Krainov. Vibrational Spectra of Polyatomic Molecules (in Russian). Moscow: Nauka; 1970. 421 p.

[25] Smith BK. Infrared Spectral interpretation, a systematical approach. Boca Raton: CRC Press; 1999. 266 p. DOI:10.1201/9780203750841

[26] Gribov LA, Orville-Thomas WJ. Theory and Methods of Calculation of Molecular Spectra. Chichester, New York, Brisbane, Toronto, Singapore: Wiley; 1988. 636 p.

[27] Tuomikoski P. On the Fermi Resonance and Isotope Effect in the Infrared Spectrum of Carbon Tetrachlorid. J. Chem. Phys. 1955;23:2083-2084. DOI:10.1063/1.1740670

[28] Chakraborty T, Rai SN. Comparative study of infrared and Raman spectra of CCl4 in vapour and condensed phases: Effect of LO–TO splitting resulting from hetero-isotopic TD–TD interactions Spectrochimica Acta A. 2006;65:406-413. DOI: 10.1016/j.saa.2005.11.021

[29] Grinvald I, Kalagaev I, Petukhov A, Vorotyntsev A, Kapustin R. Self-association of carbon tetrachloride in gas and condensed phase Struct. Chem. 2019;30:1659-1664. DOI:10.1007/s11224-019-01349-2

[30] Grinvald II, Kalagaev IYu, Petukhov AN, Grushevskaya AI, Kapustin RV, Vorotyntsev IV. Association of Haloforms in Condensed and Gas Phases. IR Spectroscopy and DFT Calculations. J. Struct. Chem. 2018;59:313-320. DOI:10.1134/S0022476618020087

[31] Zundel G, Hydration Structure and Intermolecular Interaction in Polyelectrolytes. Angewandte Chemie. Edition in English. 1969;8:499-509. DOI:10.1002/anie.196904991

[32] Bell RP. The proton in chemistry. (Softcover reprint of the original 1st ed. 1973, edition April 27, 2013), Heidelberg, New York: Springer; 2013. 317 p. DOI:10.1007/978-1-4757-1592-7

[33] Grinvald II, Kalagaev IYu, Petukhov AN, Kapustin RV, IR Manifestation of Non-Covalent Interaction in Organic Liquids, Russ. J. Phys. Chem. A. 2019;93:69-73. DOI:10.1134/s0036024419130107

[34] A Gavezzotti AL. Presti LL. Dynamic simulation of liquid molecular nanoclusters: structure, stability and quantification of internal (pseudo) symmetries New J. Chem., 2019;43:2077-2084 DOI:10.1039/c8nj05825c

[35] Durov VA, Tereshin OG. Modeling of Supramolecular Ordering and Physicochemical Properties in Cyclohexane-Ethanol Mixtures J. Phys. Chem. B. 2006;110:8441-8450. DOI:10.1021/jp056541y

[36] J. Samios, Durov VA, Novel approaches to the structure and dynamics of liquids: experiments, theories and simulations, Dordrecht: Springer; 2004. 546p.

[37] Durov VA. Modeling of supramolecular ordering in mixtures: Structure, dynamics and properties. J. Mol. Liq. 2003;103-104:41-82. DOI:10.1016/s0167-7322(02)00127-7

[38] Vreven T, Morokuma K, Farkas, Schlegel HB, Frisch MJ, Geometry Optimization with QM/MM, ONIOM, and Other Combined Methods. I. Microiterations and Constraints. J Comput Chem. 2003;24:760-769. DOI: 10.1002/jcc.10156

[39] Reveles JU, Köster AM, Geometry optimization in density functional methods. J. Comput. Chem. 2004;25:1109-1116. DOI:10.1002/jcc.20034

[40] Meyer R, Hausera AW. Geometry optimization using Gaussian process regression in internal coordinate systems. J. Chem. Phys. 2020;152: 084112. DOI:10.1063/1.5144603

[41] Rezac J, Hobza P. Benchmark Calculations of Interaction Energies in Noncovalent Complexes and Their Applications. Chem. Rev. 2016;116:5038-5071. DOI:10.1021/acs.chemrev.5b00526

[42] Goerigk L, Hansen A, Bauer C, Ehrlich S, Najibi A, Grimme SA. A look at the density functional theory zoo with the advanced GMTKN55 database for general main group thermochemistry, kinetics and noncovalent interactions. Phys.Chem. Chem.Phys. 2017;19:32184-32215. DOI:10.1039/c7cp04913g

[43] Burns LA., Faver JC, Zheng Z., Marshall MS, Smith DGA, Vanommeslaeghe K. MacKerell AD. Jr., Merz KM Jr., Sherrill CD. The BioFragment Database (BFDb): An open-data platform for computational chemistry analysis of noncovalent interactions. J. Chem. Phys. 2017;147:161727. DOI:10.1063/1. 5001028

[44] Kříž K, Nováček M, Řezáč J. Non-Covalent Interactions Atlas Benchmark Data Sets 3: Repulsive Contacts. 2020; 16:6305–6316. Theory Comput. J. Chem. DOI:10.1021/acs.jctc.0c01341

[45] Politzer P. Book Review of Halogen Bonding: Fundamentals and Applications. Structure and Bonding, 126. J. Am. Chem. Soc. 2008;130:10446-10447

[46] Bakhmutov VI. Dihydrogen Bonds. New York: Wiley; 2008. 239 p. DOI:10.1002/9780470226759

[47] Boese AD. Density Functional Theory and Hydrogen Bonds: Are We There Yet?. ChemPhysChem.

2015;16:978-985. DOI:10.1002/cphc.201402786

[48] Rezac J. Non-Covalent Interactions Atlas Benchmark Data Sets: Hydrogen Bonding, J. Chem. Theory Comput. 2020;16:2355-2368. DOI:10.1021/acs.jctc.0c00715

[49] Řezác J. Non-Covalent Interactions Atlas Benchmark Data Sets 2: Hydrogen Bonding in an Extended Chemical Space J. Chem. Theory Comput. 2020; 16:6305–6316. DOI:10.1021/acs.jctc.0c00715

[50] Kříž K., Nováček M, Řezáč J. Non-Covalent Interactions Atlas Benchmark Data Sets 3: Repulsive Contacts, J. Chem. Theory Comput. 2021;17:1548–1561. DOI:10.1021/acs.jctc.0c01341

[51] S. Ghosh S, Wategaonkar S. C–H···Y (Y=N, O, π) Hydrogen Bond: A Unique Unconventional Hydrogen Bond. J. Indian Inst. Sci. 2020;100:101-125. DOI:10.1007/s41745-019-00145-5

[52] I. I. Greenwald, I. Yu. Kalagaev, Water complexes with organic solvents in liquid phase. An IR spectroscopic study, Tetrahedron Lett. 2010;51:2610-2612. DOI:10.1016/j.tetlet.2010.03.023

[53] Y. C. Park, J. S. Lee. Accurate ab initio binding energies of the benzene dimer. J. Phys. Chem. A. 2006;110:5091-5095. DOI: 10.1021/jp0582888

[54] Dinadayalane TC, Leszczynski J. In the pursuit of small "red shift" of C–H stretching vibrational frequency of C–H···π interactions for benzene dimer: How to amend MP2 calculations to reproduce the experimental results. J. Chem. Phys. 2009;130:081101. DOI:10.1063/1.3085815

[55] Moraes EEd, Tonel MZ, Fagan SB, Barbosa MC, Atomistic model derived from ab initio calculations tested in Benzene–Benzene interaction potential. Physica A: Statistical Mechanics and its Applications. 2020;537:122679. DOI:10.1016/j.physa.2019.122679.

[56] Zeegers-Huyskens T, Huyskens P. Influence of proton transfer on the formation of hydrogen-bonded complexes higher stoichiometry and on their dissociation into free ions. In: Ratajczak H, Orville-Thomas WJ, editors. Molecular Interactions. Volume 2. New York: Wiley; 1981. p. 1-32.

[57] Huyskens P, Platteborze K, Zeegers-Huyskens T. Proton transfer complexes between phenols or tetrazole and 1,8-bis(dimethylamino)naphthalene. Non-ergodicity of the equilibria. J. Mol. Struct. 1997;436-437:91-102. DOI:10.1016/s0022-2860(97)00111-7.

[58] M.V. Basilevsky, M.V. Vener, Theoretical investigations of proton and hydrogen atom transfer in the condensed phase, Rus. Chem.Rev. 2003;72:1-33. DOI: 10.1070/RC2003v072n01ABEH000774

[59] Mayer JM. Understanding Hydrogen Atom Transfer: From Bond Strengths to Marcus Theory. Acc. Chem. Res. 2011;44:36-46. DOI:10.1021/ar100093z

[60] Pluth MD, Bergman RG, Raymond KN. Proton-Mediated Chemistry and Catalysis in a Self-Assembled Supramolecular Host. Acc. Chem. Res. 2009;42:1650-1659. DOI: 10.1021/ar900118t

[61] Stephenson RM, Malanowski S., Handbook of the Thermodynamics of Organic Compounds, Springer, Netherlands, Dordrecht, 1987, p. 552. DOI:10.1007/978-94-009-3173-2

Chapter 9

Superconductivity in Materials under Extreme Conditions: An *ab-initio* Prediction from Density Functional Theory

Abstract

The relation between thermodynamically stable and electronic structure preparation is one of the fundamental questions in physics, geophysics and chemistry. Since the discovery of the novel structure, this has remained as one of the main questions regarding the very foundation of elemental metals. Needless to say this has also bearings on extreme conditions physics, where again the relation between structure and performance is of direct interest. Crystal structures have been mainly at ambient conditions, i.e. at room temperature and ambient pressure. Nevertheless it was realized early that there is also a fundamental relation between volume and structure, and that this dependence could be most fruitfully studied by means of high pressure experimental techniques. From a theoretical point of view this is an ideal type of experiment, since only the volume is changed, which is a very clean variation of the external conditions. Therefore, at least in principle, the theoretical approach remains the same irrespective of the high pressure loading of the experimental sample. Theoretical modeling is needed to explain the measured data on the pressure volume relationships in crystal structures. Among those physical properties manifested itself under high pressure, superconductivity has emerged as a prominent property affected by pressure. Several candidate structure of materials are explored by *ab initio* random structure searching (AIRSS). This has been carried out in combination with density functional theory (DFT). The remarkable solution of AIRSS is possible to expect a superconductivity under high pressure. This chapter provide a systematically review of the structural prediction and superconductivity in elemental metals, i.e. lithium, strontium, scandium, arsenic.

Keywords: *ab initio* random structure searching, density functional theory, superconductor, lithium, strontium, scandium, arsenic

1. Introduction

It is a long time since Kohn and Sham pave the way to the self-consistent equation, based on the exchange and correlation effects in 1965, leading the Kohn–Sham (KS) Equation [1]. This has ignited the success of quantum physics and chemistry, specifically many-body problem, owing to the KS equation can be utilized for the ground state energy. Briefly stated, the KS equation formalism of density functional theory (DFT) described the motion of electron nuclei, which separated to be two part: the energy of electron $E_{electron}$ and the Coulomb interactions between the nuclei E_{nuclei} And what is more, the details of Ewald summations have been described extensively in Refs. [2, 3]. Apart from this, $E_{electron}$ and E_{nuclei} were performed by using the pseudopotential approximation within the KS equation. It is because of the effects of Coulomb interactions between the nuclei E_{nuclei}, as being in accordance with the the core electrons E_{core}, that the terms $E_{electron}$ and E_{core} used in the static crystal energy of materials relevant to the energies of valence electrons and pseudocores. Subsequently, the KS equation displayed the term $E_{electron}$ is from the summation of quasiparticle eigenvalues, corresponding the Kohn–Sham orbital, of occupied states.

Regarding thermodynamic properties, the Gibbs free energy is considered for the static crystal energy of materials; however, the KS equation formalism of DFT carried out at a temperature of 0 K. The Gibbs free energy therefore reduced to the Enthalpy. This, appearing at first glance to be high potential for high-pressure physics, is actually demonstrated the importance of superconductivity. According to the aforementioned theoretical findings by the KS equation formalism of DFT, resulting the exchange and correlation effects E_{xc}. Following this, Perdew et al. [4] presented a simple derivation of a simple of generalized gradient approximations (GGA) with E_{xc}. This methodology appropriated, it is well known to GGA with Perdew-Burke-Ernzerhof (PBE), for description of atoms, molecules, and solids. This is due to the fact that the GGA-PBE method give an accurate with the most energetically important. As a result of this, the role of the GGA-PBE method is key factor in achieving the ground-state energy of the static crystal materials. Herein, we preformed mainly the PBE formalism of GGA for calculations of lithium, strontium, scandium, and arsenic under high pressure.

The extensive studies of electronic structure were initiated chiefly by the KS equation formalism of DFT. In principle, one should note the quasiparticle eigenvalues of occupied states is useful for achieving the electronic band structure, density of sates, phonon dispersion. It is also interesting to note the DFT used mainly strong sides for prediction the metallicity, leading to the prediction of superconducting transition temperature. For considered the superconductivity, the PBE formalism of GGA for exchange-correlation energy is suitable for interpret the metallicity. This implied that the reliable theoretical study has quite a predictive potential, moreover, the GGA-PBE for the exchange-correlation energy give an accurate description of dynamical stability of crystal structure. One of the well-known Bardeen-Cooper-Schrieffer (BCS) theory [5] were already discussed phonon mediated superconductivity, leading to the way to vast both experimental and theoretical studies on high-pressure research. At this stage, using the KS equation formalism of DFT with the GGA-PBE for the exchange-correlation energy were used to have unique features of phonon mediated superconductivity, showing towards the evidence of superconducting materials as well.

There is alternative way to use the KS equation formalism of DFT with the GGA-PBE. It is well known to *ab initio* random structure searching (AIRSS). The AIRSS method have been described extensively in Refs. [6, 7]. Especially, the AIRSS method is useful in achieving the high-pressure research owing to it can predict novel structure under compressed conditions. The reliable theory for ground-state structure can help to interpret experimental data. In fact, there is also quite some experimental observation cannot identify the atomic position and crystal symmetry. The AIRSS method is powerful tool and it can guide further experimental studies.

High pressure physics is important for structural phase transitions in materials [8–18]. Regarding a crystal structure of materials under high pressure, it can enhance electronic properties of materials [19–21]. Nowadays, superconductivity is one of the most charming in physical properties. Many materials were predicted to be a superconducting transition temperature (T_c), such as SH_3 [22–25], LaH_{10} [26–28], YH_{10} [28–30], CeH_{10} [17, 31]. It is worth note that hydrogen (H) is a role important for promoting a T_c. For example, the case of LaH_{10} was shown that the T_c reached 250 K at 170 GPa [26]. The existence of H displayed that it can support pure element lanthanum (La), one can see that the T_c of pure element lanthanum is 5.88 K [32]. It is interesting to note that the physical property of pure elemental metal should be mentioned.

As mentioned above, a structural prediction is a key factor for achieving a T_c. We referred the original predictions regarding superconductivity in strontium (Sr), it is beginning to show that the T_c of the predicted phase increased with a increasing pressure [33]. The case of strontium is interested. At high pressure, Sr. displayed structural phase transition from a simple structure to a complex structure [34–37]. We can see that Sr. is a normal metal at ambient pressure, with a increasing pressure, Sr. is a metallicity, indicating that it is a superconductivity [33]. Moreover, Sr. is not only superconducting phase at high pressure, but also calcium (Ca) indicating possible increasing of the T_c also at high pressure [38].

A curious aspect of a T_c increased with increasing pressure. We found that Ca is one of a periodic table, indicating the highest T_c among the periodic table [38–40]. Moreover, it is not always clear whether or not that the increased pressure and T_c are increased. We note that scandium (Sc) [14] and arsenic (As) [12], shown a possible decreasing of T_c with a increasing pressure [12, 14]. Hence, the focus is on pure elemental metals are interesting. This is because that the prediction discovered to novel structure [12, 14], leading to the superconductivity at high pressure.

According to the aforementioned superconductor findings, the characteristic of electronic structure is often attributed to the T_c [14, 17, 41]. It is interesting that Lithium (Li) has the second highest T_c among the elemental metals [42–45]. The electron localization function (ELF) is one of the tools can determine the T_c [14, 17]. The nature of chemical bonding is directly shown in the ELF, it is considered to be consistent with the highest T_c; this implies that a strong bonding supports the T_c of the metals [46].

Regarding superconductivity in the metals [12, 14, 17, 41], a lattice dynamic is a key factor for consideration a stable structure. In practice, we can achieve the superconducting structure through electron–phonon coupling (EPC) [12, 14, 17, 41]. For example, recent work on LaH_{10} has shown that the quantum effect is important for the stabilization and destabilization [27]. In fact, both thermodynamically and dynamically structures have to consistent. Generally, the solution of dynamically structure is a harmonic phonon but the case of LaH_{10} shown that it displayed an anharmonic phonon. This because the EPC exhibited the destabilized structure. Hence, it is worth to note that Sr. is possible to be an anharmonic phonon in the Sr-III structure (the β-tin structure) At this point, we found that there is a discrepancy between a experimental observation and a theoretical study [33–37, 47]. Herein we review the superconductivity in the elemental metals both the experimental observation and the theoretical study under high pressure. In this review, we provide the success of the metals [12, 14] is BCS-type superconductor [12, 14, 17, 19, 41, 48–53]. Also, we hope that this review is useful for those interested readers in superconductivity in elemental metal under high pressure.

2. Methodology

In considered in the present work, we performed the first-principles calculations, based on the density functional theory, to examine the thermodynamic stability as a function of pressure. The static crystal energy of materials was considered at a temperature of 0 K. The calculation details of stable structure were determined by neglecting the entropy contributions. This is because the calculations were carried out at 0 K, indicating that the ground-state energy can confirm phase stability. Here, the KS equation formalism of DFT with the GGA-PBE for the exchange-correlation energy were used for Li, Sr., Sc, and As. For further details of the energy cutoff for plane waves and the Monkhorst–Pack k-point mesh as well as the DFT software have been described extensively in Refs. [10, 12–14]. Our works used the AIRSS technique, based on the density functional theory, to predict the novel structure. Following the AIRSS method, we calculated the enthalpies of the phases at any pressure using the simple linear approximation [7]. For each relaxed structure, the structures were simulated to be a non-symmetry and randomly placed in atomic position. During the calculations of the structures, it started to relax from bias until it reaches unbias. The shape is generating by shaking within a reasonable pressure range. It led to higher-symmetry space groups obtained in a search. The AIRSS technique is the approach in the local minima by giving the lowest enthalpy. We have studied the phonon mediated superconductivity by using isotropic Eliashberg theory, as implemented in the quantum espresso (QE) [54, 55]. Following the result of isotropic Eliashberg theory, the Allen-Dynes modify McMillan Equation [56] was used to estimated the superconducting transition temperature.

3. Result and discussion

3.1 Lithium

According to the aforementioned in the introduction, high pressure physics is useful in achieving a novel structure and superconductivity [12–14, 17, 33, 57–59]. Li is one of the challenging to find a novel structure [43, 60–63]. Since it is interesting that there is complex structures were discovered in alkali metal, i.e. sodium (Na) [64], potassium (K) [65–68], and rubidium (Rb) [69, 70]. Therefore, Li might be expected to possible to be a complex structure at high pressure. For the transitions sequence of Li, we found that the Im-3 m structure transformed into the Fm-3 m structure at pressure 8 GPa. Next, the Fm-3 m structure transformed into the R-3 m structure at pressure 39 GPa. With increasing pressure, the R-3 m structure transformed into the I-43d structure at pressure 44 GPa, then it transformed into C2mb at pressure 73 GPa. On further compression, the C2mb structure transformed into the C2cb structure at 80 GPa. Finally, it transformed into Cmca 120 GPa. It is interesting that there is no found the incommensurate host-guest structure at any pressure among such sequence [43, 60–63, 71].

Li was observed by optical spectroscopic through diamond anvil cells (DAC) [72]. The solution of the experimental study revealed that there is unknown phase above 50 GPa. Moreover, the characteristic of the high frequency band, i.e. Li-Li vibration, can interpret to be an incommensurate host-guest structure. The commensurate host-guest structure is defined by the different the number of the guest atoms in channels in along the c axis of the host structure, referring to the commensurate value c_H/c_G, also known as γ. At this point, it is interested to examine the unknown structure by following the Ref. [72].

As mentioned above, the unknown structure can be identified by a random search techniques. The random search technique is the high performance for the prediction of the materials. For elemental Li, the *ab initio* random structure searching (AIRSS) technique [6] is employed for determination the unknown structure. The remarkable result shown that Li is predicted to be the incommensurate host-guest structure above 50 GPa. Tsuppayakorn-aek et al. [13] was pointed out that structural phase transitions of Li might be considered to be different origins in two-phase transition sequences (**Figure 1**). Interestingly, one of two transition sequences can be obtained the incommensurate host-guest structure, indicating that it is energetically stabilized above 50 GPa and Li is likely to crystallize in the incommensurate host-guest structure at high temperature.

The existence of the incommensurate host-guest structure can be considered from the ELF calculation. As a possible cause of this, one might think of there is the s-pμ hybridization between the host–host atoms at 150 GPa (**Figure 2**) [13]. The study useful to point out that the possibility of the incommensurate host-guest structure is stable. Moreover, the nature of chemical bonding shown that the

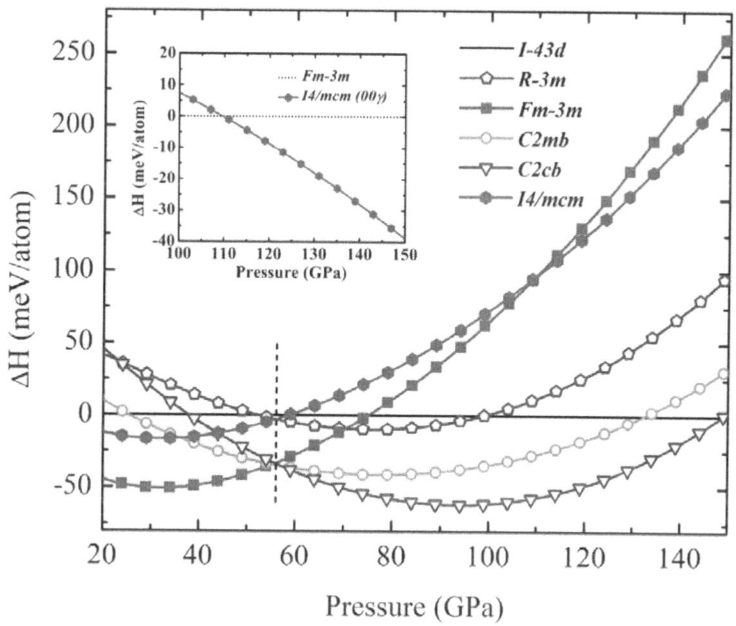

Figure 1.
The relative enthalpy of Li as a function of pressure.

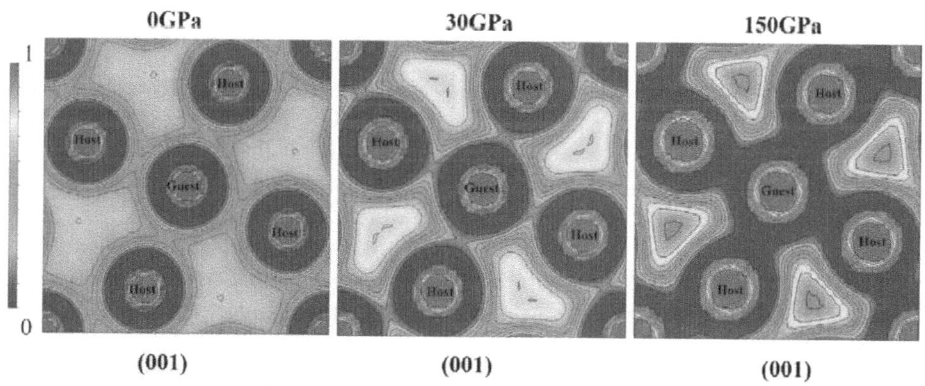

Figure 2.
The electron localization function (ELF) of the host-guest structure of Li is calculated in the (001) atomic plane.

incommensurate host-guest structure has tend to favor superconductivity at higher pressure. It is worth to note that the nature of the chemical bonding of the host–host atoms, i.e. the μ bonging, might be considered to be a superconducting phase.

3.2 Strontium

Structural phase transitions in alkaline earth metal under high pressure is interested among the periodic table. Nowadays, there are several works reported a transition sequences [33, 37, 40, 47, 58]. It is interesting to consider that a transition sequences of Ca and Sr. are similar. Ca shown that it exhibited stable structure at high temperature and low pressure through compression [58, 73]. The experimental observations [74] and the theoretical study [58, 73] reported that the simple cubic (sc) structure is stable at room temperature. At this point, the solution of theoretical study revealed that the sc structure is stable by performing a molecular dynamics (MD) calculation [73]. This is because the MD calculation can include a temperature via *ensemble*. However, the sc structure is considered by a lattice dynamics calculation [75], indicating that it is unstable structure. This is due to that fact that the sc structure is not a harmonic phase, but it is anharmonic phase [75]. Here, the sc structure is difficult to estimate the T_c by theoretical study.

In 2009, Ca was reported a novel structure at high pressure that it is the β-tin structure [58]. Here, it is worth to note that the transitions sequence of Ca is similar Sr. (the Fm-3 m structure transformed into the Im-3 m structure, then it transformed into the β-tin structure) Here, the β-tin structure is found that it is stable at high pressure and low temperature [58]. The T_c of the β-tin structure was estimated to be 5 K at 40 GPa. The case of Ca is interesting due to the d electrons are important for the estimated T_c. As a possible cause of this, one might think of the d electron is dominated near the Fermi level.

It is interesting to note that structural phase transitions in Sr. [33–37, 47]. The remarkable studies revealed that there are discrepancy between the experimental observations [34–36] and the theoretical studies [33, 37, 47]. The experimental observations were reported that the Fm-3 m structure transformed into the Im-3 m structure, then it transformed into the β-tin structure. Next, the β-tin structure transformed into the Sr-IV, finally, the Sr-IV structure transformed into the Sr-V structure. On the contrary, the theoretical studies were reported that the Fm-3 m structure transformed into the Im-3 m structure, then it transformed into the Sr-IV structure, showing that the the β-tin structure is not energetically favored over the Sr-IV structure.

In 2012, Sr. was predicted that there is a candidate structure [33]. The relative enthalpy of Sr. was reported that the Cmcm structure is thermodynamically favored over the Fm-3 m structure, the Im-3 m structure, and the β-tin structure. In addition, the Cmcm structrue was displayed that it can transform into the hcp structure as well. The Cmcm structure was investigated superconductivity, showing that the T_c of the Cmcm is estimated to be 4 K. The remarkable result manifested that the predicted T_c values are in good agreement with experiment [76, 77].

However, the discrepancy between the experimental observations and the theoretical studies were not solved yet. In 2015, the discrepancies in transition sequence between the experimental and theoretical works was explained by Tsuppayakorn-aek et al. [10]. Regarding transition sequence in Sr., it was investigated by the hybrid exchange-correlation functional, i.e. screened exchange local density approximation (sX-LDA) [78–80]. The stable structure of the β-tin was corrected by sX-LDA functional. In fact, the sX-LDA functional is important for the d electrons. At this point, it is interesting to compare the experimental observation and the theoretical study [10] by considering the energy levels in each electron configuration of isolate strontium (**Figure 3**). The solution of the energy levels indicated that the sX-LDA functional is in good agreement with the experiment [81].

The remarkable result of the Ref. [10] shown that the β-tin structure is thermodynamically favored over the hcp structure by sX-LDA functional (**Figure 4**). The Ref. [10] manifested that the Im-3 m structure transformed into the β-tin structure,

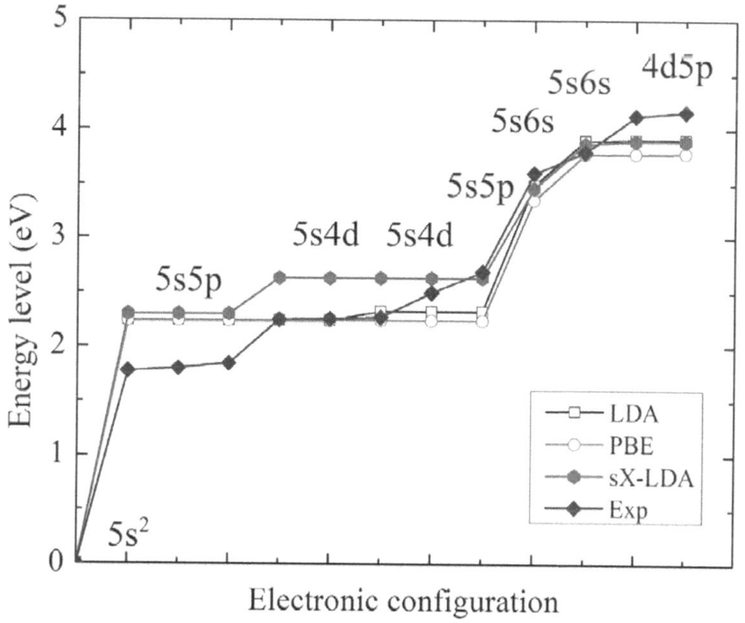

Figure 3.
The energy level each electron configuration of isolate Sr.

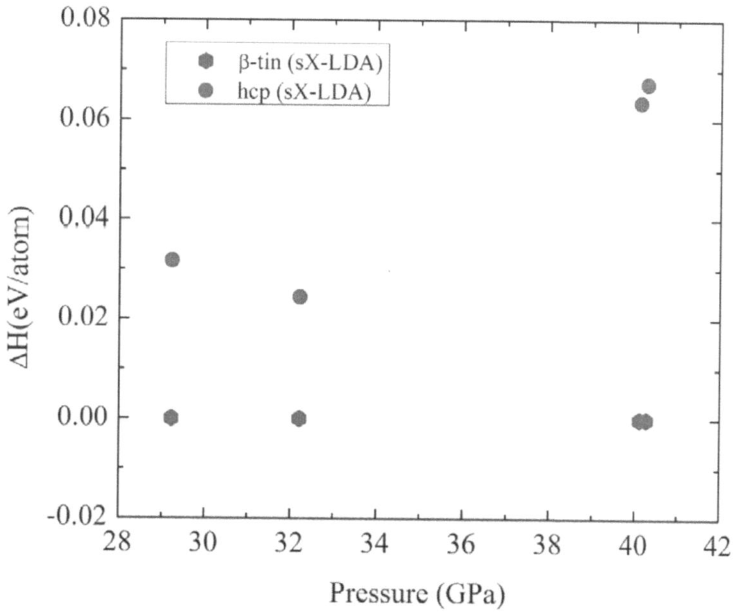

Figure 4.
The relative enthalpy of Sr. as a function of pressure by sX-LDA functional.

showing that the theoretical study is in good agreement with experimental observations [34–36].

Regarding the superconductor in the β-tin structure is interesting. Although the β-tin structure is thermodynamically stable by sX-LDA functional, it is not calculated the T_c. This because the sX-LDA functional is not implemented for the T_c calculation. However, other hybrid exchange-correlation functionals, i.e. PBE0 or HSE06, are possible for investigation the stability of the β-tin structure, leading to find the T_c.

3.3 Scandium

Structural prediction at high pressure is suitable for identifying unknown structure. Scandium (Sc) is one of d-transition metal, showing that there is an unknown structure (Sc-III) at high pressure [82]. The transition sequences is found that the hcp structure transformed into the host-guest structure [83, 84]. The host-guest structure is thermodynamically stable up to 70 GPa [85]. It is interesting to note that what is the unknown structure beyond the host-guest structure above 70 GPa. In 2018, Tsuppayakorn-aek et al. [14] was identified the unknown structure by *ab initio* random structure searching (AIRSS). The predicted structure was manifested that Sc-III is the tetragonal structure with space group $P4_12_12$. The $P4_12_12$ structure was shown that it is thermodynamiclly stable favored over the hcp structure and the host-guest structure above 93 GPa (**Figure 5**) [14]. Also, the $P4_12_12$ structure was found that it is dynamically structure at 120 GPa, as shown in (**Figure 6**). Moreover, the solution of the simulated XRD pattern [14] is in good agreement with the observed XRD pattern from the experimental study [82]. Structural phase transitions of Sc was reported that the hcp structure transformed into the host-guest structure, and then, it transformed into the $P4_12_12$ structure.

Regarding superconductor of the $P4_12_12$ structure, it was found to be the metallicity by considering density of state (DOS), leading to investigate the T_c. The $P4_12_12$ structure displayed that the estimated T_c is 8.36 K at 110 GPa. While, the

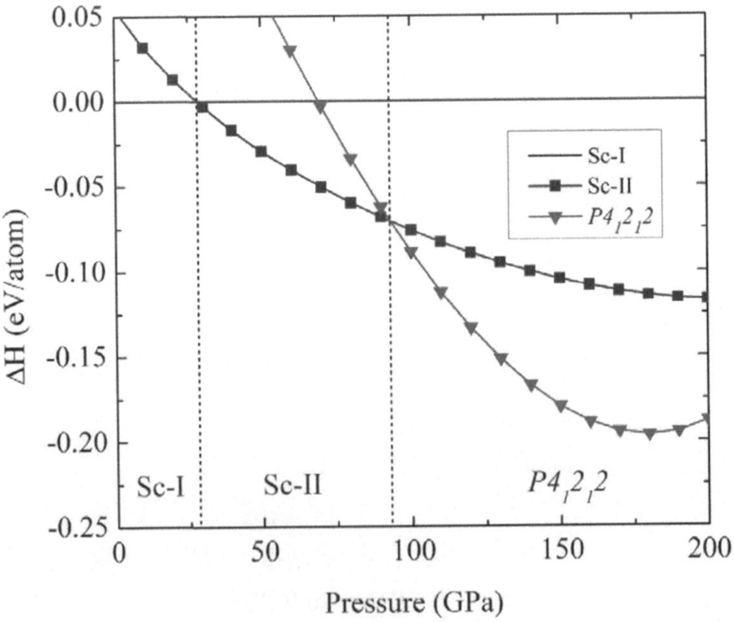

Figure 5.
The relative enthalpy of Sc as a fucntion of pressure.

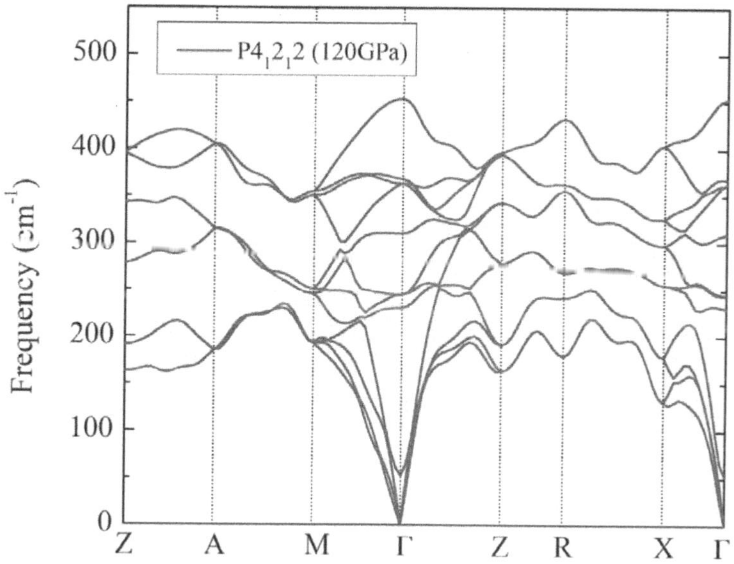

Figure 6.
The phonon dispersion of the $P4_12_12$ *structure.*

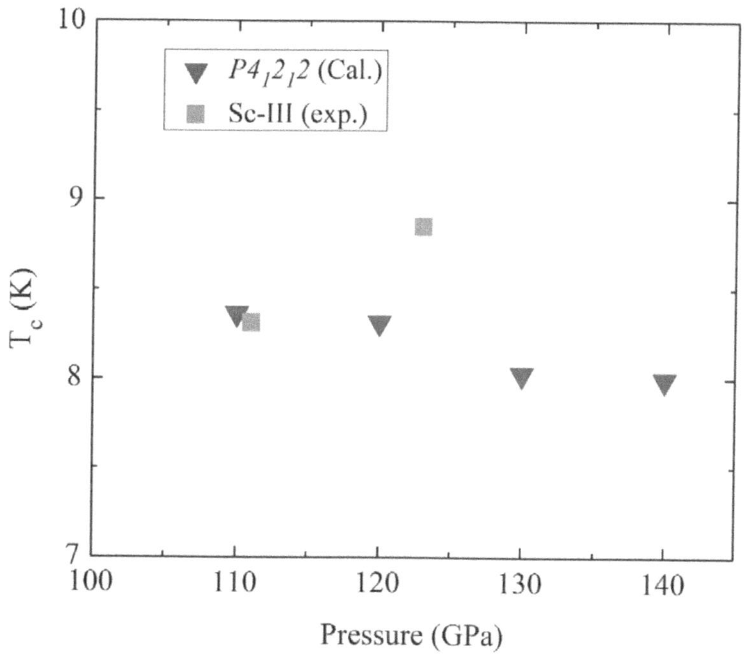

Figure 7.
The T_c *of the* $P4_12_12$ *structure compare the* T_c *of Sc-III phase.*

experimental study was reported the T_c is 8.31 K and 111 GPa [86]. Moreover, the $P4_12_12$ structure was explored the T_c above 130 GPa. Also, it was found that the T_c decreased monotonically with increasing pressure (**Figure** 7). In addition, the EPC strengths decreased with increasing pressure as well.

Tsuppayakorn-aek et al. [14] was revealed in that the T_c of the $P4_12_12$ structure decreased with increasing pressure occurred from the mechanical of the DOS. It can be easily understood by considering the partial-density of state. They were shown that the p-electron decreased with increasing pressure. In contrast, the s electron increased with increasing pressure. In addition, the decreasing of T_c value is supported by the ELF calculation. The ELF is displayed in the (110) atomic plane of the $P4_12_12$ structure, showing that the characteristic of electron state. One can see that the p-electron is accumulated between Sc atoms, indicating that the nature of the chemical bonding is the weak bonding. On increasing pressure, the p-electron transferred into the s and d electrons. This implied that the decreasing of the p-electron might affect the T_c value.

Sc is one of the group-IIIB element was shown that structural phase transformation displayed the complex to simple transition. Also, it promoted the superconducting temperature transition to be 8.36 K at 110 GPa, which it is in good agreement with the experimental observation.

3.4 Arsenic

The group-V element is one of central interest in superconductor. It is interesting to note that arsenic (As), antimony (Sb), and bismuth (Bi) share the remarkable similarity of structural and property [87, 88]. Structural of the group-V element was reported that As-III, Sb-IV, and Bi-III are the incommensurate host-guest structure [89–92]. Also, it is worth to note that the Im-3 m structure is thermodynamically stable favored over the incommensurate structure [87, 88].

Tsuppayakorn-aek et al. [12] was explored the high-pressure phase in As. This because it is interesting to find the high-pressure phase, leading to go beyond the Im-3 m structure. The structural prediction was investigated up to 300 GPa. The predicted structure was shown that the body-centered tetragonal (bct) structure with space group $I4_1/acd$ to be the stable structure at high pressure. The $I4_1/acd$ structure is energetically and dynamically stable. Also, it is thermodynamically favored over the host-guest structure. The $I4_1/acd$ structure displayed that it compete with the Im-3 m structure. Moreover, The $I4_1/acd$ structure and the Im-3 m structure are very closed in enthalpy from 100 to 300 GPa. Also, the $I4_1/acd$ structure is sub-spacegroup of the the Im-3 m structure. It is possible that the $I4_1/acd$ structure is coexistence phase with the Im-3 m structure.

Here, the $I4_1/acd$ structure was discovered to be the metallicity, indicating that it is superconducting phase. As already mentioned, the $I4_1/acd$ structure and the Im-3 m structure are wonderfully closed in enthalpy. It is interesting to investigate the superconducting phase of both of them. An important and a fundamental of the spectral function led to consider superconductor. In fact, the spectral function is associated with the electron–phonon coupling (EPC). The $I4_1/acd$ structure was regarded in superconductor, it was found that the estimated T_c is 4.2 K at 150 GPa. On increasing pressure, the T_c of the $I4_1/acd$ structure decreased with the EPC. Likewise, the T_c of the Im-3 m is likely to decrease, where a pressure increasing. It is worth to note that the $I4_1/acd$ and Im-3 m structures are very similar in the T_c [12].

The remarkable results of the T_c value were shown that the T_c of the $I4_1/acd$ structure has higher than the Im-3 m structure at 150 GPa. The reason can be considered by the spectral function (α^2F) (**Figure 8**). The contribution of the α^2F shown that the $I4_1/acd$ structure is higher than those of the Im-3 m structure around middle frequency regime (6–13 THz).

Now, it is worth to note that the $I4_1/acd$ structure hold the metallic state at 300 GPa. Tsuppayakorn-aek et al. [12] suggested that the $I4_1/acd$ structure is not favored superconductor above 300 GPa, indicating that it is likely to transform into a normal metallic state (**Figure 9**). As a possible cause of this, one might think of phase transformation [19]. Moreover, the EPC of the $I4_1/acd$ structure is very poor characterized by compression. At this point, it is possible that a novel phase might occur above 300 GPa.

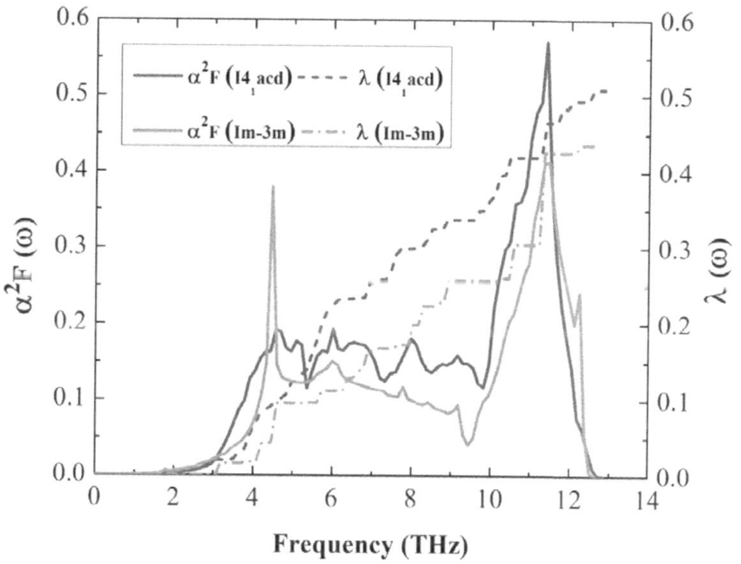

Figure 8.
The spectral function of the $I4_1/acd$ and Im-3 m structures at 150 GPa.

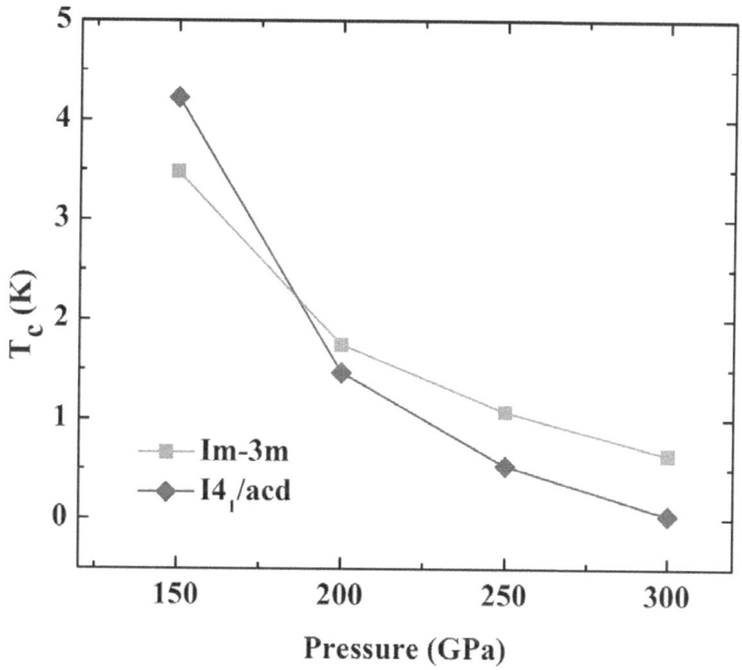

Figure 9.
The T_c of the $I4_1/acd$ and Im-3 m structures as a function of pressure.

4. Conclusion

Ab initio random structure searching is combined with density functional theory has been use to predict a candidate structure in lithium, strontium, scandium, and arsenic under high pressure. The predicted host-guest structure in lithium is expected to be superconductor, where the electron localization function is considered. The discrepancy between the experimental observations and the theoretical studies in strontium is solved by hybrid exchange-correlation functional. Moreover, the β-tin structure is worth to explore a superconductor by performing hybrid exchange-correlation functional. The role of the electron phonon coupling displays that it is crucial for scandium ans arsenic under compression. The remarkable result of the superconducting transition temperature of scandium and arsenic share to a similar character, indicating that the superconducting transition temperature of both of them is likely to decrease with increasing pressure.

References

[1] Kohn, W. and Sham, L. J. Self-Consistent Equations Including Exchange and Correlation Effects. DOI: 10.1103/PhysRev.140.A1133

[2] Yin, M. T. and Cohen, Marvin L. Theory of static structural properties, crystal stability, and phase transformations: Application to Si and Ge. DOI:10.1103/PhysRevB.26.5668

[3] Payne, M. C. and Teter, M. P. and Allan, D. C. and Arias, T. A. and Joannopoulos, J. D, Iterative minimization techniques for ab initio total-energy calculations: molecular dynamics and conjugate gradients. DOI: 10.1103/RevModPhys.64.1045

[4] Perdew, John P. and Burke, Kieron and Ernzerhof, Matthias, Generalized Gradient Approximation Made Simple. DOI:10.1103/PhysRevLett.77.3865

[5] Bardeen, J. and Cooper, L. N. and Schrieffer, J. R. Theory of Superconductivity. DOI:10.1103/PhysRev.108.1175

[6] Pickard, Chris J. and Needs, R. J. High-Pressure Phases of Silane. DOI: 10.1103/PhysRevLett.97.045504

[7] Chris J Pickard and R J Needs, Ab initio random structure searching. DOI: 10.1088/0953-8984/23/5/053201

[8] Prutthipong Tsuppayakorn-aek and Piya Phansuke and Pungtip Kaewtubtim and Rajeev Ahuja and Thiti Bovornratanaraks, Enthalpy stabilization of superconductivity in an alloying *S–P–H* system: First-principles cluster expansion study under high pressure. https://www.sciencedirect.com/science/article/pii/S0927025621000070

[9] Prayoonsak Pluengphon and Prutthipong Tsuppayakorn-aek and Burapat Inceesungvorn and Udomsilp Pinsook and Thiti Bovornratanaraks, Structural, thermodynamic, electronic, and magnetic properties of superconducting FeSe–CsCl type: Ab initio searching technique with van der Waals corrections. https://www.sciencedirect.com/science/article/pii/S0254058421004910

[10] Tsuppayakorn-aek, P. and Chaimayo, W. and Pinsook, U. and Bovornratanaraks, T. Existence of the β-tin structure in Sr: First evidence from computational approach. DOI:10.1063/1.4931810

[11] Prayoonsak Pluengphon and Thiti Bovornratanaraks, Phase stability and elastic properties of CuGaSe2 under high pressure. http://www.sciencedirect.com/science/article/pii/S0038109815002082

[12] Tsuppayakorn-aek, Prutthipong and Luo, Wei and Ahuja, Rajeev and Bovornratanaraks, Thiti, The High-Pressure Superconducting Phase of Arsenic. DOI:10.1038/s41598-018-20088-8

[13] Tsuppayakorn-aek, Prutthipong and Luo, Wei and Watcharatharapong, Teeraphat and Ahuja, Rajeev and Bovornratanaraks, Thiti, Structural prediction of host-guest structure in lithium at high pressure. DOI:10.1038/s41598-018-23473-5

[14] Tsuppayakorn-aek, Prutthipong and Luo, Wei and Pungtrakoon, Wirunti and Chuenkingkeaw, Kittana and Kaewmaraya, Thanayut and Ahuja, Rajeev and Bovornratanaraks, Thiti, The ideal commensurate value of Sc and the superconducting phase under high pressure. DOI:10.1063/1.5047251

[15] R. Manotum and R. Klinkla and U. Pinsook and K. Kotmool and P. Tsuppayakorn-aek and R. Ahuja and T. Bovornratanaraks, Effect of pressure on

the structure stability, electronic structure and band gap engineering in Zn16O1S15. http://www.sciencedirect.com/science/article/pii/S2352214318302119

[16] Jimlim, Pornmongkol and Tsuppayakorn-aek, Prutthipong and Pakornchote, Teerachote and Ektarawong, Annop and Pinsook, Udomsilp and Bovornratanaraks, Thiti, Theoretical predictions for low-temperature phases, softening of phonons and elastic stiffnesses, and electronic properties of sodium peroxide under high pressure. DOI:10.1039/C9RA03735G

[17] Prutthipong Tsuppayakorn-aek and Udomsilp Pinsook and Wei Luo and Rajeev Ahuja and Thiti Bovornratanaraks, Superconductivity of superhydride CeH10 under high pressure. DOI:10.1088%2F2053-1591%2Fababc2

[18] Kotmool, Komsilp and Tsuppayakorn-aek, Prutthipong and Kaewmaraya, Thanayut and Pinsook, Udomsilp and Ahuja, Rajeev and Bovornratanaraks, Thiti, Structural Phase Transitions, Electronic Properties, and Hardness of RuB4 under High Pressure in Comparison with FeB4 and OsB4. DOI:10.1021/acs.jpcc.0c03959

[19] Bovornratanaraks, Thiti and Tsuppayakorn-aek, Prutthipong and Luo, Wei and Ahuja, Rajeev, Ground-state structure of semiconducting and superconducting phases in xenon carbides at high pressure. DOI:10.1038/s41598-019-39176-4

[20] Prayoonsak Pluengphon and Thiti Bovornratanaraks and Prutthipong Tsuppayakorn-aek and Udomsilp Pinsook and Burapat Inceesungvorn, High-pressure phases induce H-vacancy diffusion kinetics in TM-doped MgH2: Ab initio study for hydrogen storage improvement. http://www.sciencedirect.com/science/article/pii/S0360319919322773

[21] Prayoonsak Pluengphon and Prutthipong Tsuppayakorn-aek and Burapat Inceesungvorn and Thiti Bovornratanaraks, Pressure-induced structural stability of alkali trihydrides and H2-desorption occurrence: Ab initio study for hydrogen storage improvement. http://www.sciencedirect.com/science/article/pii/S0360319920324587

[22] Drozdov, A. P. and Eremets, M. I. and Troyan, I. A. and Ksenofontov, V. and Shylin, S. I. Conventional superconductivity at 203 kelvin at high pressures in the sulfur hydride system. DOI:10.1038/nature14964

[23] Papaconstantopoulos, D. A. and Klein, B. M. and Mehl, M. J. and Pickett, W. E. Cubic H_3S around 200 GPa: An atomic hydrogen superconductor stabilized by sulfur. DOI:10.1103/PhysRevB.91.184511

[24] Guigue, Bastien and Marizy, Adrien and Loubeyre, Paul, Direct synthesis of pure H_3S from S and H elements: No evidence of the cubic superconducting phase up to 160 GPa. DOI:10.1103/PhysRevB.95.020104

[25] Goncharov, Alexander F. and Lobanov, Sergey S. and Prakapenka, Vitali B. and Greenberg, Eran, Stable high-pressure phases in the H-S system determined by chemically reacting hydrogen and sulfur. DOI:10.1103/PhysRevB.95.140101

[26] Drozdov, A. P. and Kong, P. P. and Minkov, V. S. and Besedin, S. P. and Kuzovnikov, M. A. and Mozaffari, S. and Balicas, L. and Balakirev, F. F. and Graf, D. E. and Prakapenka, V. B. and Greenberg, E. and Knyazev, D. A. and Tkacz, M. and Eremets, M. I. Superconductivity at 250 K in lanthanum hydride under high

pressures. DOI:10.1038/s41586-019-1201-8

[27] Errea, Ion and Belli, Francesco and Monacelli, Lorenzo and Sanna, Antonio and Koretsune, Takashi and Tadano, Terumasa and Bianco, Raffaello and Calandra, Matteo and Arita, Ryotaro and Mauri, Francesco and Flores-Livas, José A, Quantum crystal structure in the 250-kelvin superconducting lanthanum hydride. DOI:10.1038/s41586-020-1955-z

[28] Shipley, Alice M. and Hutcheon, Michael J. and Johnson, Mark S. and Needs, Richard J. and Pickard, Chris J. Stability and superconductivity of lanthanum and yttrium decahydrides. DOI:10.1103/PhysRevB.101.224511

[29] Liu, Hanyu and Naumov, Ivan I. and Hoffmann, Roald and Ashcroft, N. W. and Hemley, Russell J. Potential high-T_c superconducting lanthanum and yttrium hydrides at high pressure. https://www.pnas.org/content/114/27/6990

[30] Heil, Christoph and di Cataldo, Simone and Bachelet, Giovanni B. and Boeri, Lilia, Superconductivity in sodalite-like yttrium hydride clathrates. DOI:10.1103/PhysRevB.99.220502

[31] Peng, Feng and Sun, Ying and Pickard, Chris J. and Needs, Richard J. and Wu, Qiang and Ma, Yanming, Hydrogen Clathrate Structures in Rare Earth Hydrides at High Pressures: Possible Route to Room-Temperature Superconductivity. DOI:10.1103/PhysRevLett.119.107001

[32] Bağcı, S. and Tütüncü, H. M. and Duman, S. and Srivastava, G. P. Phonons and superconductivity in fcc and dhcp lanthanum. DOI:10.1103/PhysRevB.81.144507

[33] Young Kim, Duck and Srepusharawoot, Pornjuk and Pickard, Chris J. and Needs, Richard J. and Bovornratanaraks, Thiti and Ahuja, Rajeev and Pinsook, Udomsilp, Phase stability and superconductivity of strontium under pressure. DOI:10.1063/1.4742323

[34] D. R. Allan and R. J. Nelmes and M. I. McMahon and S. A. Belmonte and T. Bovornratanaraks, Structures and Transitions in Strontium

[35] McMahon, M. I. and Bovornratanaraks, T. and Allan, D. R. and Belmonte, S. A. and Nelmes, R. J. Observation of the incommensurate barium-IV structure in strontium phase V. DOI:10.1103/PhysRevB.61.3135

[36] Bovornratanaraks, T. and Allan, D. R. and Belmonte, S. A. and McMahon, M. I. and Nelmes, R. J. Complex monoclinic superstructure in Sr-IV. DOI:10.1103/PhysRevB.73.144112

[37] Phusittrakool, A. and Bovornratanaraks, T. and Ahuja, R. and Pinsook, U. High pressure structural phase transitions in Sr from ab initio calculations. DOI:10.1103/PhysRevB.77.174118

[38] Sakata, Masafumi and Nakamoto, Yuki and Shimizu, Katsuya and Matsuoka, Takahiro and Ohishi, Yasuo, Superconducting state of Ca-VII below a critical temperature of 29 K at a pressure of 216 GPa. DOI:10.1103/PhysRevB.83.220512

[39] Lei, Shi and Papaconstantopoulos, D. A. and Mehl, Michael J. Calculations of superconducting properties in yttrium and calcium under high pressure. DOI:10.1103/PhysRevB.75.024512

[40] Yao, Yansun and Tse, John S. and Song, Zhe and Klug, Dennis D. and Sun, Jian and Le Page, Yvon, Structures and superconducting properties of the high-pressure IV and V phases of calcium from first principles. DOI:10.1103/PhysRevB.78.054506

[41] Wiwittawin Sukmas and Prutthipong Tsuppayakorn-aek and Udomsilp Pinsook and Thiti Bovornratanaraks, Near-room-temperature superconductivity of Mg/Ca substituted metal hexahydride under pressure.

[42] Shimizu, Katsuya and Ishikawa, Hiroto and Takao, Daigoroh and Yagi, Takehiko and Amaya, Kiichi, Superconductivity in compressed lithium at 20 K. DOI:10.1038/nature01098

[43] Yao, Yansun and Tse, J. S. and Tanaka, K. and Marsiglio, F. and Ma, Y. Superconductivity in lithium under high pressure investigated with density functional and Eliashberg theory. DOI: 10.1103/PhysRevB.79.054524

[44] Struzhkin, Viktor V. and Eremets, Mikhail I. and Gan, Wei and Mao, Ho-kwang and Hemley, Russell J. Superconductivity in Dense Lithium. https://science.sciencemag.org/content/298/5596/1213

[45] Deemyad, Shanti and Schilling, James S. Superconducting Phase Diagram of Li Metal in Nearly Hydrostatic Pressures up to 67 GPa. DOI:10.1103/PhysRevLett.91.167001

[46] Heil, Christoph and Boeri, Lilia, Influence of bonding on superconductivity in high-pressure hydrides. DOI:10.1103/PhysRevB.92.060508

[47] P. Srepusharawoot and W. Luo and T. Bovornratanaraks and R. Ahuja and U. Pinsook, Evidence of a medium-range ordered phase and mechanical instabilities in strontium under high pressure. http://www.sciencedirect.com/science/article/pii/S0038109812001743

[48] Cui, Wenwen and Bi, Tiange and Shi, Jingming and Li, Yinwei and Liu, Hanyu and Zurek, Eva and Hemley, Russell J. Route to high-T_c superconductivity via CH_4-intercalated H_3S hydride perovskites. DOI:10.1103/PhysRevB.101.134504

[49] Ye, Xiaoqiu and Zarifi, Niloofar and Zurek, Eva and Hoffmann, Roald and Ashcroft, N. W. High Hydrides of Scandium under Pressure: Potential Superconductors. DOI:10.1021/acs.jpcc.7b12124

[50] Bi, Tiange and Miller, Daniel P. and Shamp, Andrew and Zurek, Eva, Superconducting Phases of Phosphorus Hydride Under Pressure: Stabilization by Mobile Molecular Hydrogen. https://onlinelibrary.wiley.com/doi/abs/10.1002/anie.201701660

[51] D. A. Papaconstantopoulos, Possible High-Temperature Superconductivity in Hygrogenated Fluorine. https://www.degruyter.com/view/journals/nsm/3/1/article-p29.xml

[52] Majumdar, Arnab and Tse, John S. and Yao, Yansun, Modulated Structure Calculated for Superconducting Hydrogen Sulfide. https://onlinelibrary.wiley.com/doi/abs/10.1002/anie.201704364

[53] Li, Yinwei and Hao, Jian and Liu, Hanyu and Li, Yanling and Ma, Yanming, The metallization and superconductivity of dense hydrogen sulfide. DOI:10.1063/1.4874158

[54] Baroni, Stefano and de Gironcoli, Stefano and Dal Corso, Andrea and Giannozzi, Paolo, Phonons and related crystal properties from density-functional perturbation theory. DOI: 10.1103/RevModPhys.73.515

[55] Paolo Giannozzi and Stefano Baroni and Nicola Bonini and Matteo Calandra and Roberto Car and Carlo Cavazzoni and Davide Ceresoli and Guido L Chiarotti and Matteo Cococcioni and Ismaila Dabo and Andrea Dal Corso and Stefano de Gironcoli and Stefano Fabris

and Guido Fratesi and Ralph Gebauer and Uwe Gerstmann and Christos Gougoussis and Anton Kokalj and Michele Lazzeri and Layla Martin-Samos and Nicola Marzari and Francesco Mauri and Riccardo Mazzarello and Stefano Paolini and Alfredo Pasquarello and Lorenzo Paulatto and Carlo Sbraccia and Sandro Scandolo and Gabriele Sclauzero and Ari P Seitsonen and Alexander Smogunov and Paolo Umari and Renata M Wentzcovitch, QUANTUM ESPRESSO: a modular and open-source software project for quantum simulations of materials. DOI:10.1088/0953-8984/21/39/395502

[56] Allen, P. B. and Dynes, R. C. Transition temperature of strong-coupled superconductors reanalyzed. DOI:10.1103/PhysRevB.12.905

[57] Yao, Yansun and Tse, John S. Electron-phonon coupling in the high-pressure hcp phase of xenon: A first-principles study. DOI:10.1103/PhysRevB.75.134104

[58] Yao, Yansun and Klug, Dennis D. and Sun, Jian and Martoňák, Roman, Structural Prediction and Phase Transformation Mechanisms in Calcium at High Pressure. DOI:10.1103/PhysRevLett.103.055503

[59] Yao, Yansun and Klug, Dennis D. Stable structures of tantalum at high temperature and high pressure. DOI:10.1103/PhysRevB.88.054102

[60] Ma, Yanming and Oganov, Artem R. and Xie, Yu, High-pressure structures of lithium, potassium, and rubidium predicted by an ab initio evolutionary algorithm. DOI:10.1103/PhysRevB.78.014102

[61] Pickard, Chris J. and Needs, R. J. Dense Low-Coordination Phases of Lithium. DOI:10.1103/PhysRevLett.102.146401

[62] Marqués, M. and McMahon, M. I. and Gregoryanz, E. and Hanfland, M. and Guillaume, C. L. and Pickard, C. J. and Ackland, G. J. and Nelmes, R. J. Crystal Structures of Dense Lithium: A Metal-Semiconductor-Metal Transition. DOI:10.1103/PhysRevLett.106.095502

[63] Lv, Jian and Wang, Yanchao and Zhu, Li and Ma, Yanming, Predicted Novel High-Pressure Phases of Lithium. DOI:10.1103/PhysRevLett.106.015503

[64] Lundegaard, L. F. and Gregoryanz, E. and McMahon, M. I. and Guillaume, C. and Loa, I. and Nelmes, R. J. Single-crystal studies of incommensurate Na to 1.5 Mbar. DOI:10.1103/PhysRevB.79.064105

[65] McMahon, M. I. and Nelmes, R. J. and Schwarz, U. and Syassen, K. Composite incommensurate K-III and a commensurate form: Study of a high-pressure phase of potassium. DOI:10.1103/PhysRevB.74.140102

[66] Lundegaard, L. F. and Stinton, G. W. and Zelazny, M. and Guillaume, C. L. and Proctor, J. E. and Loa, I. and Gregoryanz, E. and Nelmes, R. J. and McMahon, M. I. Observation of a reentrant phase transition in incommensurate potassium. DOI:10.1103/PhysRevB.88.054106

[67] McBride, E. E. and Munro, K. A. and Stinton, G. W. and Husband, R. J. and Briggs, R. and Liermann, H.-P. and McMahon, M. I. One-dimensional chain melting in incommensurate potassium. DOI:10.1103/PhysRevB.91.144111

[68] Zhao, Long and Zong, Hongxiang and Ding, Xiangdong and Sun, Jun and Ackland, Graeme J. Commensurate-incommensurate phase transition of dense potassium simulated by machine-learned interatomic potential. DOI:10.1103/PhysRevB.100.220101

[69] Falconi, S. and McMahon, M. I. and Lundegaard, L. F. and Hejny, C. and

Nelmes, R. J. and Hanfland, M. X-ray diffraction study of diffuse scattering in incommensurate rubidium-IV. DOI: 10.1103/PhysRevB.73.214102

[70] Loa, I. and Lundegaard, L. F. and McMahon, M. I. and Evans, S. R. and Bossak, A. and Krisch, M. Lattice Dynamics of Incommensurate Composite Rb-IV and a Realization of the Monatomic Linear Chain Model. DOI:10.1103/PhysRevLett.99.035501

[71] Matsuoka, T. and Sakata, M. and Nakamoto, Y. and Takahama, K. and Ichimaru, K. and Mukai, K. and Ohta, K. and Hirao, N. and Ohishi, Y. and Shimizu, K. Pressure-induced reentrant metallic phase in lithium. DOI:10.1103/PhysRevB.89.144103

[72] Goncharov, Alexander F. and Struzhkin, Viktor V. and Mao, Ho-kwang and Hemley, Russell J. title = Spectroscopic evidence for broken-symmetry transitions in dense lithium up to megabar pressures. DOI:10.1103/PhysRevB.71.184114

[73] Yao, Yansun and Martoňák, Roman and Patchkovskii, Serguei and Klug, Dennis D. Stability of simple cubic calcium at high pressure: A first-principles study. DOI:10.1103/PhysRevB.82.094107

[74] Li, Bing and Ding, Yang and Yang, Wenge and Wang, Lin and Zou, Bo and Shu, Jinfu and Sinogeikin, Stas and Park, Changyong and Zou, Guangtian and Mao, Ho-kwang, Calcium with the β-tin structure at high pressure and low temperature. https://www.pnas.org/content/109/41/16459

[75] Liu, Hanyu and Cui, Wenwen and Ma, Yanming, Hybrid functional study rationalizes the simple cubic phase of calcium at high pressures. DOI:10.1063/1.4765326

[76] Dunn, K. J. and Bundy, F. P. Pressure-induced superconductivity in strontium and barium. DOI:10.1103/PhysRevB.25.194

[77] Mizobata, Shigeki and Matsuoka, Takahiro and Shimizu, Katsuya, title = Pressure Dependence of the Superconductivity in Strontium. DOI: 10.1143/JPSJS.76SA.23

[78] Clark, Stewart J. and Robertson, John, Screened exchange density functional applied to solids. DOI: 10.1103/PhysRevB.82.085208

[79] Gillen, Roland and Robertson, John, Density functional theory screened-exchange approach for investigating electronical properties of graphene-related materials. DOI:10.1103/PhysRevB.82.125406

[80] Gillen, Roland and Clark, Stewart J. and Robertson, John, Nature of the electronic band gap in lanthanide oxides. DOI:10.1103/PhysRevB.87.125116

[81] Moore, Charlotte E, Atomic Energy Levels as Derived from the Analyses of Optical Spectra. Volume III. 42Mo to 57La; 72Hf to 89Ac

[82] Akahama, Yuichi and Fujihisa, Hiroshi and Kawamura, Haruki, New Helical Chain Structure for Scandium at 240 GPa. DOI:10.1103/PhysRevLett.94.195503

[83] Fujihisa, Hiroshi and Akahama, Yuichi and Kawamura, Haruki and Gotoh, Yoshito and Yamawaki, Hiroshi and Sakashita, Mami and Takeya, Satoshi and Honda, Kazumasa, Incommensurate composite crystal structure of scandium-II. DOI:10.1103/PhysRevB.72.132103

[84] McMahon, M. I. and Lundegaard, L. F. and Hejny, C. and Falconi, S. and Nelmes, R. J. Different incommensurate composite crystal structure for Sc-II. DOI:10.1103/PhysRevB.73.134102

[85] Determination of the Structural Parameters of an Incommensurate Phase from First Principles: The Case of Sc-II, Arapan, Sergiu and Skorodumova, Natalia V. and Ahuja, Rajeev. DOI: 10.1103/PhysRevLett.102.085701

[86] Debessai, M. and Hamlin, J. J. and Schilling, J. S. Comparison of the pressure dependences of T_c in the trivalent *d*-electron superconductors Sc, Y, La, and Lu up to megabar pressures. DOI:10.1103/PhysRevB.78.064519

[87] Häussermann, Ulrich and Söderberg, Karin and Norrestam, Rolf, Comparative Study of the High-Pressure Behavior of As, Sb, and Bi. DOI:10.1021/ja020832s

[88] Katzke, Hannelore and Tolédano, Pierre, Displacive mechanisms and order-parameter symmetries for the A7-incommensurate-bcc sequences of high-pressure reconstructive phase transitions in Group Va elements. DOI: 10.1103/PhysRevB.77.024109

[89] McMahon, M. I. and Degtyareva, O. and Nelmes, R. J. and van Smaalen, S. And Palatinus, L. Incommensurate modulations of Bi-III and Sb-II. DOI: 10.1103/PhysRevB.75.184114

[90] Coleman, A. L. and Stevenson, M. and McMahon, M. I. and Macleod, S. G. Phase diagram of antimony up to 31 GPa and 835 K. DOI:10.1103/ PhysRcvB.97.144107

[91] Khasanov, Rustem and Luetkens, Hubertus and Morenzoni, Elvezio and Simutis, Gediminas and Schönecker, Stephan and Östlin, Andreas and Chioncel, Liviu and Amato, Alex, Superconductivity of Bi-III phase of elemental bismuth: Insights from muon-spin rotation and density functional theory. DOI:10.1103/ PhysRevB.98.140504

[92] Kartoon, D. and Makov, G. Structural and electronic properties of the incommensurate host-guest Bi-III phase. DOI:10.1103/ PhysRevB.100.014104

Chapter 10

Unraveling Hydrogen Bonded Clustering with Water: Density Functional Theory Perspective

Abstract

Extensive density functional theory (DFT) studies have been compiled and additional investigation has been performed for several energetically favorable conformers of hydrogen bonded water clusters. The focus here is not to merely reviewing the literature on DFT investigations on water clusters but to understand the basic building blocks, structural patterns and trends in the energetics of the clusters during the cluster growth. The successive addition of water molecules to these clusters alters the hydrogen bonding pattern, that leads to modification in overall cluster geometry which is also reflected in the vibrational frequency shifts in simulated vibrational infra-red (IR) spectra.

Keywords: hydrogen bonding, water clusters, ab initio, quantum chemistry, density functional theory

1. Introduction

Water is the most ubiquitous substance on this planet and has probably received more scientific and technological attention than any other substance. *Water clusters*, in particular, the groups of water molecules held together by hydrogen bonds, have been the subject of a number of intense experimental and theoretical investigations for the past century due to their importance in chemistry [1], atmospheric chemistry [2], understanding cloud and ice formation, as well as a large number of physico-chemical and biochemical processes [3]. The exploration of the structural and binding properties of water clusters is the first step to understand the properties of bulk water, the essence of life. The difficulty in obtaining a rigorous molecular scale description of the structure of liquid and solid water is mainly due to the constantly fluctuating hydrogen-bonding network therein. This cooperative hydrogen-bonding in water molecules gives rise to the fascinating arrays of anomalous properties. Also, the cooperativity in aggregates of water molecules is particularly important towards understanding the behavior of the liquid and ionic as well as molecular hydration.

A look at the available literature reveals several monographs as well as reviews [4–9] providing detailed understanding of structures as well as binding, spectral and chemical properties of water and its related clusters. The following discussion comprises a brief review of earlier benchmarking studies on water clusters with a special reference to the structure, energetics and the spectra.

Almost six decades ago, Clementi and co-workers [10] performed benchmarking studies on the water clusters, $(H_2O)_n$, n = 2–8 within Hartree-Fock

(HF) framework. The water dimer, the smallest member of the water cluster family has been studied extensively within experimental [11–13] as well as theoretical framework [9, 14–17] ($\Delta E = 5.5 \pm 0.7$ kcal.mol^{-1} and C_s symmetry) since it forms a classical case of hydrogen bond.

According to several experimental [18–22] as well as theoretical studies, the energetically most favorable structure of the water trimer has been confirmed to be cyclic with C1 symmetry. This structure has one external hydrogen atoms on one side of the O-O-O plane and two hydrogen bonds on the other side of the plane. Also, each monomer in this case behaves as a donor as well as an acceptor. Several *ab initio* quantum chemical investigations [4, 6, 7, 14–16, 23–27] have advocated a cyclic, homodromic structure with S_4 symmetry of Water tetramer, $(H_2O)_4$ which corresponds to the global minimum. The cyclic water tetramer has the free hydrogen atoms in alternate fashion above and below the plane of the O-O-O-O ring. This structure has also been observed in IR-studies of benzene-$(H_2O)_4$ and VRT spectra of $(D_2O)_4$ and $(H_2O)_4$ [28, 29]. Pentagonal rings of $(H_2O)_5$ are quite common in clathrate hydrates and in the solvation of hydrophobic groups of small molecules as well as in proteins and in DNA molecules.

Interestingly, the most stable structure for the pentamer follows the puckered cyclic ring pattern [1, 14, 30–32] whereas the *ab initio* studies [4, 6, 15, 16, 23–27] also predicted such a ring structure. Wales [33] pointed out the existence of different ring structures that can be interconverted through low energy barrier pathways consisting of the flipping of hydrogen atoms and bifurcation mechanisms.

The hexamer of water, $(H_2O)_6$ shows a transition from cyclic to three-dimensional geometries, and could be yet another cluster which has been studied extensively by theory [4, 6, 15, 16, 24–27, 34] and experiments [22, 35, 36]. Several studies [4, 6, 15, 16, 22, 24–27, 34–36] have demonstrated that a large number of alternative three-dimensional structures, such as chair, boat, book, jaws, ring and cage, are likely to be of comparable energies. The study of $C_6H_6 \ldots (H_2O)_6$ by Pribble and Zwier [35] could be considered as the first experimental evidence for the cage structure. Later on, Liu *et al.* [36, 37] also have verified the cage structure of isolated water hexamer through their FIR-VRT spectroscopy. Zero-point vibrational energy (ZPE) seems to play an important role in deciding the preferred geometry of the hexamer. In a recent studies due to Bates and Tshumper [38] proposed that the prism structure is marginally by 0.06 kcal.mol-1 and 0.25 kcal. mol^{-1} at MP2/CBS and CCSD(T)/CBS corrected for zero-point vibrational energy (ZPE) respectively.

Size-specific IR spectra of benzene-$(H_2O)_7$ clusters [35] suggest a compact noncyclic structure for $(H_2O)_7$ whereas IR-UV and UV–UV double resonance spectra of jet-cooled phenol-$(H_2O)_7$ clusters supported a cuboid structure for the heptamer, with one corner being occupied by the phenolic oxygen atom [39]. The vibrational spectrum of pure $(H_2O)_7$ [40] indicate the existence of two isomers derived from the S_4 octamer cube by the removal of either a double-donor or a double-acceptor water molecule.

Extensive *ab initio* calculations [4, 10, 16, 41–47] suggest that, at low temperatures, $(H_2O)_8$ would stabilize into a cube with D_{2d} or S_4 symmetry and that, at higher temperatures, entropy considerations could favor the other geometries. Experiments involving pure water clusters [48] as well as hydrated molecules, such as and phenol-$(H_2O)_8$, [39] and C_6H_6-$(H_2O)_8$ [49] also support the cubic structure of $(H_2O)_8$. Maeda and Ohno [30] explored 164 local minima of $(H_2O)_8$ employing MP2/6-311++G(3df,2p)//B3LYP/6-311 + G** level. They have observed that the D_{2d} isomer was energetically most stable than the S_4 isomer of $(H_2O)_8$. Thus, it may be seen that the smaller water clusters, $(H_2O)_n$, n < 8 have been subject of intense investigation within theoretical framework.

Several possible conformers of various water clusters, $(H_2O)_n$, up to nonamer were explored employing continuum solvation model by Malloum and Conradie [31]. They used MN15 DFT framework [32] with 6-31++G(d,p) basis set. According to this study the structures of $(H_2O)_n$ in continuum solvation (IEF-PCM) framework qualitatively similar to their gas phase counterpart. However, the hydrogen bonds show elongation in continuum solvent. The authors also suggested the effectiveness of MN15/6-31++G(d,p) density functional theory model chemistry for studying water clusters Jensen *et al.* [50] explored 44 possible structures of water nonamer using extensive HF calculations using the 6-311G(d) and 6-311G(d,p) basis sets. It was observed that the energetically favorable structure of water nonamer consist of stacked pair of cyclic water tetramer and pentamer which could be thought of as an extended cube. The computational predictions are in qualitative agreement with the available experimental result [51].

About a decade ago, Shields and coworkers [52] performed a detailed investigations on $(H_2O)_n$, n = 2–10; employing a combined molecular dynamics and quantum mechanical framework. The focus of the work was on the structures and energetics of water heptamer, nonamer and decamer structures. They observed that the 30 conformers of $(H_2O)_{10}$ spanned within the short range of 2 kcal.mol^{-1} of Gibb's free energy computed at 298 K. They used least-squares fitting based extrapolation and a new approach (MP2/CBS-e) incorporating counterpoise (CP) correction to basis set superposition error. The O-H stretch spectra [51, 53] of $(H_2O)_{10}$ seem to support a butterfly structure and not a fused cage.

For larger clusters viz. $(H_2O)_n$, n > 8 there exist some notable computational studies [4–6, 8, 16, 31, 54–59] with employing different levels of theory including the Gold-Standard CCSD(T) level of theory in conjugation with complete basis set (CBS) extrapolation for some structurally important prototype clusters. Todorova and coworkers on the basis of a detailed study involving modified B3LYP, X3LYP, and PBE0 functionals concluded that the modified functionals show better agreement with the experimental structures (in terms of radial distribution function) and dynamics (in terms of self-diffusion constant) properties of liquid water.

Lee and coworkers [60] studied $(H_2O)_{11}$ (undecamer) and $(H_2O)_{12}$ (dodecamer) employing Møller–Plesset second order perturbation theory (MP_2) with TZ_2P++ basis set. The undecamer structure is though as a combination of cyclic $(H_2O)_5$ and $(H_2O)_6$ with 16 hydrogen bonds. Similar structure for $(H_2O)_{11}$ has been confirmed by Bulusu *et al.* [61] in their study at B3LYP/6-311+G(d,p) level. They also extended their study for $(H_2O)_{13}$. In case of $(H_2O)_{12}$, the isomer arising from stacking of three cyclic tetramers with 20 hydrogen bonds and patterns $(D_{2d})_2$ or $(D_{2d})(S_4)$ or $(S_4)_2$ are energetically favorable conformers [4, 46] on potential energy surface. There also exists another low-lying conformer comprising two fused hexameric units held together by 18 hydrogen bonds. The $(H_2O)_{13}$ structure as confirmed by Bulusu et al. [61] is composed of an addition of isolated water molecule to the $(H_2O)_{12}$ cluster with cuboid conformation.

CCSD(T) level study on dipole polarizability of water clusters upto $(H_2O)_{12}$ has been reported by Hammond *et al.* [62]. This study also involve performance assessment of six density functionals namely. PBE, PBE0, B3LYP, BLYP, X3LYP and PW91. The authors concluded that the density functional PBE0 with aug-cc-pVDZ yield better agreement with the results from CCSD(T) level of theory than the other density functionals.

In a benchmarking study, Bryantsev and coworkers [16] investigated some structurally important neutral and charged water clusters viz. $(H_2O)_n$, n = 2–8, and 20, $H_3O^+(H_2O)_n$, n = 1–6, and OH-$(H_2O)_n$, n = 1–6. They employed B3LYP, X3LYP, M06, M06-L and M06-2X density functionals as well as Møller-Plesset perturbation theory (MP2) and coupled-cluster theory with single, double, and perturbative triple excitations CCSD(T) levels of theory with various extended basis sets. The basis sets employed were aug-cc-pVnZ, where n = 2, 3, 4 and 5. It was concluded that Minnesota density functionals namely M06 and M06-L used with aug-cc-pV5Z basis set yield vary small mean unsigned error (MUE) of ~0.73–0.84 kcal.mol^{-1}. However, these methods are computationally very expensive even for medium-sized water clusters. In a detailed benchmarking investigation on $(H_2O)_{16}$ and $(H_2O)_{17}$, Leverentz and coworkers [54] employed 61 levels of density functionals, 12 computational methods combining DFT with molecular mechanics (MM) damped dispersion (DF-MM), seven semiempirical methods, as well as semiempirical methods with MM- damped dispersion. The results from their study were compared with those from CCSD(T)/aug-cc-pVTZ level of theory [63]. They advocated that the density functionals viz. M06-2X, M05-2X, ωB97X-D, SCC-DFTB-g are good for binding energies.

A generalized energy-based fragmentation (GEBF) approach was used by Wang and Li [64] to compute the interaction energies of $(H_2O)_{20}$ conformers at explicitly correlated version of coupled-clusters with singles, doubles and triple excitation (CCSD(T)-F_{12}) level of theory. They have also computed interaction energies of these conformers using 32 different levels of DFT-framework. Based on the detailed investigation it was concluded that the functionals ωB97X-D and M05-2X functional show good agreement in absence of empirical dispersion correction. Also, the functionals including the empirical dispersion corrections viz. LC-ωPBE-D3 and B97-D show very close agreement with the CCSDT results. The same approach was employed by Yuan and coworkers [65] to investigate relative energies of large water clusters namely $(H_2O)_n$, n = 32, and 64.The energies computed with GEBF method were compared with CCSD(T) and MP2-levels of theories combined with CBS-limit. It was concluded that the popular functionals viz. B3LYP, PBE0, and DFTB3 do not yield accurate energies whereas the functionals viz. LC-ωPBE-D3 and ωB97X-D yield better agreement of relative energies when compared with CCSD (T)/CBS-level of theory. They advocated use of at-least aug-cc-pVTZ basis set for computational studies on water clusters as well as clusters in aqueous solutions.

Mallhoun *et al.* [8] on performed a careful study of $(H_2O)_n$, n = 2–30 employing the M06-2X functional with aug-cc-pVTZ basis set. They observed that the cage structures are energetically more dominating as compared to the stacked structures comprising tetrameric and pentameric units for $(H_2O)_n$, where n = 26–30. The computed relative energies of their study are in good agreement with the earlier results with CCSD(T)/CBS level of theory. On the basis of their results, it was concluded that performance of the density functional when compared with CCSD(T)/CBS in the increasing order is: MN15 [32], ωB97X-D, M06-2X and APFD. The density functional, MN15 from Truhlar and coworkers outperforms all other density functionals.

Thus, it may be seen from above discussion that a study comprising a short review of benchmarking studies on structures, and energetics of water clusters, $(H_2O)_n$ is warranted. The present work also involves an unbiased assessment of some well-known density functionals for small prototype water clusters, as well as the understanding of basic building blocks for building bigger clusters.

2. Computational methodology

2.1 Benchmarking of the DFT framework

In order to benchmark the density functionals, the present work involves use of two *ab initio* theories, namely Hartree-Fock (HF), Moller-Plesset second order perturbation theory (MP2) and eleven different density functionals, viz. B3LYP

[66–68], B3PW91, BHandHLyp, BLYP, PBE0, M06 [69, 70], M06-2X [69, 70], M06-L [69, 70], τ-HCTHHyb [71], ωB97X [72], AND ωB97X-D [73]. These methods were used in combination with a variety of basis sets viz. 6-31G(d,p), 6-311++G(d,p), 6-311++G(2d,2p), aug-cc-pVDZ, aug1-cc-pVTZ (aug-cc-pVTZ on Oxygen atoms and cc-pVTZ on Hydrogen atoms), aug-cc-pVTZ as well as def2-TZVP for benchmarking the density functionals. Thus, the benchmarking of smaller clusters was done with 91 different model chemistries against the Gold-standard method, coupled-cluster theory with singles, doubles and triples excitation (CCSD(T)/aug-cc-pVQZ). The water clusters, $(H_2O)_2$ and $(H_2O)_3$ and the isolated water molecules were selected for benchmarking the interaction energy and structural parameters.

2.2 Structure generation

The prototype structures viz. isolated water molecule, water dimer and trimer were selected for this benchmarking. The initial structures of the energetically favorable water clusters, $(H_2O)_n$ were taken from the earlier benchmarking studies which are based on the supermolecular approach [4, 14, 27, 62]. These structures were subjected to further geometry optimization within the judiciously shortlisted DFT-framework.

2.3 Computational framework

Initial optimization of all the structures at Hartree-Fock (HF), and Moller Plesset second order perturbation theory (MP2) level with computationally decent and manageable basis set, 6-31G(d,p) was performed employing Gaussian 09 [74] and GAMESS [75]. For standardization of the results, the geometry optimization, vibrational frequency computations were carried out using Gaussian 09 suite of program with default options. However, for all the density functional theory calculations the integration grid specified was an ultrafine pruned grid with 99 radial shells and 590 angular points per shell as defined in the Gaussian 09 suite [74].

3. Results and discussion

The present work embarks on providing an extensive overview of notable investigations on water clusters, $(H_2O)_n$ within quantum chemical as well as density functional theory framework. The results of present work are summarized below.

Selection of an appropriate computational framework which is also known as model chemistry (that is a combination of a computational theory and a basis set) for a given problem is an extremely tough task due to availability of wide range of methods and variety of basis sets. The same fact is evident in the Introduction section that outlines a wide range of computational methodologies including DFT framework employed for water clusters.

In order to understand the trends in energetics and structures, some structurally important prototype clusters, viz. $(H_2O)_n$, n = 1–6, 8, 20 were selected for which some data at high-levels of theory is available. Also, for benchmarking of density functionals the smaller clusters viz. $(H_2O)_2$, and $(H_2O)_3$ were studied employing 91 model chemistries (as mentioned in the Methodology Section) against the Gold-standard method, coupled-cluster theory with singles, doubles and triples excitation (CCSD(T)/aug-cc-pVTZ) [16]. The various $(H_2O)_n$ clusters considered for reviewing are depicted in **Figure 1**. The details of interaction energy data compiled from the present investigation as well as other notable studies is compiled in **Table 1**.

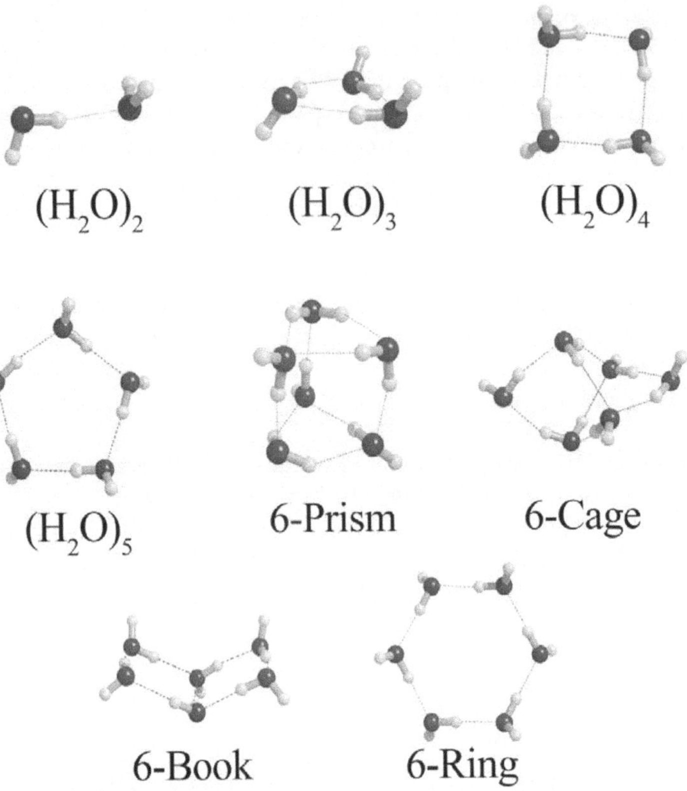

Figure 1.
Schematic representation of B3LYP/6-311++G(2d,2p) optimized structures of $(H_2O)_n$, n = 2–6. Data derived from ref. [16]. Oxygen atoms are denoted by Red spheres, Hydrogen atoms by White-Grey color. Dotted lines represent hydrogen bonds.

The mean unsigned errors for structural parameters viz. bond lengths, bond angle as well as energetics of $(H_2O)_2$ and $(H_2O)_3$ are presented in **Tables 2–4** respectively. The energies for larger clusters were adapted from other references as mentioned in the respective Tables. It may be seen from the **Table 1** that all the computational model chemistries involve an extensive aug-cc-pVTZ basis set that make use of 92 basis functions per water molecule. Thus, as an example studying a water clusters $(H_2O)_{30}$ would require 2760 basis functions, which makes the exploring and scanning of several energetically favorable conformers, rather an arduous task.

In order to explore a convenient alternative to a computationally demanding, aug-cc-pVTZ basis set, a truncated version of aug-cc-pVTZ basis set was employed which involve the use of aug-cc-pVTZ basis set on Oxygen atoms and cc-pVTZ basis set on Hydrogen atoms. This exercise reduces the number of basis functions per water molecule to 74.

It is clear from **Tables 1–4** that the decent basis set, viz. 6-31G(d,p) is good only for the initial structure optimization for larger clusters. Though the MUE for interaction energies for various quantum chemical and DFT framework are large the overall trends and structures show qualitative trends similar to the other extended basis sets.

Taking a cue from the earlier studies [16, 54, 77] as well as the basic computational insights from the present investigation it may be concluded that the density functional theory in combination with appropriate basis sets can predict interaction energies at par or even with better accuracy when compared with MP2-level of theory. On the basis of review of earlier studies and present work, the density

Clusters	Str. Code	n_H	ΔE1	ΔE2	ΔE3	ΔE4	ΔE5	ΔE6	ΔE7	ΔE8	ΔE9	ΔE10	ΔE11
$(H_2O)_2$	—	1	−5.50	−7.50	−4.92	−5.37	−4.57	−4.95	−4.74	−5.17	−4.79	−4.98	−5.01
$(H_2O)_3$	—	3	−17.10	−25.00	−15.34	−16.49	−14.35	−15.45	−15.90	−17.03	−16.0[illegible]	−15.8	−15.80
$(H_2O)_4$	—	4	−29.10	−41.70	−27.42	−28.85	−26.05	−27.76	−27.22	−28.76	−27.33	−27.6	−27.40
$(H_2O)_5$	—	5	−37.70	−53.30	−36.14	−37.86	−34.55	−36.76	−35.08	−37.32	−35.04	−36.3	−35.90
$(H_2O)_6$	Prism	9	−49.60	−72.00	−44.27	−48.59	−41.14	−44.33	−47.31	−50.04	−47.82	−45.9	−46.00
	Cage	8	−48.60	−70.70	−42.80	−48.34	−44.08	−44.60	−46.85	−49.43	−47.20	−45.8	−45.80
	Book	7	−47.75	−68.60	—	—	−44.84	−45.49	−44.94	−47.62	−45.0[illegible]	−45.6	−45.30
	Ring	6	−46.58	−64.89	—	—	−44.63	−45.51	−43.08	−46.11	−43.36	−44.8	−44.30
$(H_2O)_8$	8A	12	−76.01	−110.64	−70.37		−68.28	−71.00	−73.63	−77.34	−74.54	−72.70	−72.60
	8B	12	−75.89	−106.96	−70.37		−66.24	−70.96	−73.62	−77.34	−74.52	−72.70	
$(H_2O)_{20}$	20A	36	−216.28				−184.10	−198.02	−211.80	−221.16	−212.55	−212.60	
	20B	34	−216.04				−188.85	−202.56	−206.03	−223.71	−209.15	−217.40	
	20C	35	−213.71				−185.75	−199.59	−207.72	−220.80	−210.95	−215.00	
	20D	33	−202.49				−185.75	−198.14	−192.76	−200.63	−187.60	−200.10	
	20E	30	−187.17										

ΔE1: ΔE at RHF/6-31G(d,p) level (Present work and Ref. [4]).
ΔE2: ΔE at B3LYP/6-31G(d,p) level (Present work and Ref. [4]).
ΔE3: ΔE at B3LYP/6-311++G(2d,2p) level (Present work and Ref. [4]).
ΔE4: ΔE at MP2/6-311++G(2d,2p) level (Present work and Ref. [4]).
ΔE5: ΔE at B3LYP/aug-cc-pVTZ level (Ref. [16]).
ΔE6: ΔE at X3LYP/aug-cc-pVTZ level (Ref. [16]).
ΔE7: ΔE at M06-L/aug-cc-pVTZ level (Ref. [16]).
ΔE8: ΔE at M06-2X/aug-cc-pVTZ level (Ref. [16]).
ΔE9: ΔE at M06/aug-cc-pVTZ level (Ref. [16]).
ΔE10: ΔE at MP2/CBS level (Ref. [16]).
ΔE11: ΔE at CCSD(T)/CBS level (Ref. [16]).
nH: Number of hydrogen bonds.

Table 1.
Comparison of interaction energies (in kcal.mol^{-1}) for $(H_2O)_n$, n = 2–6, with various model chemistries. (Refer text for details.)

	B3LYP	B3PW91	BHandHLYP	BLYP	HF	M06	M06-2X	M06-L	MP2	PBE0	τ-HCTHHYB	ωB97X	ωB97X-D
6-31G(d,p)	5.71	4.28	5.15	6.29	0.76	5.38	5.83	4.22	4.33	6.13	5.82	6.94	5.78
6-311++G(d,p)	1.00	0.14	1.61	0.30	1.32	2.27	2.67	1.54	1.36	1.93	1.37	3.27	2.07
6-311++G(2d,2p)	0.42	1.33	0.17	1.07	2.54	0.69	1.12	0.25	0.36	0.51	0.02	1.78	0.65
aug-cc-pVDZ	0.84	1.77	0.14	1.58	2.78	0.27	0.63	0.72	0.24	0.04	0.38	1.41	0.21
aug-cc-pVTZ	1.25	2.20	0.59	3.64	3.25	0.18	0.53	0.49	0.98	0.21	0.95	1.00	0.23
aug-cc-pVTZ-1	0.89	1.89	0.29	2.52	3.02	0.20	0.54	0.40	1.04	0.05	0.60	1.29	0.12
def2-TZVP	0.99	0.21	1.26	0.70	1.92	1.48	2.05	1.49	1.21	1.66	1.05	2.85	1.56

Table 2.
Mean unsigned errors (MUE) of interaction energies (in kcal/Mol), computed for $(H_2O)_2$ and $(H_2O)_3$ using various model chemistries. The reference energies were taken at CCSD(T)/aug-cc-pVQZ level (ref. [14, 26]).

	B3LYP	B3PW91	BHandHLYP	BLYP	HF	M06	M06-2X	M06-L
6-31G(d,p)	0.007	0.008	0.011	0.022	0.035	0.010	0.017	0.004
6-311++G(d,p)	0.005	0.003	0.013	0.017	0.042	0.009	0.010	0.006
6-311++G(2d,2p)	0.008	0.006	0.006	0.025	0.054	0.002	0.003	0.003
aug-cc-pVDZ	0.008	0.006	0.007	0.025	0.051	0.003	0.003	0.007
aug-cc-pVTZ	0.009	0.008	0.006	0.023	0.055	0.007	0.002	0.003
aug-cc-pVTZ-1	0.009	0.008	0.007	0.023	0.056	0.003	0.001	0.002
def2-TZVP	0.006	0.006	0.008	0.023	0.048	0.004	0.002	0.005

	MP2	PBEO	τ-HCTHHYB	ωB97X	ωB97X-D
6-31G(d,p)	0.126	0.014	0.014	0.012	0.006
6-311++G(d,p)	0.004	0.012	0.012	0.014	0.010
6-311++G(2d,2p)	0.005	0.005	0.006	0.007	0.003
aug-cc-pVDZ	0.011	0.009	0.011	0.010	0.006
aug-cc-pVTZ	0.005	0.007	0.008	0.008	0.003
aug-cc-pVTZ-1	0.006	0.005	0.005	0.007	0.002
def2-TZVP	0.005	0.008	0.007	0.003	0.007

Table 3.
Mean unsigned errors (MUE) for bond lengths (in) computed for $(H_2O)_2$ and $(H_2O)_3$ using various model chemistries. The reference structural parameters were taken from the experimental data (ref. [76]).

	B3LYP	B3PW91	BHandHLYP	BLYP	HF	M06	M06-2X	M06-L
6-31G(d,p)	3.31	2.35	2.06	5.07	1.30	3.74	2.84	3.45
6-311++G(d,p)	1.64	1.48	2.36	1.04	3.71	1.72	1.58	1.25
6-311++G(2d,2p)	0.71	0.70	1.27	0.36	2.06	0.73	1.27	0.49
aug-cc-pVDZ	0.41	0.38	1.11	0.35	2.22	1.01	0.73	0.69
aug-cc-pVTZ	0.67	0.57	0.98	0.55	2.06	0.97	0.68	0.43
aug-cc-pVTZ-1	0.79	0.60	1.48	0.40	2.67	0.65	0.44	0.66
def2-TZVP	1.20	1.17	2.00	0.38	3.24	1.55	0.49	0.24

	MP2	PBEO	τ-HCTHHYB	ωB97X	ωB97X-D
6-31G(d,p)	2.66	3.01	2.94	3.30	2.79
6-311++G(d,p)	2.49	1.41	1.43	1.48	1.61
6-311++G(2d,2p)	1.21	0.32	0.32	0.72	0.50
aug-cc-pVDZ	0.37	0.14	0.22	0.36	0.43
aug-cc-pVTZ	0.46	0.31	0.38	0.42	0.35
aug-cc-pVTZ-1	0.35	0.47	0.54	0.48	0.42
def2-TZVP	0.81	0.95	1.04	1.19	1.21

Table 4.
Mean unsigned errors (MUE) for bond angles (in degrees) computed for $(H_2O)_2$ and $(H_2O)_3$ using various model chemistries. The reference structural parameters were taken from the experimental data (ref. [76]).

functionals recommended to study water clusters are: MN15 [16, 54, 77], ωB97X-D, M06, M06-L, M06-2X and the broad applicability density functional, viz. PBE0. The recommended basis set should be triple zeta-quality viz. aug-cc-pVTZ, aug1-ccp-VTZ and def2-TZVP. A comparison of the interaction energy of $(H_2O)_2$ computed with various model chemistries is given in **Table 5**.

It is clear from the **Table 5** that the truncated basis set, viz. aug1-cc-pVTZ yield interaction energies that are at par with its parent basis set viz. aug-cc-pVTZ. However, the reduction in number of basis functions results in boosting the speed of computation by a factor of ~2.0 or more for the geometry optimization as well as vibrational frequency computations of water trimer $(H_2O)_3$. This is expected to boost-up further for larger clusters. Thus, aug1-cc-pVTZ basis set is recommended for DFT-study on water clusters.

The vibrational frequency analysis of smaller clusters viz. $(H_2O)_n$, where n = 1–4 show shift as well as splitting of vibrational infra-red)IR) frequencies with successive addition of water molecules. The H-O-H bending frequency shows blue-shifts whereas stretching frequencies (symmetric and asymmetric stretching) show red-shift with successive addition of water molecules. This feature is akin to the earlier studies [4, 51, 78]. The vibrational frequency analysis for smaller clusters upto tetramer performed with PBE0/aug1-cc-pVTZ model chemistry is given in **Table 6**. Also, the successive addition is associated with increase in the spectral intensity. The increase in spectral intensity is an indication of charge separation which may be seen in terms of increase in number of hydrogen (denoted by nH in **Table 1**) with addition of water molecule to the given cluster.

The review of the present literature also reveals that the energetic stability of smaller clusters $(H_2O)_n$ with $n \leq 10$ are predominantly dependent on the total number of hydrogen bonds (HBs) in the given cluster. These hydrogen bonds are the manifestation of donor-acceptor (da) arrangement of water molecules. In case of two dimensional structures the $(H_2O)_n$ clusters with maximum number of donor-acceptor type of hydrogen bonds are energetically more favorable. Hence,

Method	Basis set	ΔE (kcal/mol)	Absolute Error (kcal/mol)
Expt.		−5.000[a]	0.000
CCSD(T)	aug-cc-pVTZ	−5.217[a]	0.217
	aug-cc-pVQZ	−5.101[a]	0.101
	aug-cc-pV5Z	−5.034[a]	0.034
M06-2X	aug-cc-pVTZ	−5.172	0.172
	aug-cc-pVTZ-1	−5.172	0.172
	def2-TZVP	−5.932	0.932
MP2	aug-cc-pVTZ	−5.181	0.181
	aug-cc-pVTZ-1	−5.037	0.037
	def2-TZVP	−5.664	0.664
PBEO	aug-cc-pVTZ	−5.219	0.219
	aug-cc-pVTZ-1	−5.007	0.007
	def2-TZVP	−5.920	0.920
wB97X-D	aug-cc-pVTZ	−5.014	0.014
	aug-cc-pVTZ-1	−5.022	0.022
	def2-TZVP	−5.854	0.854

[a]*Energy values adapted from Ref. [16].*

Table 5.
Interaction energy (in kcal.Mol^{-1}) of water dimer, $(H_2O)_2$ computed with various model chemistries.

Vibration modes assigned	H_2O	$(H_2O)_2$	$(H_2O)_3$	$(H_2O)_4$
H-O-H bending	1634.2 (78.3)	1634.6 (95.7) 1655.1 (95.8)	1645.1 (64.8) 1648.1 (98.8) 1670.1 (18.6)	1648.7 (93.9) 1663.0 (51.1)^D
O-H symmetric stretch	3864.1 (6.8)	3720.9 (357.8) 3853.7 (13.6)	3547.5 (16.5) 3617.8 (644.7) 3630.5 (597.0)	3435.6 (1634.9)^D 3479.1(21.1)
O-H asymmetric stretch	3969.7 (66.9)	3935.6 (84.7) 3955.5 (96.0)	3925.6 (95.9) 3929.6 (96.2) 3931.1 (48.3)	3924.2 (98.9) 3921.9 (90.9)^D

^{D}*Degenerate (Doublet).*

Table 6.
Vibrational frequencies (in cm^{-1}) computed at PBE0/aug1-cc-pVTZ optimized geometries for $(H_2O)n$, n = 2, 3, and 4: Results for water monomer included for comparison. Values in the parentheses indicate intensities.

the cyclic oligomers up to pentameric water clusters are energetically more favorable than their acyclic or linear chain counterparts. Also, these cyclic structures of tetramer, pentamer and hexamer form the basic building blocks for the energetically favorable larger clusters viz. $(H_2O)_n$, for n = 12, 15, 16, 20, 24, 25, 30 and so on. The hexameric ring even though energetically unfavorable as compared to prism and cage hexamers also forms the basic building blocks of higher clusters viz. $(H_2O)_n$, n = 12, 18, 20, 24, 30 and so on.

As concluded by the earlier studies [4], that one can anticipate several structures for a given water cluster $(H_2O)_n$, by changing the hydrogen bonding sequence. Hence, the number of possible energetically favorable isomers increase very fast with increase in the value of n. Considering all such isomers in a single study is beyond the scope of a single study.

4. Conclusions

The present short review on investigation of water clusters employing DFT-framework delineates the strength of DFT to study water clusters accurately and efficiently. The salient features of the present work are summarized as follows.

DFT in conjugation with appropriate triple zeta basis set can yield very accurate estimation of interaction energies and structures of water clusters. The results are at par with MP2/CBS or even CCSD(T)/CBS level of theory. The recommended density functionals for studying water clusters are MN15 [16, 54, 77], ωB97X-D, M06, M06-L, M06-2X and PBE0 where the recommended basis sets is aug-cc-pVTZ. The truncated basis set aug1-cc-pVTZ can also yield better agreement of interaction energies when compared with MP2/CBS and CCSD(T)/ CBS levels of theory. Hence, it is recommended for studying the larger clusters as well as lowering the computational exhaustiveness of the calculations. The cyclic structures of tetramer, pentamer and hexamer are the basic building blocks of larger clusters.

The performance of DFT in studying water clusters is indeed encouraging. It is expected that the initial guidelines from the present short-review can be gainfully employed to investigate larger water clusters and assessment of additional density functionals.

The investigations combined with the novel approaches like molecular tailoring approach will enable in obtaining the better understanding and accurate prediction of interaction energies within DFT-framework. Such studies are underway and will be undertaken in an independent venture.

References

[1] Pauling L. The Structure and Entropy of Ice and of Other Crystals with Some Randomness of Atomic Arrangement. *J. Am. Chem. Soc.* 1935;57(12):2680-2684.

[2] Buszek RJ, Francisco JS, Anglada JM. Water effects on atmospheric reactions. Int Rev Phys Chem. 2011;30 (3):335-369.

[3] Ball P. Water is an active matrix of life for cell and molecular biology. *Proc. Natl. Acad. Sci. U.S.A*. 2017;114(51): 13327-35.

[4] Maheshwary S, Patel N, Sathyamurthy N, Kulkarni AD, Gadre SR. Structure and Stability of Water Clusters $(H_2O)n$, n= 8–20: An Ab Initio Investigation. J Phys Chem A. 2001;105:10525.

[5] Furtado JP, Rahalkar AP, Shanker S, Bandyopadhyay P, Gadre SR. Facilitating Minima Search for Large Water Clusters at the MP2 Level via Molecular Tailoring. J Phys Chem Lett. 2012;3:2253.

[6] Gadre SR, Yeole SD, Sahu N. Quantum Chemical Investigations on Molecular Clusters. *Chem. Rev*. 2014;114 (24):12132-12173.

[7] Miró P, Cramer CJ. Water clusters to nanodrops: a tight-binding density functional study. *Phys. Chem. Chem. Phys*. 2013;15(6):1837-1843.

[8] Malloum A, Fifen JJ, Dhaouadi Z, Nana Engo SG, Conradie J. Structures, relative stability and binding energies of neutral water clusters, $(H_2O)_{2-30}$. *New. J. Chem.* 2019;43(33):13020-13037.

[9] Temelso B, Renner CR, Shields GC. Importance and Reliability of Small Basis Set CCSD(T) Corrections to MP2 Binding and Relative Energies of Water Clusters. J Chem Theory Comput. 2015; 11:1439.

[10] Kistenmacher H, Lie GC, Popkie H, Clementi E. Study of the structure of molecular complexes. VI. Dimers and small clusters of water molecules in the Hartree-Fock approximation. *J. Chem. Phys.* 1974;61(2):546-561.

[11] Dyke TR, Mack KM, Muenter JS. The structure of water dimer from molecular beam electric resonance spectroscopy. J Chem Phys. 1977;66:498.

[12] Dyke TR, Muenter JS. Microwave spectrum and structure of hydrogen bonded water dimer. *J. Chem. Phys.* 1977; 60(7):2929-2930.

[13] Bates DM, Smith JR, Janowski T, Tschumper GS. Development of a 3-Body:Many-Body Integrated Fragmentation Method for Weakly Bound Clusters and Application to Water Clusters (H_2O) n=3–10, 16, 17. J Chem Phys. 2011;135:044123.

[14] Miliordos E, Aprà E, Xantheas SS. Optimal geometries and harmonic vibrational frequencies of the global minima of water clusters $(H_2O)_n$, n = 2–6, and several hexamer local minima at the CCSD(T) level of theory. J Chem Phys. 2013;139:114302.

[15] Xantheas SS. Ab initio studies of cyclic water clusters $(H_2O)_n$, n=1–6. II. Analysis of many-body interactions. J. Chem. Phys. 1994;100(10):7523-7534.

[16] Bryantsev VS, Diallo MS, van Duin ACT, Goddard WA. Evaluation of B3LYP, X3LYP, and M06-Class Density Functionals for Predicting the Binding Energies of Neutral, Protonated, and Deprotonated Water Clusters. J. Chem. Theory. Comput. 2009;5(4): 1016-1026.

[17] Shields RM, Temelso B, Archer KA, Morrel TE, Shields GC. Accurate Predictions of Water Cluster Formation,

$(H_2O)_n$, n = 2–10. J Phys Chem A. 2010; 114:11725.

[18] Pugliano N, Saykally RJ. Measurement of quantum tunneling between chiral isomers of the cyclic water trimer. Science. 1992;257(5078):1937.

[19] Elrod MJ, Saykally RJ. Many-Body Effects in Intermolecular Forces. Chem Rev. 1994;94:1975.

[20] Liu K, Loeser JG, Elrod MJ, Host BC, Rzepiela JA, Pugliano N, et al. Dynamics of Structural Rearrangements in the Water Trimer. *J. Am. Chem. Soc.* 1994;116(8):3507-3512.

[21] Brown MG, Viant MR, McLaughlin RP, Keoshian CJ, Michael E, Cruzan JD, et al. Quantitative characterization of the water trimer torsional manifold by terahertz laser spectroscopy and theoretical analysis. II. $(H_2O)3$. *J. Chem. Phys.* 2019;111(17): 7789-800.

[22] Saykally RJ, Keutsch FN. Water Clusters: Untangling the Mysteries of the Liquid, One Molecule at a Time. Proc Natl Acad Sci USA. 2001;98:10533.

[23] Honegger E, Leutwyler S. Intramolecular vibrations of small water clusters. 1988;88(4):2582-2595.

[24] Xantheas SS. Recent Theoretical and Experimental Advances in Hydrogen-Bonded Clusters2000.

[25] Burnham CJ, Xantheas SS, Miller MA, Applegate BE, Miller RE. The Formation of Cyclic Water Complexes by Sequential Ring Insertion: Experiment and Theory. J Chem Phys. 2002;117:1109.

[26] Miliordos E, Xantheas SS. An Accurate and Efficient Computational Protocol for Obtaining The Complete Basis Set Limits of The Binding Energies of Water Clusters at The MP2 and CCSD (T) Levels of Theory: Application to $(H_2O)_m$, m = 2–6, 8, 11, 16, and 17. J Chem Phys. 2015;142:234303.

[27] Rakshit A, Bandyopadhyay P, Heindel JP, Xantheas SS. Atlas of putative minima and low-lying energy networks of water clusters n = 3–25. 2019;151(21):214307.

[28] Cruzan JD, Braly LB, Liu K, Brown MG, Loeser JG, Saykally RJ. Quantifying Hydrogen Bond Cooperativity in Water: VRT Spectroscopy of the Water Tetramer. Science. 1996;271(5245):59.

[29] Cruzan JD, Viant MR, Brown MG, Saykally RJ. Terahertz Laser Vibration–Rotation Tunneling Spectroscopy of the Water Tetramer. J. Phys. Chem. A 1997;101(48):9022-9031.

[30] Maeda S, Ohno K. Structures of Water Octamers $(H_2O)_8$: Exploration on Ab Initio Potential Energy Surfaces by the Scaled Hypersphere Search Method. J. Phys. Chem. A. 2007;111(20): 4527-4534.

[31] Malloum A, Conradie J. Structures of water clusters in the solvent phase and relative stability compared to gas phase. Polyhedron. 2021;193: 114856.

[32] Yu HS, He X, Li SL, Truhlar DG. MN15: A Kohn–Sham global-hybrid exchange–correlation density functional with broad accuracy for multi-reference and single-reference systems and noncovalent interactions. Chem. Sci. 2016;7(8):5032-5051.

[33] Wales DJ. Structure, Dynamics, and Thermodynamics of Clusters: Tales from Topographic Potential Surfaces. Science. 1996;271(5251):925-929.

[34] Headrick JM, Diken EG, Walters RS, Hammer NI, Christie RA, Cui J, et al. Spectral signatures of hydrated proton vibrations in water clusters. Science. 2005;308:1765.

[35] Pribble RN, Zwier TS. Size-Specific Infrared Spectra of Benzene-$(H_2O)n$ Clusters (n = 1 through 7): Evidence for Noncyclic $(H_2O)n$ Structures. Science. 1994;265(5168):75-79.

[36] Liu K, Brown MG, Saykally RJ. Terahertz Laser Vibration–Rotation Tunneling Spectroscopy and Dipole Moment of a Cage Form of the Water Hexamer. J. Phys. Chem. A. 1997;101 (48):8995-9010.

[37] Liu K, Brown MG, Carter C, Saykally RJ, Gregory JK, Clary DC. Characterization of a cage form of the water hexamer. Nature. 1996;381(6582): 501-503.

[38] Bates DM, Tschumper GS. CCSD(T) Complete Basis Set Limit Relative Energies for Low-Lying Water Hexamer Structures. J Phys Chem A. 2009;113:3555.

[39] Janzen C, Spangenberg D, Roth W, Kleinermanns K. Structure and vibrations of phenol$(H_2O)7,8$ studied by infrared-ultraviolet and ultraviolet-ultraviolet double-resonance spectroscopy and ab initio theory. The Journal of Chemical Physics. 1999;110 (20):9898-9907.

[40] Brudermann J, Melzer M, Buck U, Kazimirski JK, Sadlej J, Bush V. The asymmetric cage structure of $(H_2O)_7$ from a combined spectroscopic and computational study. J. Chem. Phys. 1999;110(22):10649-10652.

[41] Laasonen K, Parrinello M, Car R, Lee C, Vanderbilt D. Structures of small water clusters using gradient-corrected density functional theory. Chem. Phys. Lett. 1993;207(2):208-213.

[42] Kim KS, Dupuis M, Lie GC, Clementi E. Revisiting small clusters of water molecules. Chem. Phys. Lett. 1986;131(6):451-456.

[43] Kim K, Jordan KD, Zwier TS. Low-energy structures and vibrational frequencies of the water hexamer: Comparison with benzene-$(H_2O)_6$. J Am Chem Soc. 1994;116:11568.

[44] Kryachko ES. Water cluster approach to study hydrogen-bonded pattern in liquid water: Ab initio orientational defects in water hexamers and octamers. Intern. J. Quantum. Chem. 1998;70(4-5):831-853.

[45] Jensen JO, Krishnan PN, Burke LA. Theoretical study of water clusters: Heptamers. Chem. Phys. Lett. 1995;241 (3):253-260.

[46] Tsai CJ, Jordan KD. Theoretical study of small water clusters: low-energy fused cubic structures for $(H_2O)_n$, n = 8, 12, 16, and 20. J. Chem. Phys. 1993;97(20):5208-5210.

[47] Moore Plummer PL. Applicability of semi-empirical methods to the study of small water clusters: cubic structures for $(H_2O)_n$ (n = 8, 12, 16). J. Mol. Str.: THEOCHEM. 1997;417(1):35-47.

[48] Kim J, Mhin BJ, Lee SJ, Kim KS. Entropy-driven structures of the water octamer. Chem. Phys. Lett. 1994;219(3): 243-246.

[49] Gruenloh CJ, Carney JR, Arrington CA, Zwier TS, Fredericks SY, Jordan KD. Infrared Spectrum of a Molecular Ice Cube: The S_4 and D_2d Water Octamers in Benzene-$(Water)_8$. Science. 1997;276(5319):1678-1681.

[50] Jensen JO, Krishnan PN, Burke LA. Theoretical study of water clusters: nonamers. Chem. Phys. Lett. 1996;260 (3):499-506.

[51] Buck U, Ettischer I, Melzer M, Buch V, Sadlej J. Structure and Spectra of Three-Dimensional $(H_2O)_n$ Clusters, n = 8,9,10. Phys. Rev. Lett. 1998;80(12): 2578-2581.

[52] Shields RM, Temelso B, Archer KA, Morrell TE, Shields GC. Accurate

Predictions of Water Cluster Formation, $(H_2O)n=2-10$. J. Phys. Chem. A. 2010; 114(43):11725-11737.

[53] Lee C, Chen H, Fitzgerald G. Chemical bonding in water clusters. J. Chem. Phys. 1995;102(3):1266-1269.

[54] Leverentz HR, Qi HW, Truhlar DG. Assessing the Accuracy of Density Functional and Semiempirical Wave Function Methods for Water Nanoparticles: Comparing Binding and Relative Energies of $(H_2O)_{16}$ and $(H_2O)_{17}$ to CCSD(T) Results. J Chem Theory Comput. 2013;9:995.

[55] Qi HW, Leverentz H, Truhlar GG. Water 16-mers and Hexamers: Assessment of the Three-Body and Electrostatically Embedded Many-Body Approximations of the Correlation Energy or the Nonlocal Energy as Ways to Include Cooperative Effects. J Phys Chem A. 2013;117:4486.

[56] Friedrich J, Yu H, Leverentz HR, Bai P, Siepmann JI, Truhlar DG. Water 26-mers Drawn from Bulk Simulations: Benchmark Binding Energies for Unprecedentedly Large Water Clusters and Assessment of the Electrostatically Embedded Three-Body and Pairwise Additive Approximations. J. Phys. Chem. Lett. 2014;5(4):666-670.

[57] Sahu N, Gadre SR, Rakshit A, Bandyopadhyay P, Miliordos E, Xantheas SS. Low Energy Isomers of $(H_2O)25$ from a Hierarchical Method Based on Monte Carlo Temperature Basin Paving and Molecular Tailoring Approaches Benchmarked by MP2 Calculations. J Chem Phys. 2014;141: 164304.

[58] Sahu N, Khire SS, Gadre SR. Structures, Energetics and Vibrational Spectra of $(H_2O)_{32}$ Clusters: a Journey from Model Potentials to Correlated Theory. Mol Phys. 2015;113:2970.

[59] Sahu N, Singh G, Nandi A, Gadre SR. Toward an Accurate and Inexpensive Estimation of CCSD(T)/ CBS Binding Energies of Large Water Clusters. J. Phys. Chem. A. 2016;120 (28):5706-5714.

[60] Lee HM, Suh SB, Kim KS. Structures, energies, and vibrational spectra of water undecamer and dodecamer: An ab initio study. J. Chem. Phys. 2001;114(24):10749-10756.

[61] Bulusu S, Yoo S, Aprà E, Xantheas SS, Zeng XC. Lowest-Energy Structures of Water Clusters $(H_2O)_{11}$ and $(H_2O)_{13}$. J Phys Chem A. 2006;110: 11781.

[62] Hammond JR, Govind N, Kowalski K, Autschbach J, Xantheas SS. Accurate dipole polarizabilities for water clusters n=2–12 at the coupled-cluster level of theory and benchmarking of various density functionals. J. Chem. Phys. 2009;131 (21):214103.

[63] Yoo S, Kirov MV, Xantheas SS. Low-Energy Networks of the T-Cage $(H_2O)_{24}$ Cluster and Their Use in Constructing Periodic Unit Cells of the Structure I (sI) Hydrate Lattice. J Am Chem Soc. 2009;131:7564.

[64] Wang K, Li W, Li S. Generalized Energy-Based Fragmentation CCSD(T)-F12a Method and Application to the Relative Energies of Water Clusters $(H_2O)_{20}$. J. Chem. Phys. 2014;10(4): 1546-1553.

[65] Yuan D, Li Y, Ni Z, Pulay P, Li W, Li S. Benchmark Relative Energies for Large Water Clusters with the Generalized Energy-Based Fragmentation Method. J. Chem. Phys. 2017;13(6):2696-2704.

[66] Lee C, Yang W, Parr RG. Development of the Colle-Salvetti correlation-energy formula into a functional of the electron density. Phys Rev B: Condens Matter Mater Phys. 1988;37:785.

[67] Becke AD. Density-functional thermochemistry. III. The role of exact exchange. J. Chem. Phys. 1993;98(7): 5648-5652.

[68] Becke AD. Density-functional exchange-energy approximation with correct asymptotic behavior. Phys Rev A: At, Mol, Opt Phys. 1988;38: 3098.

[69] Zhao Y, Truhlar DG. Density Functionals with Broad Applicability in Chemistry. Accounts of Chemical Research. 2008;41(2):157-167.

[70] Zhao Y, Truhlar DG. The M06 suite of density functionals for main group thermochemistry, thermochemical kinetics, noncovalent interactions, excited states, and transition elements: two new functionals and systematic testing of four M06-class functionals and 12 other functionals. Theo. Chem. Acc 2008;120(1):215-241.

[71] Boese AD, Handy NC. New exchange-correlation density functionals: The role of the kinetic-energy density. J. Chem. Phys. 2002;116 (22):9559-9569.

[72] Chai J-D, Head-Gordon M. Systematic optimization of long-range corrected hybrid density functionals. J. Chem. Phys. 2008;128(8):084106.

[73] Chai J-D, Head-Gordon M. Long-range corrected hybrid density functionals with damped atom–atom dispersion corrections. *Phys. Chem. Chem. Phys*. 2008;10(44): 6615-6620.

[74] Frisch MJ, Trucks GW, Schlegel HB, Scuseria GE, Robb MA, Cheeseman JR, et al. Gaussian 092009.

[75] Schmidt MW, Baldridge KK, Boatz JA, Elbert ST, Gordon MS, Jensen JH, et al. General Atomic and Molecular Electronic Structure System. J Comput Chem. 1993;14:1347.

[76] NIST. NIST Web book U. S. A.: NIST; 2021 [Available from: http:// webbook.nist.gov/chemistry).

[77] Miliordos E, Xantheas SS. An accurate and efficient computational protocol for obtaining the complete basis set limits of the binding energies of water clusters at the MP2 and CCSD(T) levels of theory: Application to $(H_2O)_m$, m = 2-6, 8, 11, 16, and 17. J. Chem. Phys. 2015;142(23):234303.

[78] Lenz A, Ojamäe L. Theoretical IR Spectra for Water Clusters $(H_2O)_n$ (n = 6—22, 28, 30) and Identification of Spectral Contributions from Different H-Bond Conformations in Gaseous and Liquid Water. J. Phys. Chem. A. 2006; 110(50):13388-13393.

Section 3

Applications

Chapter 11

Applications of Current Density Functional Theory (DFT) Methods in Polymer Solar Cells

Abstract

DFT and time-dependant DFT (TD-DFT) quantum chemical calculations have become helpful for qualitative and quantitative analyses of materials at the molecular level. In this paper, we will attempt to outline successes and opportunities associated with the use of DFT and TD-DFT in OSC research. Density functional theory (DFT) has evolved as a QM method that is both rigorous and efficient enough to be employed in photovoltaic solar cell challenges in the last ten years. DFT is a prominent method for precisely and efficiently calculating molecular systems' electrical and optical characteristics at a low computational cost. The possible uses of DFT to polymer solar cells were comprehensively examined in this article. First, the foundations of DFT are examined. Following that, the precision of DFT for studying photovoltaic properties particular to polymer solar cell design is highlighted. Next, this chapter looks at how DFT is used in polymer solar cell research and its accuracy. Following that, a discussion of how DFT works and how it can investigate polymer solar cell features will be given.

Keywords: DFT, TD-DFT methods, Polymer solar cells, electrical and optical properties

1. Introduction

According to the International Energy Agency, the world utilized 109613 TWh of energy in 2014, with fossil fuels accounting for 80% of that total. The depletion of fossil fuels takes 40 years for oil, 60 years for natural gas, and 200 years for coal. The use of fossil fuels and nuclear energy directly affects the environment by emitting greenhouse gases, causing climatic changes such as global warming, which cause malaria and famine [1]. If adequately developed, these sources can diversify energy supply, while wind, biomass, and geothermal cannot fully replace fossil fuels [2]. Solar energy is unique among renewable energy sources because it delivers 711019 Kcal each year, 10,000 times more than the world consumes. Solar energy is widely used in turning sunshine into useable energy for various applications such as solar water heating, solar transportation, solar ventilation, and solar electricity [3].

Photovoltaic (PV) systems convert sunlight directly into electricity, providing a practical and straightforward option to sustain the growing energy demand. Currently, solar cell research and development focuses on (i) the maturity of silicon solar cells. However, the complex manufacturing process, high energy consumption, and high cost of classic silicon-based solar cells have slowed their progress. (ii) TFT solar cells GaAs [4]. CIGS [5], and CdTe [6] are examples of thin-film solar cells. Although thin-film solar cells are efficient and stable, gallium and indium are scarce in the crust, and cadmium is poisonous. (iii) New solar cells A new generation of solar cells include organic photovoltaics (OPV) DSSC [7] and perovskite solar cells (PSC) [4, 8, 9]. These solar cells are light and cheap. However, various issues prevent wide-scale application.

Although PSC, photoelectric conversion efficiency (PCE) has increased from 3.8 to 23.3 per cent, device stability has remained an issue limiting applications [10]. Evaporating silicon solar cells require advanced high-temperature manufacturing, high-quality silicon, and intricate energy sources. These cells are also mechanically rigid. Polymer solar cells are a low-cost alternative to silicon solar cells since they may be made by painting or printing [11]. Polymer cells are cheap, light, and flexible [12]. Since 1992, when Sariciftci et al. [13] demonstrated effective photoinduced electron transfer from polymer semiconductors to an electron acceptor, C_{60} polymer solar cells have gained popularity. Utilizing conducting polymers as electron donors could achieve 6–7.9% power conversion efficiency [14]. Conjugated polymers appeal to solar cell efficiency because their bandgap's and energy levels can be tuned chemically. The discovery of new electron-donor and electron-acceptor materials has recently increased OSCs' power conversion efficiencies (PCEs) to above 18% [15]. New materials, improved device processing methods and blend morphology [16, 17], and a better understanding of device physics [18] have contributed to OPV cell progress in the previous 30 years.

High-performance computing and optimization of computational chemistry codes have increased theoretical research in this area. Theoretical approaches are the best way to overcome practical synthesis problems and investigate cost-saving manufacturing and processing options. DFT can help solve these problems because it examines the electronic structure and spectroscopic properties of these materials. The only electronic structural approaches currently relevant to conjugated polymers have relied on density functional theory (DFT) [19, 20]. The highest occupied molecular orbital (HOMO), lowest occupied molecular orbital (LUMO), bandgap (Eg), ionization potential (IP), electron affinity (EA), charge mobility, open-circuit voltage (V_{OC}), and reorganization energies (λ) are key electronic parameters that determine the optoelectronic properties of polymer solar materials. This chapter examines DFT's usage and validity in polymer solar cell research. This will be followed by a description of how the DFT method looks at polymer solar cell characteristics. In addition, to describe the DFT and TD-DFT applications for polymer solar cells characteristics is also highlighted, as is DFT's overall utility in polymer solar cell design.

2. Computational quantum methods for polymer solar cells

Computational quantum chemistry may play an essential role in the development of OPV research and technology. In addition, the results of these computations can be used as a low-cost guide for developing and improving solar materials. Overall, the field's research can be divided into quantum science and charge transfer dynamics, new structures and creative concepts, and materials development for various applications. Understanding the functioning mechanism of polymer solar cells is the primary focus of the study. The main point of contention is how the bonded electron–hole pair breaks. The "hot exciton effect" is the most frequently recognized answer to this subject. When an electron is absorbed by one semiconductor material from another, it brings energy differences, causing the electron to get heated and acquire velocity, allowing it to escape from the confined exciton state. Even though this idea has recently gained widespread acceptance, numerous investigations have cast doubt on its validity [21].

These methods are now being used to predict component excited-state properties in light-harvesting systems. Nonetheless, this is a difficult task. Furthermore, large systems pose a significant challenge to conventional quantum chemistry techniques for excited states. Thus, calculations and theoretical models are beneficial when used in conjunction with CT experiments. Optical absorption investigations can measure exciton binding energies. Electrochemistry can be used to assess OPV components' oxidation and reduction potentials, as well as their HOMO–LUMO gaps [22]. The charge transfer and recombination kinetics can be studied using femtosecond transient absorption and time-resolved emission measurements [22]. These studies rely on well-established electron and energy transmission [23, 24] and their thermodynamic and kinetic consequences [25].

In polymer solar cells, computational quantum theories such as density functional theory (DFT) are used. TD-DFT is the essential instrument in the quantum mechanical simulation of quantum chemistry for computing excited-state characteristics and charge-transfer (CT) excitations in large systems [26]. TD-DFT can extract excitation energies, frequency-dependent response qualities, and photo-absorption spectra from molecules and materials. There are several hurdles to implementing TD-DFT computations on photovoltaic systems. Assumptions in these applications can cause catastrophic modeling errors. Many hypotheses and refinements to this basic technique have been proposed [27].

Density Functional Theory (DFT) investigates the electronic structure (principally the ground state) of many-body systems, such as atoms, molecules, and condensed phases. A many-electron system's properties can be determined using functionals, i.e. function of another function. Inorganic photovoltaic properties such as bandgap, optical absorption, intra-molecular and inter-molecular charge transfer, exciton binding energy, charge transfer integral, reorganization energy and rate of charge transfer and recombination in donor-acceptor complexes can be calculated or developed using DFT based computational methods. In practice, the effects of organic photovoltaic media on these qualities should be considered. Although the impact of the medium varies depending on the transitions involved, polarization continuum solvation models may commonly be used to account for solvation at a low computational cost [28].

The Schrödinger equation, which explains the behavior of electrons in a system, is reformulated in density functional theory (DFT) so that approximate solutions are tractable for functional materials. Hohenberg and Kohn [29] proposed the idea in 1964, claiming that all ground-state features may be expressed as an operative of the charge density that must be reduced in energy. However, rather than tackling the Schrödinger Equation [30] head-on, these theorems showed that an initial guess of the charge density might be improved iteratively.

Hohenberg, Kohn, and Sham created such a theory in density functional theory (DFT) [31], leading to two of the top ten most preferred articles of all time6 and for which Kohn received the Nobel Prize in Chemistry in 1999. The Schrödinger equation's ground-state solutions are restated in DFT to find energy as a charge density function.

The exchange-correlation functional chosen determines the precision of the DFT calculation. Although theorists may frequently improve computation accuracy by employing more intricate functionals (at a higher processing cost), there are some highly coupled electron systems that most functionals fail. Other disadvantages of traditional DFT include the small system size, the difficulty of modeling weak (van der Waals) interactions, dynamics over long periods, and non-ground state characteristics (finite temperatures or excited conditions). Larger systems can be tackled with the linear scaling approach [32], finite-temperature effects can be addressed with lattice dynamics [33] and cluster expansion [34], electronic excitations can be modeled with time-dependent DFT [35], the GW method [36], and the Bethe–Saltpeter [37], and several approaches can be used to overcome these limitations. In his assessment, Carter [38] gives a quick outline of some of these options. A DFT calculation requires the coordinates and orientations of the atoms in the material within a repeating lattice, the exchange-correlation functional, parameters and algorithms for numerical and iterative convergence, and, optionally, a method for more efficiently treating the system's core electrons (for example, through the use of pseudopotentials). DFT generates the electronic charge density, total energy, magnetic configuration, and electronic band structure.

3. DFT methods in polymer solar cells

The density-functional theory (DFT) has proven massively popular among the massive panel of current theoretical techniques. This success is mainly because no changeable inputs parameters are required, efficient numerical codes exist, and a high level of adaptability, particularly in representing semiconductor and metal ground state features. The ability of DFT approaches to handle larger systems has grown as computing power has increased. However, the present limit does not yet reach the 10000–10 million atom window involved in the active device region of PV cells. Modern semiconductor optoelectronic devices, such as quantum wells [39] and quantum dots [40], have features with a feature size of a few nanometers. Such systems are composed of various materials and alloys' complex two-dimensional (2D) and three-dimensional (3D) geometries. Nonetheless, DFT approaches can be used to get insight into the physical phenomena of individual device components, such as particular materials or tiny heterostructures. Quantitative PV design, for example, necessitates a valid prediction of electronic bandgap's band-lineups and effective masses.

DFT based on the local density approximation (LDA) [41] or the generalized gradient approximation (GGA) [42] is well recognized for failing to replicate the excited states of molecules adequately. Hybrid approaches that contain a fraction of Hartree-Fock exchange, on the other hand, may be able to avoid the band-gap problem, but their results are highly dependent on the material of interest. Even while Heyd et al. [43]. HSE06 hybrid functional is a good option for computing band gaps, band offsets, or alloy characteristics [44]. It fails to simulate the direct–indirect crossover in GaAsP alloys [45]. Many-body perturbation theory (MBPT) can also provide factual findings, mainly when using the GW technique (GW, where G stands for Green's function and W for the screened potential), which can be utilized in a perturbative scheme [46] or self-consistently [47]. The ionization potential and electron affinity of the donor and acceptor materials, respectively, are essential parameters for charge separation because they determine the relative alignment of electron and hole levels. DFT inside a super-lattice (SL) approximation can be used to estimate the drop of the interface's potential in heterostructures, resulting in a reasonable estimate of the band-lineup. The DFT potential drop at the interface can be quickly compensated by the GW eigenvalues derived for the bulk valence band states [48], but a GW treatment for a complicated stack is out of reach.

In DFT simulation, a reasonable estimation of alloy electrical characteristics is likewise a difficult task. Indeed, even typical semiconductors experience significant band-gap bowing; that is, the band-gap energy follows:

$$E_g(x) = \overline{E}_g(x) - b.x.(1-x) \quad (1)$$

Where b is a parameter for bowing, super-cell techniques are better than virtual crystal approximations for statistically random alloys. However, DFT with specific quasi-random structures (SQS) (small super-cells) that recreate mixing enthalpies and atomic correlations of extremely large super-cells can provide identical results for specific alloy compositions. Chemical mixing, strain, and atomic relaxation effects are all included in SQS models.

Furthermore, semiempirical approaches for studying mechanical or electronic properties, such as the valence force field (VFF) and the tight-binding approximation [49] or elasticity and the K.P method [50], require precise electronic parameters as input. They can be calculated using DFT or discovered through experiments. For example, the density functional perturbation theory (DFPT) [51] is used to get quantitative estimates of electromechanical tensors of bulk materials. When just by-products of a first-order perturbation computation are required, an efficient application of the "2n + 1" theorem yields second-and third-order derivatives of the total energy, provided that atomic-displacement variables are removed. Various physical responses of insulating crystals, such as elastic constants, linear piezoelectric tensors, and linear dielectric susceptibility, as well as tensor properties related to internal atomic displacements like Born charges and phonons, can be obtained using second-order derivatives [51]. Interference techniques and symmetry analysis must be used in conjunction with the DFPT method for third-order components linked to physical qualities such as nonlinear electrical susceptibilities, nonlinear elasticity, or photoelastic and electrostrictive effects [52].

4. TD-DFT methods in polymer solar cells

TD-DFT has become the workhorse of quantum chemistry for computing excited-state characteristics and charge-transfer excitations in complex networks [53]. Linear and non-linear simulations may be popular due to their scalability and variety of methodologies. However, photovoltaic TD-DFT computations pose significant problems, notably with long-range CT interactions. Furthermore, although standard computer programs make TD-DFT easy to use, it is not a "black box" technique since approximations utilized in TD-DFT tools can occasionally produce major systemic mistakes in estimated results.

Making correct CT excitation energies with TD-DFT is problematic because approximate exchange-correlation (XC) potentials lack the unique features of exact Kohn–Sham (KS) potentials [54]. Thus, a non-local, exact-exchange contribution throughout the exchange-correlation kernel solves the charge-transfer problem in TD-DFT.

Many hybrid functionals (HF) have recently performed well in benchmark tests [55]. For example, Zhao and Truhlar created a functional (M06-2X) that contains the complete non-local exact-exchange contribution [56]. Prior studies investigated comprehensive adjustments for non-hybrid local adiabatic XC potentials targeting the CT problem utilizing constraint variational density functional. According to Ziegler et al. [57], the linear response approach produces qualitative variances between the time-dependent Hartree-Fock (TD-HF) and adiabatic local potential-based TD-DFT excitation energies and also between the TD-DFT and SCF excitation energies. In the variational approach, the mix of occupied and empty orbitals is allowed above linear terms to determine transition densities. However, the recent surge of activity in the field gives reason for optimism.

Constrained DFT (CDFT) [58] is an alternative technique that uses a variational constraint approach to alleviate TD-DFT's drawbacks. To handle charge- and spin-constrained electronic states within the ground state KS DFT technique [59], CDFT was created. Implementation is computational. CDFT has been proven to be useful for long-range electron transmission [60–62]. While TD-DFT's thrilled states (valence states) are inaccessible to CDFT, many of TD-DFT's bothersome excited states are dealt with naturally in CDFT.

CDFT can be used in systems where the ground-state electron density must meet a threshold. The localization of electronic density in space indeed reduces the variation in electron number between donor and acceptor areas by two. To enforce the constraint, Lagrange multipliers are used in CDFT. To counteract self-interaction mistakes caused by approximation XC functionals, CDFT uses semi-local functionals that are denser than the density. It defines diabatic states for electron transfer kinetics and chemical reaction rates. Using a ground state DFT functional, these limitations can also compute long-range charge transfer and low-lying spin states [61, 62].

However, this paradigm has several drawbacks. In CDFT, the electron density is partitioned by nuclear populations, affecting the constraint potential's shape. So the exact electron density partitioning is unknown. As a result, CDFT can only represent a subset of electronic excitations. Therefore, CDFT may not be the best method for simulating some diabatic circumstances. Also, a time-dependent optimal potential treatment of exact exchange can help describe charge-transfer excitations [63]. However, the exchange potential is too local to correctly predict the bandgap (i.e. HOMO–LUMO gap). Semi-quantitative hybrid functionals can forecast band gaps better than orbital dependent functionals. They use correlations to communicate precisely. Somewhat of focusing on density, we can improve approximation solution XC functionals in orbitals. They can also compensate for imperfections in self-interaction, and exchange energy is naturally stated in orbitals.

Dispersion correction to KS-DFT is widely employed to handle long-range electron correlations that cause dispersion forces [64, 65]. Unfortunately, local DFT functionals overestimate dispersion forces, whereas non-local and hybrid DFT functionals underestimate them [64, 65]. Developing DFT functionals optimized for improved management of Vander Waals interactions, on the other hand, appears promising. The Vander Waals density functional (vdW-DF) [65] techniques successfully cope with London dispersion interactions. Modern dispersion-corrected DFT methods include empirical components because they function best for long-range interactions. However, standard density functionals perform well for close interactions. So any dispersion-correction method using DFT has to integrate the short and long-range asymptotic areas, which are both well understood individually.

The most widely used non-empirical method for determining dispersion energy for molecular systems is the vdW-DF approach [66]. Due to the charge transfer reliance of dispersion being incorporated via electron density, this technique naturally accommodates it. Regular adjustments also alter thickness. Despite its mathematical complexity, this approach can achieve a smooth transition between chemical binding at small distances and Vander Waals attraction at long distances.

The KS inherent DFT defect is the KS-dispersion DFT term. Adding damped inter-atomic potentials of the kind C6R6 to the KS-DFT energy appears to be another successful empirical way of accounting for dispersion [64]. This approach is substantially faster to calculate, has high numerical stability, and provides physical insight. Despite its semi-classical origin, it can supplement standard density functionals in treating long-range electron correlation. In addition, estimating supramolecular complex binding energies has been proved to be easier with this strategy.

5. Application to DFT and TD-DFT methods in polymer solar cells

Many properties of interest in OPVs can be calculated or developed using DFT based computational methods, including bandgap, optical absorption, intramolecular and intermolecular charge transfer, exciton binding energy, charge transfer integral (to quantify electronic coupling), reorganization energy, and the rate of charge transfer and recombination in D–A compositors. In addition, in natural systems, the effects of the surrounding OPV media on these properties should be addressed. Although the impact of the medium varies depending on its nature and the nature of the transitions involved, low-cost polarization continuum solvation models can be utilized to account for solvation. In such situations, the solvent is treated as a continuous with a static dielectric constant, polarizing and polarized by the solute.

The following is a discussion of the theoretical methodologies and computational techniques mentioned above to foster a greater understanding of the connection between chemical structures and the optical and electrical properties of D–A systems about the rational design of OPV devices.

5.1 Bandgap engineering

A range of experimentally observed methods utilized may or may not indicate appreciable quantities in diverse contexts. Band gaps (see infra) are essential features to consider when evaluating conducting polymers. Controlling band gaps can improve the electroluminescence of OLEDs or the light absorption efficiency of photovoltaic cells [67]. To make organic polymers with good nonlinear optical response [67] or semiconductors having high electrical conductivity [68], materials with tiny band gaps are sought.

The phrase "bandgap" has multiple meanings. An infinite periodic system's electrical structure is called a "band." Also examined are monomers of conjugated polymers and oligomers of various sizes. The term bandgap also refers to a finite method feature that converges to the infinite periodic (band-structure) limit with oligomer size. The "gap" is the difference in electronic energy levels. Such as the energy difference between the highest occupied molecular orbital (HOMO) and the lowest unoccupied one, or computed energy gaps MO or CO (LUMO) [69]. Lowest optically permissible electronic excitation energy A visible energy gap (also known as an adiabatic or vertical optical gap) is an EO. Also, adiabatic or vertical electron attachment/detachment has associated energies like electron affinity (EA), ionization potential (IP) and electronegativity (EF) = IP-EA. Orbital energies and eigenvalues in KS DFT and HF may or may not be visible. However, approximations in actual computations might lead to big mistakes (see below). Furthermore, an orbital energy gap cannot reflect both EO and EF. Although the interpretation of the orbital energy gap as the fundamental gap is accurate in principle, the multiple approximations in the functionals make this interpretation useless in practice.

Band gaps often exhibit a roughly linear dependence on 1/m, allowing extrapolation to $m \to \infty$, $1/m \to 0$. Generally, the quality of calculated results for different properties strongly relies on the physical models employed [70]. Extrapolating the infinite-periodic limit from a sequence of oligomer simulations requires prudence. Band gaps can diverge from linearity in 1/m [71]. These can also compute band gaps for polymers with unlimited chain lengths. PBCs have the advantage of not requiring further computations or extrapolation. However, PBC quantum chemistry algorithms are less functional than their molecular counterparts, which is a disadvantage.

The cheap computational cost of DFT and TD-DFT allows for the study of large systems. In contrast, the most commonly used functionals for molecules, such as generalized gradient approximations (GGA) and global hybrid GGA functionals, result in the incorrect asymptotic behavior of a potential, significant delocalization errors, and lack of derivative discontinuity, all of which negatively impact o They showed that the band gaps computed with different hybrid functionals differ significantly, indicating that present hybrid functionals do not yield proper band gaps [72–74]. The link between projected orbital energy gaps and measured band gaps absorption energies is also poor for nonhybrid density functionals. Therefore, typical hybrid and meta-hybrid functionals are not acceptable for assessing the performance of organic photovoltaics, according to Savoie et al. Other hybrid functional analyses found striking agreement with measured optical gaps [75]. Many studies suggest that DFT and TD-DFT have limited predictive potential in this critical field [74]. Using hybrid functionals to integrate eX solves some of the DFT difficulties in extended systems [76]. However, estimated transfer integrals in organic semiconductors are sensitive to eX fraction [77].

5.2 Intramolecular and intermolecular charge transfer

To explain the transport properties, the charge transfer rate between donor and acceptor moieties could be calculated. Using DFT calculations of the electronic coupling, reorganization energy, and free energy difference associated with one electron transfer from donor to acceptor at the high-temperature limit, we can determine a rate using Marcus's formula [78, 79]. These predictions are then discussed in light of DFT's intrinsic flaws, such as overestimation of actual ground state energy, failure of the basis set to represent the system, and the inability of the functional to approximate critical interactions, to name a few. When it comes to charge transfer, however, there is another problem that is often overlooked.

A system where an extra account is localized on a single molecular unit is impossible to simulate using conventional DFT. As a result, calculations are frequently performed first on the charge donor, then on the charge acceptor, with or without the charge transferred. The overall energy of the system is calculated by adding the energies of the individual components. This is true when a considerable distance separates the donor and acceptor, and the electron density distribution of one entity has no effect on the electron density distribution of the other [79]. The size of the organizations participating in the charge transfer and the quantity and location of the surplus charge should naturally establish this limit. More sophisticated (and consequently more expensive) techniques, such as charge-constrained DFT [80], have been proposed for situations where this limit can never be achieved, such as intramolecular charge transfer [79]. However, standard ground state DFT remains the approach of choice for intermolecular charge transfer in solar cells, which usually involves larger molecular assemblies.

5.3 Extion binding energy

Exciton binding energy (E_b) is an essential element in polymer electronics and fundamental polymer physics, and it has been a source of debate for a long time. For example, a big Eb is required for a light-emitting polymer so that charge recombination takes precedence. Both the semiempirical model study [81] and the DFT/LDA Bethe–Salpeter equation (BSE) or GW approaches [82] have relied heavily on theoretical research of E_b of conjugated polymers. The semiempirical model can be addressed nearly precisely for the electron correlation, but the results depend highly on the parameters, even though a qualitative comprehension has been reached [81].

Although the BSE or GW approach is first-principles, it is unclear if the Hohenberg–Kohn–Sham framework might accommodate these many-body adjustments at the Green's function level.

Furthermore, the final results differ from one another [82]. We highlight two recent advances in quantum chemistry: (i) Time-Dependent Density Functional Theory (TD-DFT) and its successful application to the lowest-lying excited states [83], and (ii) the hybrid GGA Becke three-parameter Lee–Yang–Parr (B3LYP) functional for the quantitative prediction of chemical and electronic structures [82]. In examining the excitation processes in conjugated systems, the combination of the two has proven quite effective [84]. Tretiak et al. [85] demonstrated that hybrid density functionals might simulate excitonic phenomena and provide satisfactory E_b findings. Pure local spin density approximation (LSDA), generalized gradient approximation (GGA, such as BLYP, BP86, BPBE, PBEPBE, BPW91), meta-GGA (such as PBEKCIS), and hybrid density functionals (H-GGA, such as O3LYP, B3LYP, B972, PBE1PBE) are used to optimize the ground-state geometries of the molecules at the DFT level.

5.4 Electron transfer parameters

The ET parameters are now frequently calculated using ab initio quantum chemistry methods. In classical Marcus theory [86], two critical parameters determine the temperature-dependent kinetics of electron transfer: the driving force and the reorganization energy. The activation energy, G#, is calculated as follows:

$$\Delta G^{\#} = \frac{(\lambda + \Delta G^{0})}{4\lambda} \tag{2}$$

which then can be used in the Arrhenius relationship for the rate constant

$$k_{ET} = A \exp\left(\frac{\Delta G^{\#}}{k_B T}\right) \tag{3}$$

The Boltzmann constant is denoted by k_B. Note that Eq. (3) is essentially classical, and when quantum effects are relevant, the temperature dependence of Eq. (3) breaks down [87].

Although free energies will be used in natural systems, ab initio calculations usually overlook entropy changes and instead use potential energy. When the reactant and product structures are known, G° may readily determine the difference between respective equilibrium energies. However, calculating is more complicated since it involves nonequilibrium energy.

Abs Initio algorithms have difficulties determining since it is not a ground-state attribute. In the adiabatic representation, the ground state potential energy curve is lower than the excited state potential energy curve. To compute, one must first know the energy of the product state at the reactant state's equilibrium structure. Excited-state energies are more difficult to calculate than ground-state energies. TD-DFT methods provide good excited-state energies (e.g., up to 100 atoms). The energy of long-range CT states in TD-DFT is underestimated [88, 89], limiting its use in ET reactions.

In ET research, constrained DFT has various advantages. For starters, restricted DFT makes accessing diabatic states and calculating Marcus parameters a breeze. Second, from ground-state computations, constrained DFT generates diabatic conditions. Excited-state calculations are avoided. Third, the equilibrium structures of the reactant and product states are obtained by optimizing constrained geometry. Third, the quality of diabatic potential energy curves from limited DFT is superior to adiabatic curves from DFT to optimize in adiabatic conditions because fractional charge systems are more susceptible to self-interaction errors [90]. Fourth, the localization of an unpaired electron is forced via constrained DFT. As a result, restricted DFT energy values are more precise. Limited DFT cannot be used for ET reactions involving a locally excited state as a ground-state approach. To investigate such responses, TD-DFT and limited DFT could be utilized. Our method, in particular, is before the electron source and acceptor. In systems in which the donor and acceptor are not separated, this can be a problem. As a result, our method is now the most effective for long-range ET responses, which is why it was created.

The coupling constant should be computed to obtain correct adiabatic energies from diabatic ones, making limited DFT more useful. In the adiabatic representation, the two curves create an upper and lower curve, with the energy gap at qc being twice the electronic coupling constant H_{ab}, which is a significant coupling constant in nonadiabatic dynamics. DFT overestimates H_{ab}, which could lead to erroneous RobinDay class III compound assignment. The limited DFT methodology for calculating high-quality diabatic energies could also be used to forecast exact H_{ab} values. One of the difficulties is that constrained DFT techniques do not provide the proper wave function. H_{ab} is the union of two independent wave functions with no equivalent in static density-dependent observables. As a result, some estimates are required to extract H_{ab} from constrained DFT. We are now doing an active study on this topic, and the results will be released soon. H_{ab} can also be employed with restricted DFT to study the issue of degenerate charge transfer states. It can also tackle problems similar to those that the restricted open-shell Kohn-Sham approach can solve [45].

5.5 Scharber's model-electronic properties

They are using density functional theory and Scharber's model to forecast the power conversion efficiency of organic solar cells. The scientific community has long sought improved polymers with excellent power conversion efficiency. Because polymer synthesis and device production take time, a guide would help find the best polymers. They published a simple model in 2006 that outlines how to estimate the power conversion efficiency of bulk heterojunction solar cells, and they claimed that these devices could attain 10% power conversion efficiencies. To evaluate a polymer's photovoltaic potential, Scharber's model requires knowledge about energy levels. Commonly, cyclic voltammetry is used to obtain these energy levels after polymer production. Modern theoretical tools like density functional theory come into use. These technologies can theoretically predict polymer properties before they are made.

The DFT has been extensively used to develop, explain, and predict the features of present and future organic solar cells [91]. Even though the model predicts certain qualities like open-circuit voltage (V_{OC}) and short-circuit current density (J_{SC}), one may wonder if the model estimates some attributes more precisely than others when combined with density functional theory. The dependability of theoretical computations is critical for understanding and predicting device attributes. There have been extensive research on oligomers [92] and crystals [93], but few comparisons of computations on polymers with experimental evidence. These highest values for power conversion efficiency can be derived by integrating density functional theory determined attributes with Sharber's model.

5.5.1 Scharber's model

This is equivalent to the maximum power density output of the device divided by the total power density receiving from the Air Mass 1.5 solar spectrum [94], which is 1000 W/m^2. The device's power density comprises the open-circuit voltage, short-circuit current density, and fill factor (FF). According to Scharber's model, the V_{OC} is connected to the difference between the acceptor's LUMO and the donor's HOMO. The V_{OC} is obtained by subtracting 0.3 eV from the energy level difference. This shift was discovered empirically and is linked to residual carrier binding energy and interface effects [95]. An integral of external quantum efficiency (EQE) multiplied by the number of photons from the Air Mass 1.5 sun spectrum at all frequencies. For energies below and above the donor's optical band gap (E_{opt}), the EQE is just a step function with a value of 0%. The fill factor is FF = 0.65 for all devices. Other EQE and FF assumptions can be made if desired. For example, the EQE could be determined by investigating the Kohn-Sham joint density of states, revealing the frequency-dependent absorption cross-section behavior. In this situation, the polymer layer is thick enough to absorb any photon over the optical band gap, and the film shape essentially limits the EQE. The following equations describe Scharber's mode.

$$PCE = \frac{V_{OC}J_{SC}FF}{1000W/m^2} \tag{4}$$

$$LUMO_{donor} > LUMO_{acceptor} + 0.3eV \tag{5}$$

$$eV_{OC} = LUMO_{acceptor} - LUMO_{donor} \tag{6}$$

$$E_{opt} = LUMO_{donor} - HOMO_{donor} \tag{7}$$

$$EQE(\omega) = 0.65 \times \Theta(\hbar\omega - E_{opt}) \tag{8}$$

$$J_{SC} = \int EQE(\omega) \times \#photons_{AirMass1.5}(\omega)d\omega \tag{9}$$

As seen in Eq. (5), this model implies a 0.3 eV energy difference between the donor and acceptor Lumos to enable effective charge transfer. This LUMO offset should not be confused with the 0.3 eV empirical shifts for Eq. (6). So E_{opt} = 0.6 eV is the maximum value for eV_{OC}.

5.6 Photoabsorption spectrum

Time-dependent DFT has surpassed all other methods for calculating organic compounds' excitation energies and optical characteristics in the last decade. Visual features like absorption spectra and optical band gap can be used to validate structures further. To better understand the electronic transitions of polymer monomers, used TD-DFT/CAMB3LYP/with varied basis set levels to perform quantum calculations on electronic absorption spectra in the gaseous phase and solvent. Aside from the bandgap, the computational prediction of whole spectrum excitation energies and cross-sections above the bandgap is equally crucial to solar cell efficiency. The excited electron tends to decay toward the conduction level (or the LUMO level) before being injected into the anode due to vibronic (molecular dyes) or photonic (solid dyes) contact with the environment, resulting in thermalization loss of the cell efficiency. The choice of time-dependent TD-DFT can be critical in accurately reproducing absorption, especially when using donor-acceptor dyes with charge-transfer excitations, as in the current study, where range corrected functionals become a viable option. The band maximum (λ_{max}) is an apparent essential feature of the absorption spectra.

These are significant for optical properties of polymers in polymer solar cell applications, ranging from TD-DFT methodologies to predicted absorption wavelengths (λ_{max}), oscillator strengths (f), and vertical excitation energies (E).

6. Concluding remarks

New materials have emerged as an appealing and profitable replacement to inorganic semiconductors due to the desire to manufacture more cost-effective electronic devices using simple manufacturing processes. Organic materials, such as conjugated polymers, tiny organic molecules, and self-assembling organic semiconductors, have recently piqued interest due to their ability to fabricate flexible, lightweight, semi-transparent, and large-area devices. As a result, DFT based computational computations have become a valuable method for investigating materials in OSC research. The relevance of DFT calculations in the knowledge of the structure-properties relationship in the design of new polymer solar cells has been highlighted in this paper. Furthermore, we discovered that DFT's improved power in understanding and forecasting features of polymer solar cells is caused by increased computational capacity and the emergence of robust and diverse computational methodologies. We have explored how theoretical calculations based on the KohnSham energy levels of density functional theory combined with Scharber's model can be utilized to locate viable photovoltaic polymers in this paper.

Within the generalized Kohn–Sham formalism that offered proper excitation energies, density functional theory and its time-dependent extensions made substantial progress along the range separated hybrid functionals. DFT has been effectively used to explain and predict molecular geometries, electronic structure, frontier molecular orbital (FMO) energy levels, absorption spectra, and intramolecular charge transfer for known conjugated organic compounds (ICT). These characteristics significantly impact the open-circuit voltage (V_{OC}), short-circuit current (J_{SC}), charge separation at the donor/acceptor interface, and solar photon absorption.

References

[1] Sovacool BK. National context drives concerns. *Nat Energy*. 2018;3(10): 820-821.

[2] Li B, Wang L, Kang B, Wang P, Qiu Y. Review of recent progress in solid-state dye-sensitized solar cells. *Sol Energy Mater Sol Cells*. 2006;90(5):549-573.

[3] Asif M, Muneer T. Energy supply, its demand and security issues for developed and emerging economies. *Renew Sustain Energy Rev*. 2007;11(7):1388-1413.

[4] Ragoussi ME, Torres T. New generation solar cells: concepts, trends and perspectives. *Chem Commun (Camb)*. 2015;51(19):3957-3972.

[5] Li W, Yang RL, Wang D. CdTe solar cell performance under high-intensity light irradiance. *Sol Energy Mater Sol Cells*. 2014;123:249-254

[6] Andersen TR, Dam HF, Hösel M; et al. Scalable, ambient atmosphere roll-to-roll manufacture of encapsulated large area, flexible organic tandem solar cell modules. *Energy Environ Sci*. 2014;7 (9):2925-2933.

[7] Wu YZ, Xie FX, Chen H et al. Thermally stable MAPbI3 perovskite solar cells with efficiency of 19.19% and area over 1 cm^2 achieved by additive engineering. *Adv Mater*. 2017;29(28).

[8] Xiao YM, Wu JH, Lin JY et al. A dual function of high performance counter-electrode for stable quasi-solid-state dye-sensitized solar cells. *J Power Sources*. 2013;241:373-378.

[9] Kojima A, Teshima K, Shirai Y, Miyasaka T. Organometal halide perovskites as visible-light sensitizers for photovoltaic cells. *J Am Chem Soc*. 2009;131(17):6050-6051.

[10] Xie FX, Chen CC, Wu YZ et al. Vertical recrystallization for highly efficient and stable formamidinium-based inverted-structure perovskite solar cells. *Energy Environ Sci*. 2017;10 (9):1942-1949.

[11] Qiao Q, Xie Y, McLeskey JT. Organic/inorganic polymer solar cells using a buffer layer from all-water-solution processing. *J Phys Chem C*. 2008;112(26):9912-9916.

[12] Darling SB. Block copolymers for photovoltaics. *Energy Environ Sci*. 2009; 2(12):1266-1273.

[13] Sariciftci NS, Smilowitz L, Heeger AJ, Wudl F. Photoinduced electron transfer from a conducting polymer to buckminsterfullerene. *Science*. 1992;258(5087):1474-1476.

[14] Chen HY, Hou J, Zhang S et al. Polymer solar cells with enhanced open-circuit voltage and efficiency. *Nature Photon*. 2009;3(11):649-653.

[15] Liu Q, Jiang YH, Jin K et al. 18% efficiency organic solar cells. *Sci Bull*. 2020;65(4):272-275.

[16] Li W, Chen M, Cai J et al. Molecular order control of non-fullerene acceptors for high-efficiency polymer solar cells. *Joule*. 2019;3(3):819-833.

[17] Meng L, Zhang Y, Wan X et al. Organic and solutionprocessed tandem solar cells with 17.3% efficiency. *Science*. 2018;361(6407):1094-1098.

[18] Yao H, Cui Y, Qian D et al. 14.7% Efficiency organic photovoltaic cells enabled by active materials with a large electrostatic potential difference. *J Am Chem Soc*. 2019;141(19):7743-7750.

[19] Vaschetto ME, Monkman AP, Springborg M. First-principles studies of some conducting polymers: PPP, PPy, PPV, PPyV, and PANI. *J Mol Struct*. 1999;468(3):181-191.

[20] Scharber MC, Wü hlbacher D, Koppe M et al. Influence of the Bridging Atom on the Performance of a Low-Bandgap Bulk Heterojunction Solar Cell. Adv. Mater. 2010, 22, 367–370.

[21] Lee JM, Kwon BH, Il Park HI et al. Exciton dissociation and charge-transport enhancement in organic solar cells with quantum-dot/N-doped CNT hybrid nanomaterials. *Adv Mater*. 2013; 25(14):2011-2017.

[22] D'Souza F, Chitta R, Ohkubo K et al. Corrole-fullerene dyads: formation of long-lived charge-separated states in nonpolar solvents. *J Am Chem Soc*. 2008; 130(43):14263-14272.

[23] Olaya-Castro A, Scholes GD. Energy transfer from Förster–Dexter theory to quantum coherent light-harvesting. *Int Rev Phys Chem*. 2011;30(1):49-77.

[24] Beljonne D, Curutchet C, Scholes GD, Silbey RJ. Beyond Förster resonance energy transfer in biological and nanoscale systems. *J Phys Chem B*. 2009;113(19):6583-6599.

[25] Veldman D, Meskers SCJ, Janssen RAJ. The Energy of charge-transfer states in electron donor–acceptor blends: insight into the energy losses in organic solar cells. *Adv Funct Mater*. 2009;19(12):1939-1948.

[26] Autschbach J. Charge-transfer excitations and time-dependent density functional theory: problems and some proposed solutions. *ChemPhysChem*. 2009;10(11):1757-1760.

[27] Zhao Y, Truhlar DG. Density functionals with broad applicability in chemistry. *Acc Chem Res*. 2008;41(2): 157-167.

[28] Difley S, Van Voorhis T. Exciton/charge-transfer electronic couplings in organic semiconductors. *J Chem Theor Comput*. 2011;7(3): 594-601.

[29] Hohenberg P, Kohn W. Inhomogeneous electron gas. *Physiol Rev*. 1964;136(3B):B864-B871. This study established the theoretical basis of density functional theory.

[30] Schrödinger E E. An undulatory theory of the mechanics of atoms and molecules. *Phys Rev*;28(6):1049-1070.

[31] Kohn W, Sham LJ. Self-consistent equations including exchange and correlation effects. *Phys Rev*. 1965;140 (4A):A1133-A1138.

[32] Fonseca Guerra C, Snijders JG, Te Velde G, Baerends EJ. Towards an order-N DFT method. *Theor Chem Acc*. 1998;99(6):391-403.

[33] Baroni S, Gironcoli SD, Corso AD, Giannozzi P. Phonons and related crystal properties from density functional perturbation theory. *Rev Mod Phys*. 2001;73(515).

[34] Sanchez, Sanchez JM, Ducastelle F, Gratias D. Generalized cluster description of multicomponent systems. *Phys A*. 1984;128(1-2):334-350.

[35] Petersilka M, Gossmann UJ, Gross EK. Excitation energies from time-dependent density functional theory. *Phys Rev Lett*. 1996;76(8):1212-1215.

[36] Hedin L. New method for calculating the one-particle Green's function with application to the electron-gas problem. *Phys Rev*. 1965;139 (3A):A796-A823.

[37] Salpeter EE, Bethe HA. A Relativistic Equation for Bound-State Problems. *Phys Rev*. 1951;84(6):1232-1242.

[38] Carter EA. Challenges in modeling materials properties without experimental input. *Science*. 2008;321 (5890):800-803.

[39] Guézo M, Loualiche S, Even J et al. Ultrashort, nonlinear, optical time

response of Fe-doped InGaAs/InP multiple quantum wells in 1.55-μm range. *Appl Phys Lett*. 2003;82(11): 1670-1672.

[40] Cornet C, Labbé C, Folliot H et al. Time-resolved pump probe of 1.55μm InAs/InP quantum dots under high resonant excitation. *Appl Phys Lett*. 2006;88(17).

[41] Perdew JP, Wang Y. Accurate and simple analytic representation of the electron-gas correlation energy. *Phys Rev B Condens Matter*. 1992;45(23).

[42] Perdew JP, Burke K, Ernzerhof M. Generalized gradient approximation made simple. *Phys Rev Lett*. 1996;77(18): 3865-3868.

[43] Heyd J, Scuseria GE, Ernzerhof M. Hybrid functionals based on a screened Coulomb potential. *J Chem Phys*. 2003; 118(18).

[44] Wadehra A, Nicklas JW, Wilkins JW. Band offsets of semiconductor heterostructures: A hybrid density functional study. *Appl Phys Lett*. 2010;97(9).

[45] Filatov M, Shaik S. Spin-restricted density functional approach to the open-shell problem. *Chem Phys Lett*. 1998;288(5-6):689-697.

[46] Surh MP, Louie SG, Cohen ML. Quasiparticle energies for cubic BN, BP, and BAs. *Phys Rev B Condens Matter*. 1991;43(11):9126-9132.

[47] Vidal J, Trani F, Bruneval F, Marques MAL, Botti S. Efects of electronic and lattice polarization on the band structure of delafossite transparent conductive oxides. *Phys Rev Lett*. 2010; 104(13).

[48] Shaltaf R, Rignanese G-M, Gonze X, Giustino F, Pasquarello A. Band ofsets at the Si/SiO2 interface from manybody perturbation theory. *Phys Rev Lett*. 2008; 100(18).

[49] Robert C, Perrin M, Cornet C, Even J, Jancu JM. Atomistic calculations of Ga(NAsP)/GaP(N) quantum wells on silicon substrate: band structure and optical gain. *Appl Phys Lett*. 2012;100 (11)

[50] Cornet C, Schliwa A, Even J et al. Electronic and optical properties of InAsInP quantum dots on InP. *Phys Rev B*. 2006;100 and InP(311) B substrates: theory and experiment:74(3).

[51] Gonze X, Amadon B, Anglade P-M et al. ABINIT: irstprinciples approach to material and nanosystem properties. *Comput Phys Commun*. 2009;180(12): 2582-2615.

[52] Pedesseau L, Katan C, Even J. On the entanglement of electrostriction and non-linear piezoelectricity in noncentrosymmetric materials. *Appl Phys Lett*. 2012;100(3):article ID 031903.

[53] Burke K, Werschnik J, Gross EKU. Time-dependent density functional theory: past, present, and future. *J Chem Phys*. 2005;123(6).

[54] Elliott P, Furche F, Burke K. fromtime-dependentdensity functional theory, in. Rev Comp Chem.. *Excited S*. Chichester, United Kingdom: John Wiley & Sons, Inc. 2009:91-165.

[55] Grimme S, Neese F. Double-hybrid density functional theory for excited electronic states of molecules. *J Chem Phys*. 2007;127(15):154116.

[56] Zhao Y, Truhlar DG. The M06 suite of density functionals for main group thermochemistry, thermochemical kinetics, noncovalent interactions, excited states, and transition elements: two new functionals and systematic testing of four M06-class functionals and 12 other functionals. *Theor Chem Acc*. 2008;120(1-3):215-241.

[57] Ziegler T, Krykunov M. On the calculation of charge transfer transitions with standard density functionals using constrained variational density functional theory. *J Chem Phys*. 2010;133 (7):074104-074111.

[58] Kaduk B, Kowalczyk T, Van Voorhis T. Constrained density functionaltheory. *Chem Rev*. 2012;112 (1):321-370.

[59] Wu Q, Van Voorhis T. Direct calculation of electron transfer parameters through constrained density functional theory. *J Phys Chem A*. 2006; 110(29):9212-9218.

[60] VanVoorhis T, Kowalczyk T, Kaduk B, Wang LP, Cheng CL, Wu Q. The diabatic picture of electron transfer, reaction barriers, and molecular dynamics. *Annu Rev Phys Chem*. 2010; 61:149-170.

[61] Kowalczyk T, Yost SR, Voorhis TV. Assessment ofthe Delta SCF density functional theory approach for electronic excitations in organic dyes. *J Chem Phys*. 2011;134(5): 054128-054136.

[62] Yeganeh S, Voorhis TV. Triplet excitation energy transfer with constrained density functional theory. *J Phys Chem C*. 2010;114(48): 20756-20763.

[63] Kümmel S, Kronik L. Orbital-dependent density functionals: theory and applications. *Rev Mod Phys*. 2008;80 (1):3-60.

[64] Jurecka P, Cerný J, Hobza P, Salahub DR. Density functional theory augmented with an empirical dispersion term. Interaction energies and geometries of 80 noncovalent complexes compared with ab initio quantum mechanics calculations. *J Comp Chem*. 2007;28(2):555-569.

[65] Grimme S. Density functional theory with London dispersion corrections. *WIREs Comp Mol Sci*. 2011; 1(2):211-228.

[66] Lee K, Murray ÉD, Kong L, Lundqvist BI, Langreth DC. Higher-accuracy van der Waals density functional. *Phys Rev B*. 2010;82(8).

[67] Gierschner J, Cornil J, Egelhaaf H-J. Optical bandgaps of π-conjugated Organic Materials at the Polymer Limit: experiment and Theory. *Adv Mater*. 2007;19(2):173-191.

[68] Jones BA, Facchetti A, Wasielewski MR, Marks TJ. Tuning orbital energetics in arylene diimide semiconductors. materials design for ambient stability of n-type charge transport. *J Am Chem Soc*. 2007;129(49): 15259-15278.

[69] Brédas J-L, Cornil J, Beljonne D, dos Santos DA, Shuai Z. Excited-state electronic structure of conjugated oligomers and polymers: A quantum-chemical approach to optical phenomena. *Acc Chem Res*. 1999;32(3): 267-276.

[70] Torras J, Casanovas J, Alemán C. Reviewing Extrapolation Procedures of the Electronic Properties on the π-conjugated Polymer Limit. *J Phys Chem A*. 2012;116(28):7571-7583.

[71] Zade SS, Bendikov M. From oligomers to polymer: convergence in the HOMO-LUMO gaps of conjugated oligomers. *Org Lett*. 2006;8(23): 5243-5246.

[72] Tozer DJ. Relationship between long-range charge-transfer excitation energy error and integer discontinuity in Kohn–Sham theory. *J Chem Phys*. 2003;119(24):12697-12699.

[73] Jain M, Chelikowsky JR, Louie SG. Reliabilty of hybrid functinal in predicting banggaps *Phys Rev Lett*. 2011; 107: 216806-216811.

[74] Savoie BM, Jackson NE, Marks TJ, Ratner MA. Reassessing the use of one-electron energetics in the design and characterization of organic photovoltaics. *Phys Chem Chem Phys*. 2013;15(13):4538-4547.

[75] Yang S, Olishevski P, Kertesz M. Bandgap calculations for conjugated polymers. *Synth Met*. 2004;141(1-2): 171-177.

[76] Salzner U. Theoretical investigation of excited states of oligothiophenes and of their monocations. *J Chem Theor Comput*. 2007;3(3):1143-1157.

[77] Sutton C, Sears JS, Coropceanu V. Bredas JL. Understanding the Density Functional Dependence of DFT-Calculated Electronic Couplings in Organic Semiconductors. J. Phys. Chem. Lett. 2013, 4, 919–924.

[78] Vaissier V, Mosconi E, Moia D et al. Effect of molecular fluctuations on hole diffusion within dye monolayers. *Chem Mater*. 2014;26(16):4731-4740.

[79] Wu Q, Van Voorhis T. Extracting electron transfer coupling elements from constrained density functional theory. *J Chem Phys*. 2006;125(16): 164109-164105.

[80] Wu Q, Van Voorhis T. Direct optimization method to study constrained systems within density-functional theory. *Phys Rev A*. 2005;72 (2): 024502-024505.

[81] Yaron D, Moore EE, Shuai Z. Br'edas. *J Chem Phys*. 1988;108(7451):81. (h) Tiago ML, Rohlfing M, Louie SG. Bound excitons and optical properties of bulk trans -polyacetylene. *Phys Rev B*. 2004;70(19): 193204-193208.

[82] Lee C, Yang W, Parr RG. Development of the Colle-Salvetti correlation-energy formula into a functional of the electron density. *Phys Rev B Condens Matter*. 1988;37(2):785-789.

[83] Yang L, Feng JK, Ren AM, Sun JZ. The electronic structure and optical properties of carbazole-based conjugated oligomers and polymers: A theoretical investigation. *Polymer*. 2006; 47(4):1397-1404.

[84] Pogantsch A, Heimel G, Zojer E. Quantitative prediction of optical excitations in conjugated organic oligomers: A density functional theory study. *J Chem Phys*. 2002;117(12): 5921-5928.

[85] Igumenshchev KI, Tretiak S, Chernyak VY. Excitonic effects in a time-dependent density functional theory. *J Chem Phys*. 2007;127(11):114902.

[86] Hammes-Schiffer S. Theoretical perspectives on proton-coupled electron transfer reactions. *Acc Chem Res*. 2001; 34(4):273-281.

[87] Grätzel M.Photoelectrochemical cells Nature(London) 2001, 414: 338-344.

[88] Burke K, Werschnik J, Gross EK. Time-dependent density functional theory: past, present, and future. *J Chem Phys*. 2005;123(6):62206.

[89] Dreuw A, Weisman JL, Head-Gordon M. Long-range charge-transfer excited states in time-dependent density functional theory require non-local exchange. *J Chem Phys*. 2003;119(6): 2943-2946.

[90] Zhang Y, Yang W. A challenge for density functionals: self-interaction error increases for systems with a noninteger number of electrons. *J Chem Phys*. 1998;109(7):2604-2608.

[91] Ku J, Lansac Y, Jang YH. Time-dependent density functional theory study on Benzothiadiazole-based low-band-gap fused-ring copolymers for organic solar Cell Applications. *J Phys Chem C*. 2011;115(43): 21508-21516.

[92] Pandey L, Risko C, Norton JE, Brédas J. Donor–Acceptor copolymers of relevance for organic photovoltaics: A theoretical investigation of the impact of chemical structure modifications on the electronic and optical properties. *Macromolecules*. 2012;45(16):6405-6414.

[93] Sokolov AN, Atahan-Evrenk S, Mondal R`C.; Zoombelt. A.P.; Bao, Z.; et al. From Computational Discovery to Experimental Characterization of a High Hole Mobility Organic Crystal. *Nat Commun*. 2002;2:437.

[94] Air mass ASTM G-173 data available at. http://rredc.nrel.gov/solar/spectra/am1.5/ Accessed April 9, 2013.

[95] Brédas JL, Beljonne D, Coropceanu V, Cornil J. Charge-´ Transfer and Energy-Transfer Processes in π-conjugated Oligomers and Polymers: A Molecular Picture. *Chem Rev*. 2004;104(11):4971-5004.

Chapter 12

Applications of Density Functional Theory on Heavy Metal Sensor and Hydrogen Evolution Reaction (HER)

Abstract

A great effort has been devoted to develop the numerical methods to solve Schrödinger equation for atoms and molecules which help to reveal the physico-chemical process and properties of various known/unknown materials. Designing the efficient probe to sense the heavy metals is a crucial process in chemistry. And, during this energy crisis, to find the effective conversion materials for water splitting is an important approach. The density functional theory (DFT) is a powerful tool to identify such materials and made great achievements in the field of heavy metal chemosensor and photocatalysis. Particularly, DFT helps to design the chemosensor for the effective sensor applications. The universe is moving towards the exhaustion of fossil fuels in a decade and so on, DFT plays a vital role to find the green energetic alternative to fossil fuel which is the Hydrogen energy. This book chapter will focus on the application of DFT deliberately on the heavy metal sensors and hydrogen evolution reaction.

Keywords: DFT, HER, hydrogen evolution reaction, heavy metal sensor

1. Introduction

Since pronounced by Dirac in 1929 that "The underlying physical laws necessary for the mathematical theory of a large part of physics and the whole of chemistry are thus completely known, and the difficulty is only that the exact application of these laws lead to equations much too complicated to be solved" [1]. So, a great effort has been devoted to develop the numerical methods to solve Schrödinger equation for atoms and molecules which help to reveal the physico-chemical process and properties. The density functional theory has become an important tool for physicst, chemist, and material scientist. Over the past three decades, DFT has been developed successfully challenging traditional wavefunction-based methods for large scale quantum chemistry calculations. DFT has become an important method and suitable alternative to *ab initio* method as well as cheaper in terms of computational cost. The well-developed modern DFT is applicable to quantum as well as classical systems based on the theorems of Hohenberg and Kohn [2]. The system is having N-particles allowing interactions with a given interparticle interaction, the total energy is completely derived by specification of external field φ(r) using Hamiltonian and ground-state wave function. Many electron wave functions j(r1, r2, rN) could be obtained by solving Schrödinger equation provided all the necessary information about the system. The density of single particle possibly acquired which on performing integration over any one of the single directions of coordinates of N-1 electrons,

$$\rho(r) = N\int....\int\psi\,(r_1, r_2, r_3......r_N)\,\psi(r_1, r_2, r_3......r_N)dr_2......dr_N \tag{1}$$

where,

$$\int\rho(r)\,dr = N \tag{2}$$

In other words, the functional of φ(r) gives the ground state energy of the system. During the development of DFT, Hohenberg and Kohn sham (HK) shown that the correspondence between external field φ(r) and the single-particle density ρ (r) and its consequence lead to the total ground state energy with the functional of ρ(r) using the following equation,

$$E[\rho] = E_0[\rho] + \int dr\,\varphi(r)\rho(r) \tag{3}$$

Moreover, Hohenberg and Kohn proved and called a second theorem which provides an energy variational principle. Further, they showed the trial density which satisfies $\int\rho(r)dr = N$

$$E[\bar{\rho}] \geq E_g \tag{4}$$

Here, E_g is the ground state energy. In Eq. (4), the Left Hand Side (LHS) and Right Hand Side (RHS) attain equal when $\bar{\rho}$ (r) is the true ground state single particle density.

In case, E_0() were known for an interacting electron of a given system, then the Eqs. (3) and (4) allows to calculate the ground state energy and density of electron of any multi-electron system in a given arbitrary external field. Though, HK approach produce the total energy calculation but it does not provide any prescription for its determination. So, it is ultimate goal of a researcher to establish and develop accurate approximate functionals. Of course, in later 1990s, a number of functionals was developed to produce the experimental observations. A familiar and few density functionals are hybrid density functional B3LYP [3–5] introduced by M.J.Frisch in 1994, further gradient-corrected correlation functional: Perdew-Burke-Ernzerhof (PBE) [6] was introduced by Ernzerhof, M. in 1996, the Global Hybrid Meta-GGAs Minnesota [7–11] functionals such as M05, M06-HF, M06, M06-2X, M08-HX, M08-SO, revM06, MN15-L introduced by Donald G. Truhlar during 2005 to 2016, dispersion corrected functionals DFT-D [12], DFT-D3 [13], then the First GGA functionals were introduced by Stefan Grimme in the period of 2006–2014, moreover the another dispersion corrected method ωB97XD [14] was introduced by M. Head-Gordon in 2008. The functionals played an important role to predict the properties of unknown materials and also explain the post-process analysis of experimental results. Selection of the method is the key in computational chemistry to achieve appropriate results.

In the past 30 years, the evolution and success of DFT for various chemical applications, emerged as the most popular electronic structure method for the chemists and plays a vital role in computational chemistry. To parameterize electronic structure theory methods and to give the guidance to chemist, nearly five different databases evolved so far *viz*, GMTKN, MGCDB84 [15], Minnesota2015B [16], DP284 [17], and W4–17 [18]. In addition to that the W4–17-RE [19], and MN-RE [20] are two newly developed databases for reaction energies. In 2018, Peverati P et al., comprised the above mentioned databases and published it in the name of ACCDB [21]-which includes data from 16 different research groups, for a total of 44,931 unique reference data points. Now a days, data points and databases plays an important role to proceed the research further to get clear picture on history. This kind of databases will be the standard reference for the future research. In computational chemistry, the various DFT methods have been implemented and studied in several systems including chemosensor [22, 23], hydrogen evolution reaction (HER) [24–26], oxygen reduction reaction (ORR) [27, 28], oxygen evolution reaction (OER) [29], molecular machines [30, 31], DNA mutation [32–34], selective etching [35, 36], atomic layer deposition (ALD) [37, 38] etc. Although a number of applications of DFT were reported, this book chapter precisely focused on heavy metal sensor and hydrogen evolution reaction.

2. Heavy metal sensor

With the increasing population globally, there is a demand to provide clean and safe drinking water to them. However, contaminants in water at several countries/cities restrict the water supply. The contaminants most likely the heavy metals, occurring due to anthropogenic, agricultural, mining, industrial revolution etc. Commonly identified heavy metals are Hg^{2+}, Pb^{2+}, Cu^{2+}, Cd^{2+}, Cu^{2+}, which are although essential for living being but these heavy metals are toxic at higher concentrations [39]. So, it is a need to find the heavy metal sensor and this chapter will specifically focus on those sensors.

A series of 2,2`-Bipyridyl functionalized Iridium(III) complexes were synthesized by Zhao et al. [40] enabled sensing behavior towards Zn^{2+}, Cd^{2+}, Cu^{2+} ions. The modification of binding sites led to conversion of intra-ligand charge transfer (ILCT) to Ligand–Ligand Charge Transfer (LLCT) and Metal–Ligand Charge Transfer (MLCT) transition upon binding with metal ions. DFT calculations using gradient corrected correlation functional PBE1PBE and 6-31G(d,p) basis set were employed to confirm the transition mechanism using HOMO and LUMO energies of the Iridium(III) complexes with heavy metal ions. Kim et al. [41] synthesized two triphenylamine-based dyes as fluorescent probes for effective sensing of Hg^{2+} ions. The fluorescence emission band was decreased and increased on addition of Hg^{2+} ions due to metal to ligand charge transfer (MLCT) mechanism. The optimization of the dye structures was carried out at DFT level with exchange correlation functional of LDA based on PWC set. The LUMO energy levels of mono-benzoxazole and di-benzoxazole were found to be −4.61 eV and −4.74 eV, respectively. The HOMO energy levels of mono-benzoxazole and di-benzoxazole were found to be −2.51 and −2.82 eV, respectively. In this way, Anand et al. [42] reported the aminoquinoline fluorescent probe for selective and sensitive detection of Pb^{2+} and Al^{3+} ions with significant fluorescence enhancement. The fluorescence signaling was due to the inhibition of photo-induced electron transfer process and restriction of C=N isomerization. The fluorescent turn-on mechanism was understood by carrying out DFT calculations at B3LYP level. The HOMO, HOMO-1 and LUMO energies of the probe were −5.18 eV, −5.67 eV and −2.29 eV, respectively. On complexation with Pb^{2+}, the HOMO, HOMO-1, HOMO-2 and LUMO energies were −6.18 eV, −6.42 eV, −6.95 eV and −3.69 eV respectively and for Al^{3+}, it was found to be −6.03 eV, −6.93 eV, −7.35 eV and −3.54 eV respectively. Gonzalaz et al. [43] reported the Tripodal pyrenyl-triazole probe and demonstrated citrate ion recognition with significant fluorescence enhancement. DFT calculations using B3LYP-D3/def2-TZV basis set in gas and solvent phase revealed the stable geometry of the tripodal probe as nesting type, whereas all triazole CH units bind with citrate ions. The probe also showed a remarkable chromogenic behavior from pale to deep orange seen by naked-eye which was further confirmed by DFT calculations. In 2018, Liao et al. [44] have designed and developed $NiCo_2O_4$ nanoplatelets modified glass carbon electrodes for detection of Pb^{2+} ions. The DFT computations were explored to calculate the adsorption bond energies of the $NiCo_2O_4$ nanoparticles with Pb^{2+} heavy metal ions. Pb has the largest adsorption energy of −2.78 eV with the nanoparticles and was in-line with the electrochemical experimental values. The modified glass electrodes were analyzed in real water samples and found that detection of Pb^{2+} ions was successful compared with detection of Hg^{2+} and Cu^{2+} ions. Particularly for Cu^{2+} ion sensor, Sun et al. [45] reported 1,8-diaminonaphthalene derived probe as effective fluorescence chemosensor towards Cu^{2+} ions. The fluorescence quenching was experienced due to the inhibition of intramolecular charge transfer (ICT) process and chelation enhanced fluorescence quenching (CHEQ) of paramagnetic Cu^{2+} ions. The TD-DFT calculations were performed using B3LYP/6–31(d) and LANL2DZ basis sets produced the HOMO-LUMO band gap of copper complex as 2.15 eV and the probe as 2.82 eV. This reduction in the band gap confirms the interaction of Cu^{2+} ions with the receptor. The adsorption studies explored using cetyltrimethyl ammonium bromide on Poly-L-Lysine coated Ag nanoparticles were employed at B3LYP level using 6-31 g(d) and LANL2DZ basis sets reported by Moudgil et al. [46]. The binding energy of the organic core with silver nanoparticles was found to be −0.75 eV indicates the chemisorption of pi-bonding orbitals on the Ag surface. The binding energy was observed to be reduced further 1 eV while interacting with Hg^{2+} ion which confirms the aggregation mechanism and a charge transfer through amine group. Thiocarbazone based probe was reported by Mahajan et al. [47] as multi-ions sensor for Ni^{2+}, Cu^{2+}, Co^{2+} and Cd^{2+}. The calculated stabilization energy by DFT and the observed result from the fluorescent spectra produced the same order for metal ions binding with probe. The order of the metal ion sensing was found to be $Cu^{2+} > Ni^{2+} > Co^{2+} > Cd^{2+}$. Changes in the bathochromic shift in absorption spectra and fluorescence quenching for these metal ions were due to the interaction of sulfur and hydroxyl O atoms of the probe. The binding interaction was supported by TD-DFT calculations at B3LYP level using LANL2DZ basis set. The calculated HOMO-LUMO band gap values were decreased for the metal ion-probe complex than that of probe confirms the complex formation. The real time experiment was explored with river water samples. The imidazole based receptor was reported by Khan et al. [48] to sense picric acid (donor-acceptor) for detection of toxic nitrobenzene through static quenching, Dexter electron transfer and Forster resonance energy transfer quenching mechanisms. The HOMO and LUMO orbital energies were obtained at B3LYP/6-31 g(d) level as −3.00 eV and −6.91 eV for the probe and the HOMO and LUMO energies were found to be −3.15 eV and −7.64 eV upon binding with nitrobenzene. These orbital energy values confirmed the transfer of electron from LUMO of the probe to the LUMO of nitrobenzene which is supported by the quenching mechanism. The charge transfer between the probe and the metal ion was confirmed by electrostatic potential calculation of the binding site. Moreover, the HOMO-LUMO energy gap of the probe-Co^{2+} complex was reduced to 1.86 eV well adhering to the charge transfer process. Kumar et al. [49] reported a dicarboxylic acid based receptor as turn-on fluorescent probe for detection of Hg^{2+} ions. The binding of Hg^{2+} ion takes place through the D-pi-A system by -COOH and $-NH_2$

functional groups through intra-molecular charge transfer (ICT) mechanism. DFT calculations confirmed the ICT mechanism by analyzing the HOMO-LUMO energy differences. The band gap was found to be decreased for the Hg^{2+} ion complex rather than the unbound probe which confirms the host-guest interaction. Chiral CdSe nanoplatelets – L/D-cysteine as sensitive optical probe for recognition of Pb^{2+} ions was reported by Wang et al. [50]. The TD-DFT calculations at B3LYP level using LANL2DZ basis set was carried out. The results of frontier molecule orbital calculations of the HOMO and LUMO orbitals showed overlapping between the cysteine capped CdSe nanocluster and the organic core. Moreover, the reason for the transformation into chirality was observed to be the orbital coupling effect in the nanocluster. The Pb^{2+} was recognized by L-Cysteine capped CdSe nanoplatelets. The merocyanine dye based receptor [51] crystallized in monoclinic P21/n space group showed pi-pi stacking between merocyanine units. The B3LYP method was employed to optimize the geometries of the free receptor and mercuric complex. For optimization, 6-31 g(d) and LANL2DZ basis sets were used for non-metal and metal atoms, respectively. The change in the HOMO-LUMO band gap for the mercury complex than the receptor confirms the selective sensing of Hg^{2+} ion with phenolic and methoxy oxygen moities. Modified calix[4] arene based probe was reported by Anandababu et al. [52] selectively detected Zn^{2+} ions with 240-fold fluorescence enhancement due to the imine C=N isomerization. In turn, Cu^{2+} and Hg^{2+} ions showed fluorescence quenching due to its paramagnetic and spin-orbit coupling properties. DFT studies revealed the HOMO-LUMO energy difference of unbound probe as 1.07 eV and [probe-Zn^{2+}] has a decrement to 0.15 eV, confirming the strong complexation. The luminescent metal organic frameworks (LMOFs) as nanocages to recognize Fe^{3+} and Cu^{2+} ions in trace quantity was designed and synthesized by Li et al. [53]. The fluorescent quenching activity of the nanocages was due to their weak interaction with the metal ions at N-rich sites. DFT calculations confirmed the uncoordinated N atoms in the nanocage interact weakly with the metal ions and hence the fluorescence was quenched. The other chemosensor based on Nitrogenous carbon dots impregnated on natural microcline nanostructures demonstrated a dual fluorometric response towards Fe^{3+} and Cr^{6+} metal ions was reported by Bardhan et al. [54]. The fluorescence turn OFF response was attributed by photo-induced electron transfer mechanism and was confirmed by TD-DFT calculations. The B3LYP level and 6-31 g(d,p) with RIJCOSX basis sets were utilized and revealed that the microcline structure holds the carbon dots which in turn bind with the metal ions. The possible binding site is the nitrogen atoms present in the C-dots deduced by its electrostatic potential surface calculation. The HOMO-LUMO band gap was reduced by 2.49 eV in the metal ion complexes. Carbohydrazide based colorimetric sensors for detection of multiple ions were reported by Tekuri et al. [55]. The receptors showed selectivity towards Cd^{2+}, Hg^{2+}, Pb^{2+}, Cu^{2+} and AsO_2^- ions with low detection limit. Using B3LYP functional and non-relativistic effective core potential (ECP) with LANL2DZ and 6 31 g(d,p) basis sets confirmed the binding of receptor with different metal ions as per hard and soft (Lewis) acids and bases (HSAB) theory. The hard -OH group and soft C=N group binds with complementary metal ions. The HOMO-LUMO band gap for the [receptor(s)-metal ion(s)] complexes showed a reduction in their values confirmed the binding of the respective metal ions. The novel silsesquioxane cage – porphyrin nanocomposites reportedly [56] a dual chemosensor towards fluoride and cyanide ions particularly these ions bind with the Si atoms in the cage resulted in turn on fluorescence. The microporous surface of the polymeric cage with extended conjugation was confirmed by non-local density functional theory using slit shape modeling. The extended conjugation gives a red shift in the absorption band and was confirmed by DFT calculation specifically the dropping in band gap value of the nanocomposites. In turn, the fluorescence emission was quenched in the presence of Hg^{2+} and Cu^{2+} ions due to the restricted photo-induced electron transfer process. An eco-friendly garlic extracted alliin based silver nanoparticles was synthesized by Paw et al. [57]. This AgNPs selectively detects Hg^{2+} and Sn^{2+} ions with easily detectable color change from brown to colorless. The structure was optimized by M06-2X level with LANL2DZ and 6-31 g(d) basis sets. Ag^0 with alliin stabilizes the nanocluster with the stabilization energy of 18.4 kcal mol^{-1}. On interaction with Hg^{2+} and Sn^{2+} ions, alliin separated out from AgNPs with stabilization energies of 7.6 kcal mol^{-1} and 8.3 kcal mol^{-1}, respectively. A novel N-doped diaza derivative of ovalene as nanographenes was reported by Jin et al. [58]. The N atom in the zigzag edges of ovalene favored the interaction with protons, Cu^{2+} and Fe^{2+} ions. The HOMO-LUMO band gap with 2.20 eV and 2.13 eV for diaza and ovalene nanographenes were calculated at DFT level of theory in gas phase. The N-bearing diaza nanographene showed fluorescence quenching in the presence of Cu^{2+} and Fe^{2+} ions due to intermolecular charge transfer. The aromaticity was reported by nucleus-independent chemical shift (NICS) calculations using GIAO at B3LYP/6-31G(d) level. The result shows that the NMR chemical shift has also been computed and confirmed.

3. Hydrogen evolution reaction

The Hydrogen evolution reaction ($2H^+ + 2e^- \rightarrow H_2$) is the cathodic half-cell reaction in acid-based electrolyzers. To understand the mechanism of electrocatalytic/photo-catalytic processes the density functional theory has been utilized as well as to predict and design the new catalyst for water splitting [59]. Production of hydrogen in efficient manner from water-splitting is an underpinning science to realize the hydrogen economy. The Janus nanoparticles which have the two different faces each consisting of different chemistry, size, morphology, material and further one face have the hydrophilic and another face have the hydrophobic nature introduced by de Gennes in his noble lecture [60]. In 2019, Chuan Zhao et al. [61] observed the higher hydrogen evolution reaction using Janus nanoparticle catalyst with a nickel–iron oxide interface and multi-site functionality. This nanoparticles have also been compared with benchmark platinum on carbon catalyst. The structure orientation during the hydrogen evolution reaction revealed by DFT calculations that Ni–O–Fe bridge at Ni-γ-Fe_2O_3 interface modifies the Gibbs free energy of the adsorption of the intermediate H atoms promote the HER. Moreover, the DFT result shows that the H atom adsorb on top site of O atoms in γ-Fe_2O_3 (311) or in fcc site of Ni(111) with the ΔG_H* of −0.62 and −0.31 eV, respectively. The result of negative ΔG_H shown to be responsible for higher HER. The study displays that the Ni(111) fcc responsible for higher rate adsorption of hydrogen as well as the good amount HER. Yong K. et al. [62] investigated the NiCoP and vanadium doped NiCoP material based on the results of crystal structure, XRD, TEM and XPS. At two different places, the Co and V were replaced by Ni and produced the most stable material. The water has allowed to interact with the surface of NiCoP. Right after the dissociation of water, the produced OH and H are placed at the surface of Ni bridge and Ni hollow sites, respectively. During the process, Vanadium was doped in NiCoP and has observed to be the increased value of adsorption energy of OH and decreased water dissociation energy by 0.05 eV on the doped system. At the same time, the adsorption energy for hydrogen on the surface observed to be lower which is advantageous to desorption for hydrogen molecule from the surface and the desorption energy observed to be decreased by 0.09 eV. The DFT results well supported and further insights to the experimental observation of lower over potential and Tafel slope of NiCoVP as lower compared to NiCoP material. Dong et al. [63] used DFT tool to understand the role of sulfur vacancies on Co_9S_8 and Co_3S_4 in dissociation of water and HER. The author used the Vienna *ab initio* simulation package (VASP) for the calculation using generalized gradient approximation (GGA) using the

Perdew-Burke-Ernzerhof approach for the exchange–correlation term with the inclusion of correction as implemented in the method of Stephen Grimme. The adsorption energy for the H atom on Co_9S_8 and Co_3S_4 as observed to be −2.22 and −1.99 eV, respectively, investigated by Ford et al. The adsorption energy shows that both the surfaces are equally potential for the water splitting process. To understand the reaction kinetics, activation energy for H_2 dissociation has been calculated using Nudged Elastic Band method on Co_3S_4 and Co_9S_8. The result shows that product of the reaction is exothermic. Moreover, the activation energy for H_2 dissociation has observed to be 0.4 eV and 1.1 eV on Co_3S_4 and Co_9S_8, respectively. Also, the PDOS result shows the variation during the time of creation of the sulfur vacancy formation. From the PDOS result and low coordination of octahedral on Co_9S_8 surface, it can be concluded that the Co_9S_8 possibly act as a best catalyst for water splitting. While using Ni_2Mo_3N [64] as electrocatalyst for hydrogen evolution reaction, the DFT studies shown the coordination of four for N-Mo, responsible for higher adsorption energy of hydrogen and responsible for HER. The adsorption energy of hydrogen atom at Ni and Mo site of Ni_2Mo_3N are calculated to be −0.47 and −0.15 eV, respectively. The strong adsorption energy is expected to be the lower yield of HER. So, the active site for this surface is N rather than Ni or Mo. So, the calculated hydrogen adsorption energy was found to be in the range of −0.21 to 0.38 eV. In metal free electrocatalyst, to understand and show the interlayer electronic-coupling effect between g-C_3N_4 and N-graphene, the DFT calculations have been carried out particularly density of states have been computed. In the structure of C_3N_4@NG hybrid shows the downshifting of valence and conduction bands resulting to Fermi level crosses the conduction band of g-C_3N_4 responsible and significant enhancement of electrocatalytic HER. Moreover, DGH for g-C_3N_4 and NG were observed to be −0.54 and 0.57 eV which shows the strong and weak, respectively, adsorption of H on the surfaces. So, both the chemical surfaces are unfavorable for HER. However, while coupling both g-C_3N_4 and NG have yielded good and enhanced HER activity. To understand the noble-metal-free nature of catalyst on HER, recently, noble-metal-free and earth-abundant electrocatalysts as noble-metal-free core-shell catalyst, MoS_2/Ni_3S_2 on Ni foam has been designed, synthesized and tested for HER. The synergistic effect of MoS_2 and Ni_3S_2 combination shown to be enhanced HER. The activation energy for H_2 dissociation on MoS_2 has observed as without barrier and negative reaction energy of 1.36 eV. It shows the Mo is the most responsible surface for HER and have the high potential values to explore by the chemist in the near future with various synergism. In this way, in 2018 Ternary Ni-S-Se Nanorod Arrays shown to be good catalyst supported by DFT for HER. The success of this selenium doped electrocatalytic performance towards HER as Se 3d orbitals were bonded to 3d orbitals of Ni and near Fermi level of s p orbitals, and was observed to be significant electron transfer between nickel and selenium atoms. The excellent performance of the catalyst due to the synergistic effect, 3D core structure, electronic modification of Se into nickel sulfide. To understand the solvent effect [65], a single or double layer water molecules are framed around the catalyst and investigated the water splitting process through simulation. The thermodynamic barrier estimated to be 0.6 eV for the H spill-over process, and it shows that the kinetics of HER was enhanced at Ni-RGO [66] synergistic point with the new active sites for the discharge of water. In 2015, Srinivasadesikan V et al. [67] demonstrated the DFT applications to find the role of Li on Lithium decorated surface of $TiO_2(101)$ anatase surface. The results observed to be the maximum of 13 Lithium adsorbed on the surface and furthermore addition of Lithium incorporated to sub-surface of TiO_2 anatase. The barrier energy for the H_2 dissociation on Li adsorbed TiO_2 surface has calculated to be 39.8 kcal/mol and for five Lithium decorated case, it has calculated to be 37.8 kcal/mol. The result of Lithium effect enormously reduced the barrier H_2 dissociation on bare TiO_2 surface which was reported [68]. Later, to understand the Ni effect on the way of doping and decorated metal atoms on the TiO_2 surface, the DFT based first principle calculation has been carried out. Nickel doping and 3Ni metal atoms decorated on TiO_2 surface [69] has been investigated. Upon understanding the successful catalyst by theory, the same has been synthesized in the lab and shown good HER. With the arrival of new peak in between valence band and conduction band in density of states calculation on KSCN activation on NiO/TiO_2 [70] anatase surface and upon reduction of band gap further confirms the better effect of catalyst towards HER and the same has been confirmed by experiment.

With the above discussion, one can able to understand the power of DFT on various applications. Moreover, the number of unknown science will be explored by using the density functional theory in computational chemistry. Solving the chemical problems with understanding of physics through mathematical equation will be explored for new challenges in science.

References

[1] Dirac, PAM Proc. Roy. Soc. Ser. A, 1929, 123, 714

[2] Hohenberg, P.; Kohn, W. Phys. Rev. 1964, 13

[3] Becke A.D. J. Chem. Phys. 1993, 98, 5648 5652

[4] Stephens, P.J.; Devlin, F.J.; Chabalowski, C.F.; Frisch, M.L. J. Phys. Chem. 1994, 98, 11623-11627

[5] Stephens, P.J.; Devlin F. J.; Ashvar, C.S.; Chabalowski, C.F.; Frisch, M.J. Faraday Discuss, 1994, 99, 103-119

[6] Perdew, J.P.; Burke, K.; and Ernzerhof, M. Phys. Rev. Lett. 1996, 77, 3865.

[7] Zhao, Y.; Schultz, N. E.; Truhlar, D. G. J. Chem. Phys. 2005, 123, 161103.

[8] Zhao, Y.; Truhlar, D. Theor. Chem. Acc. 2008, 120, 215-241.

[9] Zhao, Y.; Truhlar, D. G. J. Phys. Chem. A 2006, 110, 13126-13130

[10] Zhao, Y.; Truhlar, D. G. J. Chem. Theory Comput. 2008, 4, 1849-1868.

[11] Yu, H. S.; He, X.; Truhlar, D. G. J. Chem. Theory Comput. 2016, 12, 1280-1293.

[12] Grimme, S.; Antony, J.; Ehrlich, S.; Helge, K. J. Chem. Phys. 2010, 132, 154104.

[13] Moellmann, J.; Grimme, S. J. Phys. Chem. C, 2014, 118, 7615-7621.

[14] Chai, J.; Head-Gordon, M. Phys. Chem. Chem. Phys. 2008, 10, 6615.

[15] Mardirossian, N.; and Head-Gordon, M. Mol. Phys., 2017, 115, 2315-2372.; Goerigk, L.; Hansen, A.; Bauer, C.; Ehrlich, S.; Najibi, A.; Grimme, S. Phys. Chem. Chem. Phys. 2017, 19, 32184-32215.

[16] Yu, H.S.; He X.; Truhlar, D.G.; J. Chem. Theory Comput., 2016, 12, 1280-1293.; Yu, H.S.; He, X.; Li S.L.; Truhlar, D.G. Chem. Sci., 2016, 7, 5032-5051.

[17] Hait D.; Head-Gordon, M. J. Chem. Theory Comput., 2018, 14, 1969-1981

[18] Karton, A.; Sylvetsky, N.; Martin, J.M.L. J. Comput. Chem., 2017, 38, 2063-2075

[19] Morgante, P.; Pevarati, R. Journal of Computational Chemistry 2019, 40, 839-848

[20] 20.Goerigk, L.; Hansen, A.; Bauer, C.; Ehrlich, S.; Najibi, A.; Grimme, S. Phys. Chem. Chem. Phys., 2017, 19, 32184-32215

[21] Morgante, P.; Pevarati, R. Journal of Computational Chemistry 2019, 40, 839-848.

[22] Sun, X.; Gao, A.; Zhang, H. Scientific Reports, 2020, 10, 711

[23] Kumar, S.K.A.; Vijayakrishna, K.; Sivaramakrishna, A.; Brahmmananda Rao C.V.S; Sivaraman, N.; Sahoo, S.K. Inorganic chemistry, 2018, 57, 15270-15279

[24] Huang, B.; Hart, J.N. Phys. Chem. Chem. Phys. 2020, 22, 1727-1737.

[25] Li, H.; Xu, S.; Wang, M. Chen, Z.; Ji, F.; Cheng, K.; Gao, Z.; Ding, Z.; Yang, W. J. Mater. Chem. A, 2020, 8, 17987-17997

[26] Li, H.; Han, Y.; Zhao, H.; Nat Commun, 2020, 11, 5437

[27] Li, Y.; Zhai, X.; Liu, Y.; Wei, H.; Ma, J.; Chen, M.; Liu, X.; Zhang, W.; Wang, G.; Ren, F.; Wei, S. (2020)

[28] Deng, C.; Li, F.; Tang, Q. J. Phys. Chem. C 2019, 123, 44, 27116-27123

[29] Seh, Z.W.; Kibsgaard, J.; Dickens, C.F.; Chorkendor, I.B.; Nørskov, J.K.; Jaramillo, T.F. Science 2017, 355, 4998

[30] Crucio, M.; Nicoli, F.; Palrinieri, E.; Folis, E.; Tabacchi, G.; Cavallo, L.; Silvi, S.; Baroncini, M.; Credi, A. J. Am. Chem. Soc. 2021, 143, 21, 8046-8055

[31] Arumugamperumal, R.; Shellaiah, M.; Lai, Y.K.; Venkatesan, P.; Raghunath, P.; Wu, S.P.; Lin, M.C.; Sun, K.W.; Chung, W.S.; Lin, H.C. Journal of Materials Chemistry C 2021, 9, 3215-3228

[32] Xue, J., Guo, X., Wang, X. et al. Sci Rep 2020, 10, 9671

[33] Townsend, P.A.; Grayson, M.N. Chem. Res. Toxicol. 2021, 34, 179-188

[34] Srinivasadesikan, V.; Sahu, P.K.; Lee, S.L. J. Phys. Chem. B 2011, 116, 11173-11179

[35] Park, S.; Jung, H.; Min, K.A.; Kim, J.; Han, B. Applied Surface Science, 2021, 551, 149376

[36] Wang, Y.; Zhang, H.; Han, Y.; Liu, P.; Yao, X.; Zhao, H. Chem. Commun., 2011, 47, 2829-2831

[37] Elliott, S.D.; Dey, G.; Maimiti, Y. J. Chem. Phys. 2017, 146, 052822

[38] Shirazi, M.; Kessels, W.M.M.; Bol, A.A. Phys. Chem. Chem. Phys., 2018, 20, 16861-16875

[39] McDonald, R.I.; Weber, K.; Padowski, J.; Flörke, M.; Schneider, C.; Green. P.A. et al Glob. Environ. Change, 2014, 27, 96-105.

[40] Zhao, N.; Wu, Y.H.; Wen, H.M.; Zhang, X.; Chen, Z.N. Organometallics 2009, 28, 5603-5611

[41] Kim, B.S.; Kim, Y.S.; Kim, S.H.; Son, Y.A. J. Nanosci. Nanotechnol. 2010, 10, 7730-7734.

[42] Anand, T.; Sivaraman, G.; Mahesh, A.; Chellappa, D. Analytica Chimica Acta 2015, 853, 596-601

[43] Gonzalez, M.C.; Oton, F.; Espinosa, A.; Tarraga, A.; Molina, P. Org. Biomol. Chem. 2015, 13, 1429

[44] Liao, J.; Zhang, J.; Wang, C.Z.; Lin, S. Analytica Chimica Acta 2018, 1022, 37-44

[45] Sun, T.; Li, Y.; Niu, Q.; Li, T.; Liu, Y. Spectrochimica Acta Part A: Molecular and Biomolecular Spectroscopy 2018, 195, 142-147.

[46] Moudgil, L.; Jaiswal, J.; Mittal, A.; Saini, G.S.S.; Singh, G.; Kaura, A. Journal of Molecular Liquids 2019, 276, 910-918

[47] Mahajan, P.G.; Dige, N.C.; Vanajre, B.D.; Kamaraj, E.; Seo, S.Y. Journal of Photochemistry and Photobiology A: Chemistry 2019, 385, 112089

[48] Khan, I.M.; Shakya, S. ACS Omega 2019, 4, 9983–9995

[49] Kumar, A.; Ananthakrishnan, R.; Jana, G.; Chattaraj, P.K.; Nayak, S.; Ghosh, S.K. Chemistry Select 2019, 4, 1-11

[50] Wang, X.; Hao, J.; Cheng, J.; Li, J.; Miao, J.; Li, R.; Li, Y.; Li, J.; Liu, Y.; Zhu, X.; Liu, Y.; Sun, X.W.; Tang, Z.; Delville, M.H.; He, T.; Chen, R. Nanoscale 2019, 11, 9327

[51] Kumar, A.; Kumar, A.; Sahoo, P.R.; Kumar, S. J. Mol. Struc. 2020, 1206, 127702

[52] Anandababu, A.; Anandan, S.; Syed, A.; Marraiki, N.; Ashokkumar, M. Inorganica Chimica Acta 2021, 516, 120133

[53] Li, Y.W.; Li, J.; Wan, X.Y.; Sheng, D.F.; Yan, H.; Zhang, S.S.; Ma, H.Y.; Wang, S.N.; Li, D. Ch.; Gao, Z.Y.; Dou, J.M.; Sun, D.; Inorg. Chem. 2021, 60, 671–668

[54] Bardhan, S.; Roy, S.; Chanda, D. Kr.; Ghosh, S.; Mondal, D.; Das, S.; Das, S. Dalton Trans., 2020, 49, 10554

[55] Tekuri, V.; Mohan, M.; Trivedi, D.R. Chemistry Select 2020, 5, 5289-5299

[56] Majumdar, R.; Wannasiri, C.; Sukwattanasinitt, M.; Ervithayasuporn, V. Polym. Chem., 2021, 12, 3391

[57] Paw, R.; Hazarika, M.; Boruah, P.K.; Kalita, A.J.; Guha, A.K.; Das, M.R.; Tamuly, C. RSC Adv., 2021, 11, 14700

[58] Jin, E.; Yang, Q.; Ju, C.W.; Chen, Q.; Landfester, K.; Bonn, M.; Mullen, K.; Liu, X.; Narita, A. J. Am. Chem. Soc. 2021, 143, 10403–10412

[59] Nemiwal, M, Gosu, V.; Zhang, T.C.; Kumar, D. Int. J. Hydrogen Energy, 2021, 46, 10216-10238

[60] P.-G. de Gennes, Angew. Chem. Int. Ed. Engl. 31(7), 842-845 (1992)

[61] Suryanto, B.H.R.; Wang, Y.; Hocking, R.K.; Adamson, W.; Zhao, C. nature Comm. | (2019) 10:5599

[62] Jeung, Y.; Jung H.; Kim, D.; Roh, H.; Lie, C.; Han, J.; Yong, K.; J. Mater. Chem. A, 2021, 9, 12203-12213

[63] Fronzi, M.; Assadi, M.H.N.; Ford, M.J.; ACS Omega 2018, 3, 12215-12228

[64] Park, S.H.; Jo, T.H.; Lee, M.H.; Kawashima, K.; Mullins, C.B.; Lim, H.K.; Youn, D.H. J. Mater. Chem. A, 2021, 9, 4945

[65] J.A. Gauthier, C.F. Dickens, L.D. Chen, A.D. Doyle, J.K. Nørskov, J. Phys. Chem. C121 (2017) 11455

[66] Chanda, D.; Hnat, J.; Dobrota, A.S.; Pasti, I.A.; Paidar, M.; Bouzek, K. Phys. Chem. Chem. Phys., 2015, 17, 26864-26874

[67] V. Srinivasadesikan; P. Raghunath; M. C. Lin J Mol Model (2015) 21:142

[68] Raghunath P, Huang WF, Lin MC (2013) J Chem Phys 138:154705

[69] Lin, C.K.; Chuang, C.C.; Raghunath, P.; Srinivasadesikan, V.; Wang, T.T.; Lin, M.C. Chemical Physics Letters, 2017, 667, 278-283

[70] Wanga, T.T.; Chiang, C.L.; Lin, Y.C.; Srinivasadesikan, V.; Lin, M.C.; Lin, Y.G.; Applied Surface Science 511 (2020) 145548.

Chapter 13

Density Functional Theory Study of the Solvent Effects on Electronic Transition Energies of Porphyrins

Abstract

We have calculated the solvent effects on the ground state and the lowest triplet state absorption spectra of *meso*-tetraphenylporphyrin (TPP), meso-tetrakis(*p*-sulfonatophenyl)porphyrin (TSPP) and their diprotonated forms (H_4TPP and H_4TSPP) in thirty-nine different solvent using time-dependent-DFT density functional theory (TD-DFT) coupled with CPCM method. The results of the calculations show that the Q-bands and Soret-bands (or B-bands) in the absorption spectra of these compounds substantially change as function of solvent dielectric constant (ε) up to 20.493 (acetone), but become stabile in high polar solvents with dielectric constants $\varepsilon > 20$. The relative shifts in the B-bands are more significant than that in the Q-bands. The magnitude of the shifts in the spectral position of the Q and B bands are in the following order: H_4TSPP > H_4TPP > TPP > TSPP for the B-bands and H_4TSPP > H_4TPP > TSPP > TPP for the Q-bands. We also have determined that the energy-gaps between the B/Q-bands and their nearest triplet states are also solvent dependent for $\varepsilon < \sim 20.493$.

Keywords: porphyrin, TPP/TSPP, absorption, solvent effect, DFT

1. Introduction

Porphyrins are tetrapyrrolic macrocycles with conjugated electronic systems that exist in nature and have a large number of applications in several fields of research from biomedicine to materials science. The ubiquity of porphyrins in natural systems and their subtle yet important biological and chemical functions have inspired scientists to explore the unique structure/dynamics characteristics of compounds of this family, and to endeavor to imitate their properties in synthetic molecular analogs that display efficient use of solar energy [1–5] and could be used as active elements in molecular electronic devices [6, 7]. Furthermore, in recent times, porphyrin-like molecular systems have drawn a great deal of interest due to the use of therapeutic drugs, photosensitizers in photodynamic therapy of cancer [8], several applications in the treatment of nonmalignant conditions such as psoriasis, blocked arteries and pathological and bacterial infections [9], as well as in HIV research [10]. As known, biological effects of porphyrins generally derive from their photophysical and physicochemical properties, such as the formation of molecular aggregates and axial ligation that give rise to significant modifications in absorption spectra, quantum yields, and fluorescence and triplet state lifetimes [11–15].

Along with rapid development in computer technology and computational quantum chemical methods, molecular properties have been intensively investigated in gas-phase and solvent-phase. Computational quantum chemistry is a powerful tool for understanding real-world chemical problems. Several molecular properties can be obtained by solving quantum mechanical equations. When the results of calculations are in agreement with their experimental results, we can interpret the experimental results more reasonably. Computational studies are not only performed to provide an understanding of experimental data, for instance the position and source of spectroscopic peaks, but also may be used to explore reaction mechanisms and molecular environment effect on the molecular properties.

Photophysical and photochemical properties of porphyrin and its derivatives have been widely studied by experimentalists and theoreticians [16–19]. Dipole allowed electronic excitation spectra of porphyrin compounds exhibit two major absorption bands in which are called as the Q-band in the visible region and the Soret band (or B-band) in the near UV region [20]; both of which have been the subject for a number of quantum chemical studies [21–29]. It has been experimentally reported that when porphyrin molecules are excited in the Soret- or B-band region, internal conversion (IC) takes place from the B-band to the Q-band, which is then followed by a fluorescence from the Q-band to the ground state S_0; i.e., $S_0 + h\nu_0 \rightarrow S_2$(B-band) $\rightarrow S_1$(Q-band) $\rightarrow S_0 + h\nu$. During relaxation of the S_1 state, a fraction of molecules also relaxes to the triplet T_1 via intersystem crossing (ISC).

The electrostatic interaction between a molecular system and its surrounding environment leads to change in geometric and spectroscopic properties. For instance, Jun Takeda and Mitsuo Sato [30] have experimentally studied solvent effect on the absorption spectra of meso-tetraphenylporphyrin (TPP) and dodecaphenylporphyrin (H_2DPP) in thirty-seven different neat solvent. The authors reported that the solvent leads to red shifts in Q-bands and B-band (Soret-band) in their absorption spectra, and that red shifts in H_2DPP are greater than those in TPP due to the nonplanarity of the H_2DPP macrocycle. The authors concluded that these red shifts in the absorption spectra of both compounds are caused by the hydrogen-bonding interactions of pyrrole NH protons and pyrrolenine nitrogen lone pairs with solvent.

Li Ye [31] used steady-state and time-resolved spectroscopic techniques to investigate photophysical properties of several metalloporphyrins have been examined in several different solvents. The author reported that the absorption spectrum of the Cu($TPPCl_8$) in nonpolar solvents does not display a shift in the spectral position of the absorption bands and there is no evidence found for charge transfer (CT) transitions in the visible or near UV regions, however, in the polar solvent, blue shifts are observed in the absorption spectrum of the Cu($TPPCl_8$) an intramolecular CT band takes place in absorption spectra. The author concluded that the activation free energy of the charge-transfer transition decreases with increasing outer reorganizational energy owing to increasing solvent polarity.

In order to take the solvent effect on the molecular properties into account in calculations, a number of implicit (continuum) theoretical models have been developed in the last decade, such as the polarizable continuum model (PCM), the dielectric PCM (DPCM), conductor-like PCM (CPCM), integral equation formalism PCM (IEFPCM), and the conductor-like screening model (COSMO)); see refs. [32, 33] for more details. Density functional theory (DFT) and Time-dependent DFT (TD-DFT) coupled with one of the PCM methods, which is becoming a general routine in most quantum chemical software packages like Gaussian, can be used to compute solvent effect on the geometric and spectroscopic properties [34–37]. Results from such calculations has been shown to be successful in supporting analyses of experimental data with useful insights for better understanding of photophysical and photochemical pathways in solution.

L. Edwards et al. [38] have experimentally studied solvent effect on the spectral positions of dipole allowed singlet-singlet transitions ($S_0 \rightarrow S_n$) of the free-base porphyrin (FBP). Their experimental results shows that the Q-bands at 626 and 512 nm and the Soret band at 372 nm in gas-phase spectrum were respectively significantly shifted to 614, 519 and 391 nm in ethanol.

As far as we know, there is no theoretical study systematically investigating the solvent effect on porphyrin compounds in the literature, except our previous work [39] where we have studied solvent effect on the absorption spectra of the free-base porphyrin (FBP) and diprotonated form (H_4FBP) in the ground state and the lowest triplet state by using TD-DFT/CPCM techniques in thirty-nine different solvent (with solvent dielectric constant (ε) changes from 1 to 181.56). The results from calculations show that dipole allowed electronic transitions in each absorption spectrum change as function of increasing solvent dielectric constant up to about ε = 20.493 and remain almost constant with further increment of the solvent dielectric constant. We have also studied meso- substitution effect on the absorption spectra of the porphyrin (parent porphyrin) such as the meso-tetraphenylporphyrin (TPP), mesotetrakis (p-sulfonatophenyl) porphyrin (TSPP) and their diprotonated forms (H_4TPP and H_4TSPP), where the results from the calculations have shown that the meso-substituted functional groups result in a significant read shift in the electronic transition energies in the porphyrin absorption spectrum [19, 40].

This *work is a continuation* of our previous studies as mentioned above. In this present work we used TD-DFT/CPCM method to investigated solvent effect on the singlet-singlet and triplet-triplet electronic spectra of the TPP, TSPP and their protonated derivatives (H_4TPP and H_4TSPP).

2. Calculation section

The singlet-singlet and triplet-triplet electronic absorption spectra of the porphyrins and their derivatives in the gas phase and thirty-eight solvents were calculated using the Gaussian 09 software package [41]. Geometries for the ground and the lowest triplet states in each solvent were optimized using unrestricted density functional theory (at B3LYP level) [42, 43] with the 6-311G(d,p) basis set for C and H atoms and 6-311 + G(d.p) basis set [44] used for N, O and S atoms. Solvent effects were taken into account using self-consistent reaction field (SCRF) calculations [45], with the conductor-like polarizable continuum model (CPCM) [46–48] as contained in the Gaussian 09 software package. Type of solvent used in calculations are listed in **Table 1**. All compounds in both gaseous and solvents phases were optimized to minima on their ground and lowest triplet state potential energy surfaces (PESs) that is verified by the absence of imaginary frequencies in calculated vibrational spectra.

Time-dependent-DFT (at TD-B3LYP level) coupled with CPCM solvation method was used to calculate the first 24 singlet-singlet vertical electronic transitions ($S_0 \rightarrow S_n$) and 25 triplet-triplet vertical electronic transitions ($T_1 \rightarrow T_m$) in the gas and solvent phases. The GaussSum 0.8 freeware program [49] was used to check outputs and to generate computed absorption spectra from the output file of the Gaussian 09 software.

Solvent	ε	Solvent	ε	Solvent	ε
Gas	1.000	Bromoform	4.249	Acetone	20.493
Argon	1.430	Chloroform	4.711	1-Nitropropane	23.730
Krypton	1.519	Diiodomethane	5.320	Ethanol	24.852
Xenon	1.706	Chlorobenzene	5.697	Nitroethane	28.290
Heptane	1.911	Aniline	6.888	Methanol	32.613
Cyclohexane	2.017	Tetrahydrofuran	7.426	Nitrobenzene	34.809
Carbontetrachloride	2.228	Ethylmethanoate	8.331	Nitromethane	36.562
Benzene	2.271	Dichloromethane	8.930	1,2-Ethanediol	40.245
Toluene	2.374	Dichloroethane	10.125	Dimethylsulfoxide	46.826
Carbondisulfide	2.611	Benzylalcohol	12.457	Formicacid	51.100
Dibutylether	3.047	Cyclohexanone	15.619	Water	78.355
Propanoicacid	3.440	Cyclopentanol	16.989	Formamide	108.940
Diphenylether	3.730	Propanal	18.500	n-Methylformamide-mixture	181.560

Table 1.
The list of the solvents with their dielectric constants (ε) used in this work.

Additionally, we obtained a best fit to the calculated relative shifts in the Q and B bands positions by using the following equation: $f(\varepsilon)=\sum_{n}^{5}C_n\left[\frac{(\varepsilon-1)}{(\varepsilon)}\right]^n$, where ε is the dielectric constant of the solvent and C_n is a constant. We would like to point out that $\frac{(\varepsilon-1)}{\varepsilon}$ in the fitting equations is part of the solvent correction in the CPCM method.

3. Results and calculations

3.1 Solvent dependence of TPP

It has been reported in literature that the measured absorption spectra of the TPP considerably depend on the dielectric constant of the solvent. For instance, the experimentally measured absorption spectrum the TPP exhibited the Soret-band at 398 nm in a pulsed supersonic expansion of helium [50], which is considerably shifted to 420 nm in toluene (ε = 2.374) [51], 419 nm in benzene (ε = 2.271) [52], 417 nm in *tetrahydrofuran* (ε = 7.426) and 415 nm in acetone (ε = 20.493) [51]. The Q(0-0) bands have been measured at 529 and 640 nm in a pulsed supersonic expansion of helium [50], which are also significantly shifted to 548 and 647 nm in benzene [52]. Furthermore, the lowest triplet state (T_1) of TPP at 77 K in methyl-cyclohexane [53] and diethyl ether/petroleum ether/isopropyl alcohol (in the ratio of 5/5/2) [54] was observed at 865 nm and 859 nm, respectively. The triplet-triplet absorption transitions of TPP in toluene were experimentally observed at 780, 690, 430, 405 and 390 nm [55].

It is worthy to note that the geometric structures of the TPP molecule both in the ground state and the lowest triplet state belong to the C_{2v} symmetry point group. The calculated solvent effect on the electronic spectrum of the TPP are summarized below.

Singlet TPP: Figure 1A provides the calculated absorption spectra of FBP in the different solutions used. The shifts in spectral positions of the Q, B (Soret) and L bands caused by the solvent are given in **Figure 1B** as functions of the dielectric constant of the solvent (ε).

The calculations indicate that the solvent effect on the Q1 and Q2 bands of the TPP are inconsequential. For instance, the Q1 band at 17398 cm^{-1} (574.8 nm) and Q2-band at 18566 cm^{-1} (538.6 nm) in gas-phase spectrum are 40 and 90 cm^{-1} red shifted to 17358 (576 nm) and 18473 cm^{-1} (541.4 nm), respectively, with increase in ε from 1.00 to 5.32; which then increased and become almost constant at around 17384 and 18506 cm^{-1} for ε > 20.493. However, the dependence of the calculated Soret bands (B1 and B2-bands) of the TPP significantly depend on solvent polarity. For example the B1 band at 25405 cm^{-1} (393.6 nm) and B2 band 26271 cm^{-1} (380.6 nm) in the gas phase spectrum initially decreased to 23890 (418.6 nm) and 24160 cm^{-1} (414.4 nm), respectively, with increase of ε from 1 to 5.32, and then blue-shifted to 23890 cm^{-1} (418.6 nm) and 24131 cm^{-1} (414.4 nm), respectively, for ε = 20.493 (acetone). With further increase in the dielectric constant (ε > 20.493), these bands positions become nearly stable at around 24574 (406.9, B1) and 24919 cm^{-1} (401.3 nm, B2) within 120 cm^{-1} fluctuations.

Similar to the modification in the positions of the B-bands, the spectral position of the L-band at 28862 cm^{-1} (346.5 nm) in the gas-phase (ε = 1.00) is first red-shifted to 28212 cm^{-1} (354.5 nm) in region of ε = 1 to 5.324, then blue-shifted to 29067 cm^{-1} (334 nm) for ε = 20.493, and then remains unchanged at 28334 cm^{-1} (352.9 nm) within ±20 cm^{-1} variation with the further increase in ε (**Figure 1**).

Triplet TPP: The lowest triplet state T_1 of the TPP in both gas-phase and solvent phase was estimated from calculated global energy difference between the singlet ground state and the lowest triplet state, i.e., E(T_1) = E(the global energy of the lowest triplet state)- E(the global energy of the ground state). The energy level of the T_1 state of the TPP in the gas phase lies about 11406 cm^{-1} (876.8 nm) above the ground state energy (S_0). The estimated solvent effect on the energy gap between S_0-T_1 states is inconsequential, only about 20 cm^{-1} red-shifted with increasing solvent polarity (see **Figure 2B**).

The calculated triplet-triplet electronic transitions up to 25000 cm^{-1} (400 nm) are given in **Figure 2(A and B)**. The calculations indicate that the triplet states

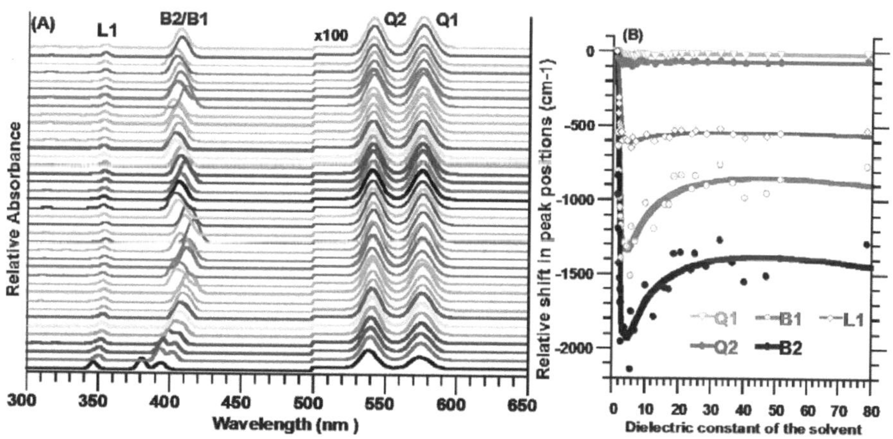

Figure 1.
(A) The calculated solvent-dependence of the singlet-singlet electronic absorption spectrum of the TPP in the thirty-nine different environments, where the dielectric constant of the molecular environment increases from bottom (ε = 1.00) to top (ε = 181.560); (B): The shift in the positions of the Q1, Q2, B1, B2 and L1 bands reference to their corresponding values in gas-phase spectrum as function of dielectric constant of the solvent (ε).

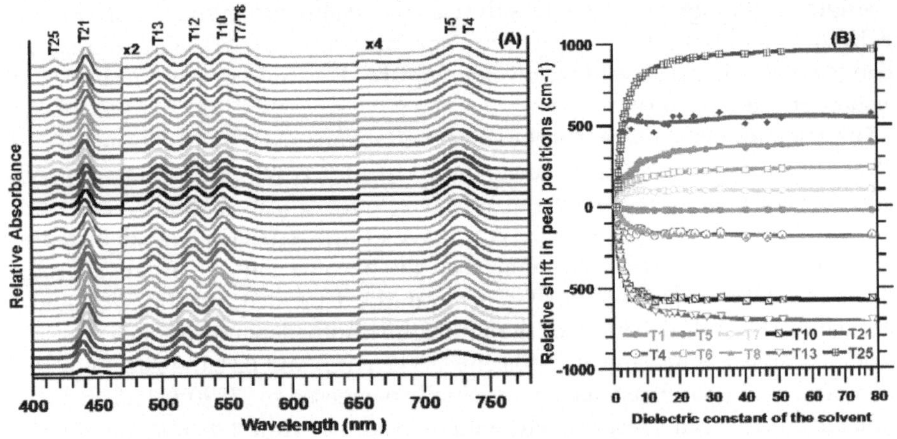

Figure 2.
(A) The calculated solvent-dependence of the triplet-triplet electronic absorption spectrum of the TPP in the thirty-nine different environments, where the dielectric constant of the molecular environment increases from bottom (ε = 1.00) to top (ε = 181.560); (B): The shift in the energy level of the T_n state reference to their corresponding value in gas-phase spectrum as function of dielectric constant of the solvent (ε). The shift in the energy level of the T_1 state: $\Delta E(T_1,\varepsilon) = E(T_1,\varepsilon)-E(S_0,\varepsilon)$.

at 18805 cm^{-1} (531.8 nm; labeled as T_{10}) and 20722 cm^{-1} (482.6 nm, T_{13}) in the gas-phase spectrum of the TPP are gradually red-shifted to 18232 and 20041 cm^{-1} (or 548.5 and 499.0 nm) as a function of ε up to 28.29), respectively, and become stable within a few wavenumbers variation with the further increment in solvent dielectric constant (ε). On contrary to red-shifts in the energy level of the T_{10} and T_{13} states, the T_{21} (at 22130 cm^{-1}/451.9 nm) and T_{25} (22926 cm^{-1}/436.2 nm) in the gas-phase are blue-shift to 22666 cm^{-1} (441.2 nm) and 23849 cm^{-1} (419.3 nm), respectively, as function of ε up to 28.29. Additionally, the solvent leads to blue shifts (around 400 cm^{-1}) and red shifts (about 200 cm^{-1}) in the calculated energy levels of other triplet states as seen in see **Figure 2B**.

The İntersystem crossing (ISC) between singlet and triplet excited states of a molecular system is very important for many purposes in photochemistry. Therefore, we also investigated solvent effect on the energy barrier between singlet and triplet state, ΔE(ISC) = E(triplet)-E(singlet), where the ISC may take place.

The results from calculations show that the energy gap between Q2-band and T_3 triplet state regularly decreases to −4 cm^{-1} (ε = 10.125) from −330 cm^{-1} (in gas-phase) with increasing solvent polarity and then slowly increases up to 60 cm^{-1} with the increase solvent polarity up to ε = 78.355. The gap between B2 band and T_4 triplet state rapidly decreases from −972 cm^{-1} in the gas phase to −84 cm^{-1} in Xenon (ε = 1.706), increases from 132 cm^{-1} (ε = 1.911) to a maximum value of 972 cm^{-1} (ε = 5.32), then follow an exponentially decrease to 110 cm^{-1} (ε = 78.355). The triplet state T_7 lies about 191 cm^{-1} above the L1-band in gas-phase, which start to decrease to 130 cm^{-1} with increase in the ε.

3.2 Solvent dependence of the H_4TPP

The UV–vis absorption spectrum of the diprotonated TPP (H_4TPP) exhibited a Soret band (B-band) at 22272 cm^{-1} (449 nm) in *acidic* chloroform solutions [56], 22831 cm^{-1} (438 nm) in acidic dichloromethane solution [57], 23419 cm^{-1} (427 nm) in acidified THF solution [58], and 22422 cm^{-1} (446 nm) *and* 22272 cm^{-1} (*449* nm) in acidic dichloromethane solution containing chloride *anion* (Cl^-) and bromide anion (Br^-), respectively [59]. The Q(0,0)-band was observed at 14837 cm^{-1} (674 nm) in acidic chloroform solutions [56], 15337 cm^{-1} (652 nm) in acidic dichloromethane solution [57], 15432 cm^{-1} (648 nm) in acidic (H_2SO_4) dichloromethane solution, and 15129 cm^{-1} (661 nm) *in* acidic (H_2SO_4) dichloromethane solution containing chloride *anion* (Cl^-) [59]. These experimental results reveal that the solvent's polarity has significantly effects on the absorption spectrum of diprotonated TPP (H_4TPP). The calculated solvent effect on the absorption spectra of the H4TPP in thirty nine different solutions can summarized as follows.

Singlet H_4TPP: On contrary to red-shifts of the bands in the absorption spectrum TPP, the solvent leads to substantial blue-shifts of the absorption bands in the protonated TPP (H_4TPP) spectrum as seen in **Figure 3(A** and **B)**. For instance, the calculated Q (14761 cm^{-1}/ 677.5 nm), B (20583 cm^{-1}/485.8 nm), L1 (21803 cm^{-1}/458.7 nm), L2 (24095 cm^{-1}/415.0 nm) and M (28465 cm^{-1}/351.3 nm) bands in gas phase spectrum are progressively blue shifted to 15352, 22981, 25069, 25731, and 29355 cm^{-1} (or 651.4, 435.2, 404.4, 398.9 and 388.6 nm), respectively, as function of solvent dielectric constant (ε) from 1.00 to 20.493. With further increase in ε, these electronic bands remain almost constant within a few ten of wavenumbers fluctuation. These calculated results are consistent with the experimentally observed dependence of the absorption spectrum of H4TPP on molecular environments as discussed above.

Triplet H_4TPP: While **Figure 4(A)** provides the calculated electronic spectra of the triplet H_4TPP in the solvents used in this work, **Figure 4B** provides the shifts in the peak positions as function of the solvent dielectric constant relative to their corresponding positions in the gas phase spectrum. The results from calculations show that the solvent gives rise to change in dipole allowed triplet-triplet vertical electronic transition energies in the H_4TPP spectrum as well energy level of the lowest triplet state (T_1) as a function of solvent dielectric constant only in the region of solvent dielectric constant from ε = 1 to 28.493 and, with further increase of solvent dielectric constant, remain unchanged within a few wavenumber variations, see **Figure 4A** and **B**. Therefore, we only provide the maximum shifts when molecular environment changes from the gas phase to acetone (ε = 28.493). It is worthy to point that the shift in the energy level of the lowest triplet state (T_1) was estimated from the equation: $\Delta E(T_1,\varepsilon) = E(T_1,\varepsilon)-E(S_0,\varepsilon)$.

The predicted T_1 lowest triplet state gradually blue shifts from 8232 cm^{-1} (1215 nm) in gas phase to 8767 cm^{-1} (1141 nm) in acetone solvent medium. The T_4

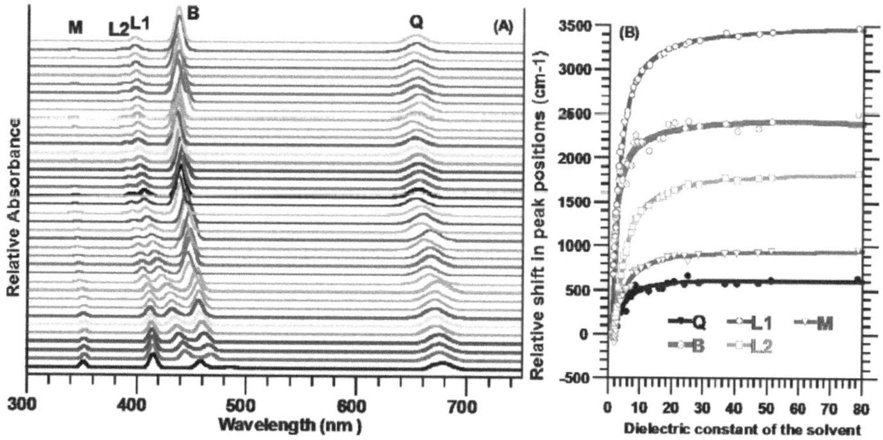

Figure 3.
(A) The calculated solvent-dependence of the singlet-singlet electronic absorption spectrum of the H_4TPP in the thirty-nine different environments, where the dielectric constant of the molecular environment increases from bottom (ε = 1.00) to top (ε = 181.560); (B): The red-shifts in the absorption band positions as function of dielectric constant of the solvent (ε), reference to their corresponding values in gas-phase spectrum.

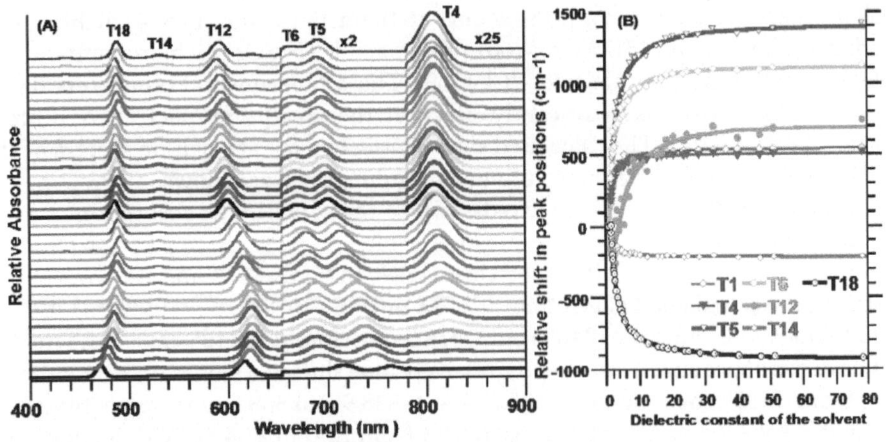

Figure 4.
(A) The calculated solvent-dependence of the triplet-triplet electronic absorption spectrum of the H_4TPP in the thirty-nine different environments, where the dielectric constant of the molecular environment increases from bottom (ε = 1.00) to top (ε = 181.560); (B): The shift in the energy level of the T_n state reference to their corresponding value in gas-phase spectrum as function of dielectric constant of the solvent (ε). The shift in the energy level of the T_1 state is obtained using the equation: $\Delta E(T_1,\varepsilon) = E(T_1,\varepsilon)-E(S_0,\varepsilon)$.

(at 11900 cm^{-1}/840.3 nm), T_5 (13093 cm^{-1}/763.8 nm), T_6 (13955 cm^{-1}/716.6 nm), T_7 (13291 cm^{-1}/752.4 nm), and T_{12} (16236 cm^{-1}/615.9 nm) in the gas phase spectrum are 500, 1344, 1083, 2176, and 631 cm^{-1} blue shifted in spectrum of H_4TPP in acetone, respectively. However, the T_{14} (19024 cm^{-1}/525.7 nm) and T_{18} (21607 cm^{-1}/462.8 nm) are 210 and 890 cm^{-1} red shifted in acetone as solvent, respectively, see **Figure 4A** and **B**.

The SCF corrected triplet states showed there are four ISC pathway below 30000 cm^{-1}, which also are solvent dependent. The solvent dependence of the energy gap between the singlet and triplet excited states, $\Delta E(T\text{-}S; \varepsilon) = E(triplet; \varepsilon)-E(singlet; \varepsilon)$, decreases with increase of solvent dielectric constant up to acetone and remain almost constant with further increase of ε. For instance, when molecular environment changes from gas-phase to acetone medium, the computed energy gap ($\Delta E(T\text{-}S; \varepsilon)$) between the closest singlet and triplet state changes from −2602 to −2458 cm^{-1} (T_2-Q), from 3269 to 2138 cm^{-1} (T_3-Q), from 743 to 208 cm^{-1} (T_5-B), from 348 to 65 cm^{-1} (T_{11}-L1), from 373 to 105 cm^{-1} (T_{12}-L2), from 1076 to 5 cm^{-1} (T_{17}-M) and from 1375 to 151 cm^{-1} (T_{18}-M).

3.3 Solvent dependence of TSPP

Akins *et al.* [60] and Zhang *et al.* [61] also have measured the UV–vis spectra of the free-base TSPP and the H_4TSPP (diprotonated- or dianionic-TSPP). The author reported that the absorption spectrum of the TSPP exhibited Q(0-0)-bands at about 517 (± 2) and 640 (± 3) nm and the B-band at ~412 nm, the H_4TSPP spectrum revealed a weak broad Q(0-0)-band at 645 nm and an intense B-band at 432 nm. Furthermore, the UV–vis spectra of the singlet TSPP and H_4TSPP in the region of 420-460 nm indicated that the maximum peak position of the B-band in the ethanol, methanol, and DMSO solvents is red shifted with respect to water [62].

The excited-state dynamics of the TSPP has been experimentally studied by several research groups. The lowest triplet state (T_1) of the TSPP has been observed at 862 nm [63] and the energy gap between the Q1 band and the T_1 state in the TSPP, $\Delta E(T_1\text{-}Q1)$, is measured to be about 4000 cm^{-1} (48 kJ mol^{-1}) [64]. Moreover, the measured triplet–triplet absorption spectrum of the TSPP (at pH = 7) exhibited a strong peak between 440 and 450 nm with a three weak transitions: at 524 ± 2 nm, a broad peak between ca. 550-575 nm and third one at 624 nm [65]. For the H_4TSPP (at pH = 3), their triplet-triplet absorption spectrum in the range of 450-700 nm displayed two relatively strong and one weak peaks. Their estimated peak positions: at ~500 ± 2 nm with a shoulder (at about 530 nm) and 650 ± 5 nm; a weak one at about 596 ± 2 nm. Jirsa and *et al.* also observed a strong triplet band at 444 nm in the TSPP absorption spectrum [66].

Our calculated dipole allowed singlet-singlet and triplet-triplet electronic transition energies of the TSPP and H_4TSPP are in good agreement with these experimentally observed absorption spectra discussed above. The results from calculations show that the spectral position of the calculated absorption bands gradually change as a function of solvent dielectric constant only in the region of ε = 1 to about 20.493 (acetone) and remain unchanged with further increase of ε. Thus, we only discuss the shifts in the band position when molecular environment changes from gas phase to acetone medium and the results may be summarized as follows.

Singlet TSPP: The solvent-dependence of dipole allowed singlet-singlet electronic transition energies of TSPP is given in **Figure 5A**. **Figure 5B** shows the shifts in band positions caused by solvent, which is similar to the shifts in the TPP absorption spectra. As seen in **Figure 5A** and **B**, the bands at 17064 cm^{-1} (Q1, 586.0 nm) and 18178 cm^{-1} (Q2, 550.1 nm) in the gas phase spectrum of singlet TSPP are 576 and 541 cm^{-1} blue-shifted to 17365 cm^{-1} (575.9 nm) and 18467 cm^{-1} (541.5 nm), respectively; however, the Soret-bands B1 at 24692 cm^{-1} (405.0 nm) and B2 at 25060 cm^{-1} (399.0 nm) in gas phase spectrum are substantially shifted as function of increasing ε up to about 20.493, such as are first 958 and 1112 cm^{-1} red shifted to 23734 cm^{-1} (421.3 nm) and 23948 cm^{-1} (417.6 nm) with increase of solvent dielectric constant ε from 1.0 to 5.32, and which are then turned to increase up to 24429 cm^{-1} (409.3 nm) and 24758 cm^{-1} (403.9 nm) with increase of ε from 5.32 to about 20.493. With further increasing value of ε, they remain constant within ±120 cm^{-1} fluctuation. The bands at 26319 cm^{-1} (labeled as L1 at 380.0 nm) in the gas phase spectrum is shifted to 28117 cm^{-1} (355.7 nm) in the same region.

Triplet TSPP: The solvent-dependence of the triplet-triplet vertical electronic transition energies of the TSPP is similar to these in the triplet-TPP spectrum when molecular environment changes from gas phase to solvent phase. For instance, the SCF-corrected the lowest triplet state at 10922 cm^{-1} (915.6 nm) exhibits a exponentially blue shift up to around 11426 cm^{-1} (875.2 nm) as function of the ε (up to 20.493), which then

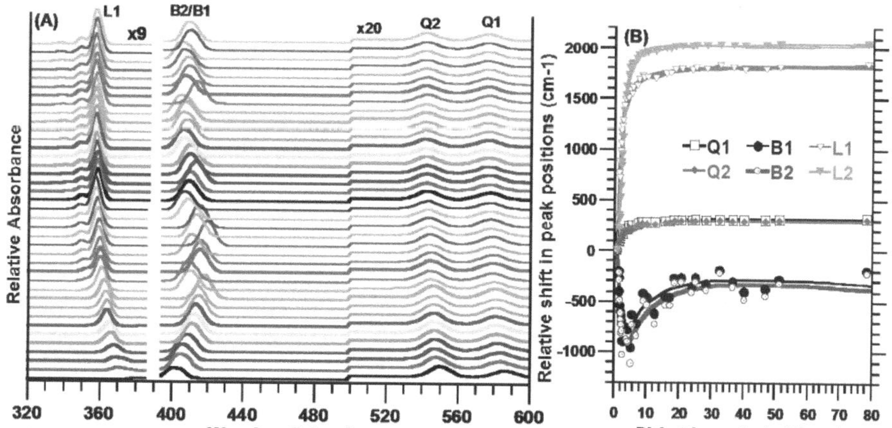

Figure 5.
(A) The calculated solvent-dependence of the singlet-singlet electronic absorption spectrum of the TSPP in the thirty-nine different environments, where the dielectric constant of the molecular environment increases from bottom (ε = 1.00) to top (ε = 181.560); (B): The red-shifts in the absorption band positions as function of dielectric constant of the solvent (ε), reference to their corresponding values in gas-phase spectrum.

remains almost constant within 20 cm^{-1} with the further increase in ε. This predicted value of the T_1 (lowest triplet state) in the solvent with its dielectric constant value $\varepsilon \geq 20.493$ is consistent with its experimental value of 862 nm as mentioned above.

As seen in **Figure 6A** and **B**, the calculated dependence of the dipole triplet-triplet vertical dipole allowed electronic transitions ($T_1 \rightarrow T_n$) energies also exhibits a exponentially red/blue-shifts as function of the ε. The triplet absorption spectrum displays the maximum shifts in acetone as solvent. For example, while the T_7 at 14980 cm^{-1} (667.5 nm), T_8 at 15282 cm^{-1} (654.4 nm) and T_{17} at 18757 cm^{-1} (533.1 nm) in the gas phase spectrum of the TSPP are respectively 2415, 2177 and 902 cm^{-1} blue-shifted (to 17295 cm^{-1} (574.9 nm), 17459 cm^{-1} (572.8 nm) and 19659 cm^{-1} (508.7 nm)) in acetone medium, the T_{24} at 22000 cm^{-1} (454.5 nm) in gas phase is 1036 cm^{-1} red-shifted to 20964 cm^{-1} (477.0 nm).

In the same region, T_4, T_5, T_9 and T_{12} states displayed a relatively weak solvent dependence such as while the triplet states T_4 (13248 cm^{-1}/754.8 nm) and T_{12} (18236 cm^{-1}/548.4 nm) display a maximum blue-shift of about 500 cm^{-1}; the T_5 (14022 cm^{-1}/713.2 nm) and T_9 (18230 cm^{-1}/548.6 nm) triplet states exhibit about 400 cm^{-1} red-shift (**Figure 6B**).

For a possible inter system crossing (ISC) process, we also examined the solvent effect on the energy-gap between the closest singlet and the triplet states, ΔE(S-T, ISC) = E(T) - E(S). The singlet states Q2, B2 and L2 singlet states are almost overlapping with the T_3 (triplet-triplet forbidden state), T_4 and T_8 triplet states, respectively. For instance, with changes solve ε = 1 to 20.493, ΔE(S-T, ISC) the energy-gap changes from: 234 to −40 cm^{-1} between the Q2 and T_3 state; −2108 to −1829 cm^{-1} (Q1 and T_2); −3322 to −2931 cm^{-1} (Q2 and T_2); 234 to −40 cm^{-1} (Q2 and T_3); −522 to 510 cm^{-1} (B1 and T_4); −890 to 181 cm^{-1} (B2 and T_4); −115 to 767 cm^{-1} (L1 and T_8); and – 576 to 87 cm^{-1} (L2 and T_8) with going from ε = 1 to 20.493. With the further increase in ε, they remain nearly unchanged within a few ten wavenumber variations.

3.4 Solvent dependence of the H_4TSPP

Singlet H_4TSPP: The results from calculations shows that the solvent effect on the dipole allowed vertical electronic transition energies of the H_4TSPP is similar to

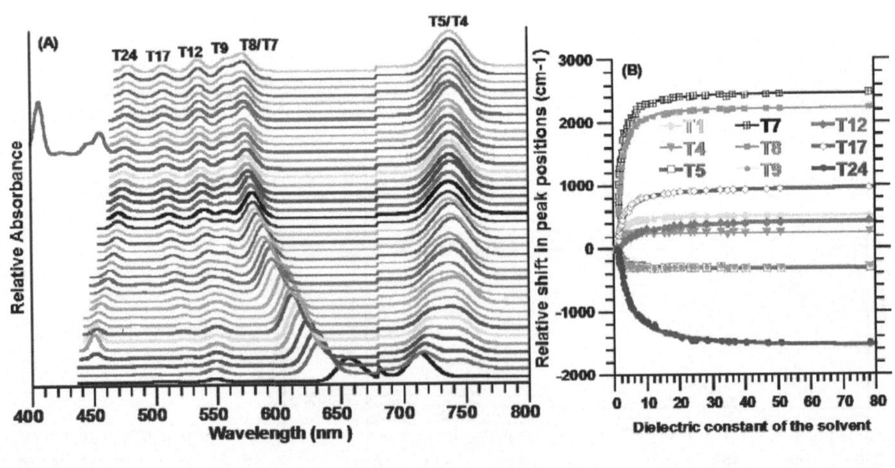

Figure 6.
(A) The calculated solvent-dependence of the triplet-triplet electronic absorption spectrum of the TSPP in the thirty nine different environments, where the dielectric constant of the molecular environment increases from bottom (ε = 1.00) to top (ε = 181.560); (B): The shift in the energy level of the T_n state reference to their corresponding value in gas-phase spectrum as function of dielectric constant of the solvent (ε). The shift in the energy level of the T_1 state is obtained using the equation: $\Delta E(T_1,\varepsilon) = E(T_1,\varepsilon)-E(S_0,\varepsilon)$.

these in the singlet-TSPP spectrum, but exhibit more stronger blue-shift. As seen in **Figure 7A** and **B**, with increasing value of ε from 1.00 to about 20.493, the nearly degenerated Q-bands predicted at 10692/10737 cm^{-1} (935.3/932.4 nm) in gas-phase exhibit extremely blue-shift with the increase of ε such as are respectively blue-shift to 15173/15188 cm^{-1} (659.1/658.4 nm) in acetone as solvent with ε = 20.493 and to 155304/15317 cm^{-1} (or 653.4/652.9 nm) in water (ε = 78.355).

Likewise, the nearly degenerated B-bands at 12720/12796 cm^{-1} (or 786.1/781.5 nm) in gas phase spectrum are respectively 9839/9755 cm^{-1} blue shifted to 22559/22551 cm^{-1} (443.3/443.4 nm) in acetone (ε =20.493) and to 22755/22751 cm^{-1} (439.5/439.5 nm) in water (ε =78.355). Relatively very weak bands with relatively weak intensity (labeled as L-bans) at 14211/14189 cm^{-1} (703.7/704.8 nm) are substantially blue-shifted (9139/9108 cm^{-1}) to 23350/23297 cm^{-1} (428.3/429.2 nm) in acetone and to 23907/23861 cm^{-1} (528.2/528.2 nm) in water, respectively.

In literature, we could not find an experimentally observed electronic spectrum of H_4TSPP in the gas-phase or in a less polar solvent environment to compare with our calculated results. In high polar or acidic solutions, the observed B- and Q-bands in the electronic spectra of H_4TSPP have been published by several researchers as discussed in the TSPP section, which are in agreement with the results presented here. Furthermore, results also shows that the stability of the electronic spectrum of the diprotonated TSPP (H_4TSPP) rapidly increases with increasing polarity or dielectric constant of the solvent (**Figure 7A** and **B**).

Triplet H_4TSPP: The calculations predicted the lowest triplet state (T_1) at 5046 cm^{-1} (1982 nm) in gas phase shifts toward shorter wavelength region with increasing value of ε from 1 to 28.29, and stay almost unchanged around 880 cm^{-1} (1136 nm) with the further increase in ε. The calculated triplet-triplet electronic transition energies of the H_4TSPP exhibit strong solvent-dependence in the range of solvent dielectric constant ε from 1.00 to 28.29. As seen in **Figure 8A** and **B**, the dipole allowed vertical electronic transition energies from the T_1 state to T_4, T_9, T_{11} and T_{12} triplet states are substantially blue shifted to 12047, 13554, 14359, and 15959 cm^{-1} (or 830.1, 737.8, 696.5 and 626.6 nm) from 6332, 8707, 12784 and 11947 cm^{-1} (or 1579, 1149, 782 and 837 nm) in the gas phase, respectively.

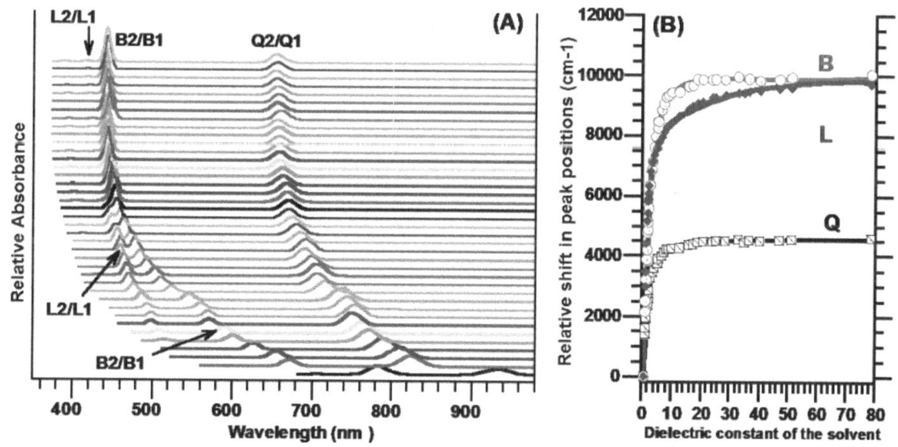

Figure 7.
(A) The calculated solvent-dependence of the singlet-singlet electronic absorption spectrum of the H_4TSPP in the thirty-nine different environments, where the dielectric constant of the molecular environment increases from bottom (ε = 1.00) to top (ε = 181.560); (B): The red-shifts in the absorption band positions as function of dielectric constant of the solvent (ε), reference to their corresponding values in gas-phase spectrum.

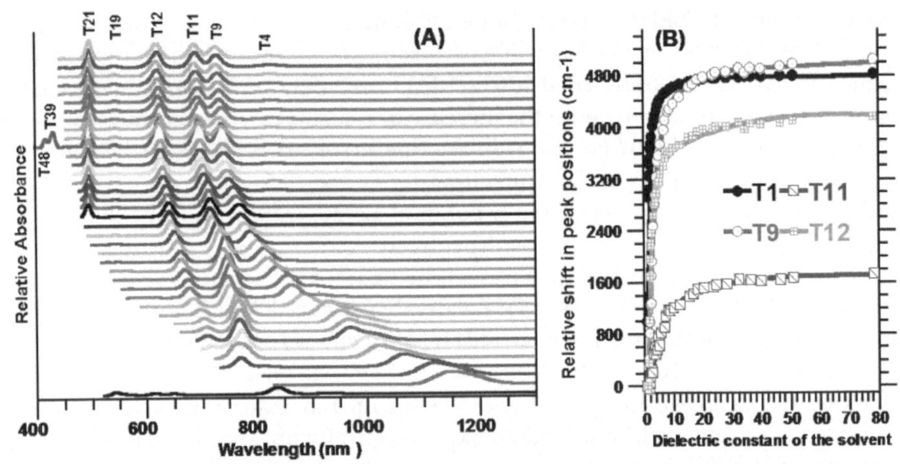

Figure 8.
(A) The calculated solvent-dependence of the triplet-triplet electronic absorption spectrum of the H_4TSPP in the thirty-nine different environments, where the dielectric constant of the molecular environment increases from bottom (ε = 1.00) to top (ε = 181.560); (B): The shift in the energy level of the T_n state reference to their corresponding value in gas-phase spectrum as function of dielectric constant of the solvent (ε). The shift in the energy level of the T_1 state is obtained using the equation: $\Delta E(T_1,\varepsilon) = E(T_1,\varepsilon)-E(S_0,\varepsilon)$.

The solvent-dependency of the smallest energy gap between the singlet and SCF corrected-triplet states, ΔE(singlet-triplet, ε), are estimated to be around 2300 cm^{-1} (T4-Q states) and about 300 ± 150 cm^{-1} (T9-B) and (T11-L).

4. Conclusion

Results from calculated electronic spectra of the TPP, TSPP and their diprotonated structures in gas-phase and thirty eight different solvent showed that solvent gives rise to a blue/red shifts in the singlet-singlet and triplet-triplet dipole allowed vertical electronic transition energies as a function of solvent dielectric constant only in the region of ε = 1 to about 20.493 (acetone), but no significant spectral shifts found for the solvent dielectric constant ε ≥ about 20.493.

In the low energy region (Q-band region), the Q-bands of the TPP first exhibit a red-shift (of about 70 cm^{-1}) with increasing solvent dielectric constant up to ε = 5.32 (diiodomethane) and then to a blue shift with increasing of the ε (from 5.32 to 20.493), However, the Q-bands of the TSPP display only a blue shift (as much as 295 cm^{-1}) with increasing of solvent dielectric constant up to 20.493. The Q-bands in both H_4TPP and H_4TSPP spectra exhibits a rapid blue-shift with increasing of solvent dielectric constant, but the shifts in the H_4TSPP is much stronger than that in the H_4TPP spectrum; such as 4480 cm^{-1} in the H_4TSPP and 490 cm^{-1} in the H_4TPP in acetone as solvent.

In the high energy region, the solvent effect on the Soret band (B-band) of the TPP is stronger than on that of the TSPP. The B-band of both molecules first display a red-shift in range from 1 to 5.32 and then turn to blue shift up to acetone (ε = 20.493). In both H_4TPP and H_4TSPP compounds, the B-band exhibit a strong blue-shift with increase of ε, but the shift in the B-band position of H_4TSPP is much stronger than that of the H_4TPP. With the further increase of the solvent dielectric constant (ε), spectral position of the Q- and B-bands become almost stabile within a few-tens wavenumber ranges. Furthermore, the intensity of the B-band slightly increases with increase of solvent dielectric constant.

The solvent results in both red- and blue-shifts in the triplet-triplet transition energies in the TPP, H_4TPP and TSPP spectra in the calculated spectral region, but only blue-shift takes place in the H_4TSPP. Furthermore, the calculations also showed that the ISC process many not only take place between the Q-bands and the nearest triplet states, but also it is possible between the B- and L-bands and the nearest triplet states because these singlet states almost overlapping with the higher triplet states. Therefore, the ISC may occurs in the singlet state B (or B-band) via surface touching, based on the competition between the IC (from the B- to Q-bands) and the ISC process. This energy gaps between Q/B-bands and the nearest triplet states is also solvent dependent for $\varepsilon < 20$.

The results of calculations also indicate that solvent effect on the nonplanar macrocycle conformation of porphyrin (H_4TPP/H_4TSPP) is more significant than that on the planar macrocycle conformation of porphyrin (TPP/TSPP). This observation suggest that the nonplanar macrocycle conformation increase the electrostatic/electronic interaction between the four hydrogen atoms at the macrocycle core and molecular environment.

The results from calculations suggest that the polarized porphyrin molecule may interact strongly with increasing solvent polarity and lead to modification of dipole allowed vertical electronic transition energies with increasing solvent polarity arising from partial charge transfer between solvent and the porphyrin macrocycle and/or the intramolecular charge transfer. Moreover, in terms of the electronic configuration of the molecular orbitals of the porphyrin, the additional charge partially may occupy a nonbonding and/or antibonding orbitals that brings about the weakening of the chemical bonding in porphyrin. Also, the partial charge transfer from an occupied nonbonding (atomic orbital of the N atom) and/or antibonding orbital results in the strengthening of the chemical bonding in porphyrin. The change in occupancy of the bonding/nonbonding orbitals affect the electronic excited state, which gives rise to shift in the electronic absorption spectrum of the porphyrin molecule.

As a result, the solvent-dependence of the spectroscopic features of porphyrin can be used to monitor micro environmental changes of porphyrin-like compounds incorporated in biological systems and nanoparticles, which also may be appropriate for study and monitoring changes of the chemical environment in different solutions and interactions in biological systems, as well as deal with nonspecific *adsorption on nanomaterials* and their orientations on the surface, which is very important for the *Surface-Enhanced Resonance Raman Scattering* (SERRS) and Aggregation-Enhanced Raman Scattering (AERS) in different solvent. Furthermore, such strongly solvent-dependent electronic bands can be used as a marker of the environmental dielectric constant.

References

[1] Dolphin, D. (Ed.) The Porphyrins; *Academic Press*: New York; Vol. 1-3, **1978/1979**

[2] M. Gouterman (Ed.). Porphyrins: Excited States and Dynamics. ACS Symp. 321, *American Chem. Soc.*: Washington D. C., **1986**.

[3] Ojima, I. (Ed.) Catalytic Asymmetric Synthesis; Vett Publisher Inc.: New York, **1993**.

[4] Kadish, K. M.; Smith, K. M.; Guilard (Eds.) R. The Porphyrin Handbook; *Academic Press*: New York, Vols. 1-10, **2000**.

[5] Kadish, K. M.; Smith, K. M.; Guilard, R. Handbook of Porphyrin. *Science* **2011**, 14, 249

[6] Holten, D.; Bocian, D. F.; Lindsey, J. S. Probing Electronic Communication in Covalently Linked Multiporphyrin Arrays. A Guide to the Rational Design of Molecular Photonic Devices. *Acc. Chem. Res.*, **2002**, 35 (1), 57-69

[7] Hopfield, J.J., Onuchic, J.N. & Beratan, D.N. A molecular shift register based on electron transfer. Science **1988**, 241, 817-820.

[8] Bonnett, R. Photosensitizers of the porphyrin and phthalocyanine series for photodynamic theory. Chem. Soc. Rev. **1995**, 24, 19-33.

[9] Andrade, S. M.; Costa, S. M. B. Spectroscopic Studies on the Interaction of a Water Soluble Porphyrin and Two Drug Carrier Proteins. Biophysical Journal **2002**, 82, 1607-1619.

[10] Ben-Hur, E.; Horowitz, B. Advances in photochemical approaches for blood sterilization. Photochem. Photobiol. **1995**, 62, 383-388.

[11] Uhehara, K.; Hioki, Y,; Mimuro. M. The chlorophyll *a* aggregate absorbing near 685 nm is selectively formed in aqueous tetrahydrofuran. *Photochem. Photobiol.* **1993**, 58, 127-132.

[12] Tominaga, T. T.; Yushmanov, V. E; Borissevitch, I. E.; Imasato, H; Tabak. M. Aggregation phenomena in the complexes of iron tetraphenylporphine sulfonate with bovine serum albumin. *J. Inorg. Biochem*. **1997**.

[13] Togashi, D. M., and S. M. B. Costa. Absorption, fluorescence and transient triplet-triplet absorption spectra of zinc tetramethylpyridylporphyrin in reverse micelles and microemulsions of aerosol OT–(AOT). Phys. Chem. Chem. Phys. **2000**, 2, 5437-5444.

[14] Dolphin (Ed.), D. The Porphyrins; *Academic Press*, New York; Vol. 1-3, **1978/1979**.

[15] Ferreira, G. C (Ed.); Kadish, K. M.; Smith, K. M.; Guilard, R. Handbook of Porphyrin Science (Volumes 26 – 30) *World Scientific Publishing Company*, **2013**.

[16] Tsvirko, M.P.: Solovjev, K.N.; Gradyushko, A.T.; Dvornikov, S.S. Phosphorescence of porphyrin free bases and their complexes with light metals. *Opt. Spectrosc*. **1975**, 38, 400-404.

[17] Volker, S.; van der Waals, J.H. Laser-induced photochemical isomerization of free base porphyrin in an *n*-octane crystal at 4·2 K. Mol. Phys. **1976**, 32, 1703-1718.

[18] Loboda, O. Quantum-chemical studies on Porphyrins, Fullerenes and Carbon Nanostructures; Springer Science & Business Media, **2012**.

[19] Aydin, M. (2014). Comparative study of the structural and vibroelectronic properties of porphyrin and its derivatives. Molecules. **2014**, *19*, 20988-21021.

[20] Gouterman, M. Spectra of porphyrins. J. Mol. Spectrosc. **1961**, 6, 138-163.

[21] Parusel, A.B.J.; Grimme, S. DFT/MRCI calculations on the excited states of porphyrin, hydroporphyrins, etrazaporphyrins and metalloporphyrins J. Porphyrins Phtalocyanines, **2001** 5, 225-232.

[22] Nakatsuji, H.; Hasegawa, J.; Hada, M. Excited and ionized states of free base porphin studied by the symmetry adapted cluster-configuration interaction (SAC-CI) method. J. Chem. Phys. **1996**, 104, 2321.

[23] Hasegawa, J.; Ozeki, Y.; Hada, M.; Nakatsuji, H. Theoretical Study of the Excited States of Chlorin, Bacteriochlorin, Pheophytin *a*, and Chlorophyll *a* by the SAC/SAC–CI Method. *J. Phys. Chem*. B. **1998**, 102, 1320-1326.

[24] Serrano-Andres, L.; Merchan, M.; Rubio, M.; Roos, B.O. Interpretation of the electronic absorption spectrum of free base porphin by using multiconfigurational second-order perturbation theory. Chem. Phys. Lett. 1998, 295, 195-205

[25] Sundholm, D. Density functional theory calculations of the visible spectrum of chlorophyll *a*. Chem. Phys. Lett. 1999, 302, 480-484.

[26] Sundholm, D. Interpretation of the electronic absorption spectrum of free-base porphin using time-dependent density-functional theory. Phys. Chem. Chem. Phys. **2000**, 2, 2275-2281.

[27] Sundholm, D. Density functional theory study of the electronic absorption spectrum of Mg-porphyrin and Mg-etioporphyrin-I. Chem. Phys. Lett. **2000**, 317, 392-399.

[28] Nguyen, K.A.; Day, P.N.; Pachter, R. Triplet Excited States of Free-Base Porphin and Its β-Octahalogenated Derivatives. J. Phys. Chem. A. **2000**, 104, 4748-4754.

[29] Aydin, M. DFT and Raman spectroscopy of porphyrin derivatives: Tetraphenylporphine (TPP). Vib. Spectrosc. **2013**, 68, 141–152.

[30] Jun Takeda and Mitsuo Sato. Unusual solvent effect on absorption spectra of nonplanar dodecaphenylporphyrin caused by hydrogen-bonding interactions. Chemistry Letters 1995, 971-972.

[31] Li, Y. Solvent effects on photophysical properties of copper and zinc porphyrins. Chin. Sci. Bull. 53 (2008) 3615-3619.

[32] J. Tomasi, B. Mennucci, and R. Cammi. Quantum Mechanical Continuum Solvation Models. Chem. Rev. 105 (2005) 2999-3093.

[33] M. Milovanovic. Experimental and theoretical approaches coupled with thermochemistry of reactions in solution and role of non-covalent interactions. PhD. Thesis, Doctoral School of Chemical Science, University of Strasburg, Faculty of Chemistry, 2018, France.

[34] J. Tomasi and M. Persico. Molecular interactions in solution: An overview of methods based on continuous distributions of the solvent. Chem Rev. 94 (1994) 2027-2094.

[35] I. Izumi, Y. Atshi and K. Takayoshi. Solvent effect for ruthenium porphyrin. Sci China Phys Mech Astron. 53(6) (2010) 1005-1012.

[36] S. H. Saraf and R. Ghiasi. Effect of the Solvent Polarity on the Optical Properties in the (OC)5Cr=(OEt)(Ph) Complex: A Quantum Chemical Study. Russian Journal of Physical Chemistry A, 2020, Vol. 94, No. 5, pp. 1047-1052.

[37] S. Kamrava, R. Ghiasi and A. Marjani. Structure, electronic properties and slippage of cyclopentadienyl and indenyl ligands in the (η^5-C_5H_5) (η^3-C_5H_5) W$(CO)_2$ and (η^5-C_9H_7) (η^3-C_9H_7) W$(CO)_2$ complexes: A C-PCM investigation. Journal of Molecular Liquids 329 (2021) 115535.

[38] L. Edwards, D.H. Dolphin, M. Gouterman, A.D. Adler, Porphyrins XVII. Vapor absorption spectra and redox reactions: Tetraphenylporphins and porphin. J. Mol. Spectrosc. 38 (1971) 16-32.

[39] M. Aydin and D. L. Akins. DFT studies on solvent dependence of electronic absorption spectra of free-base and protonated porphyrin. Computational and Theoretical Chemistry 1132 (2018) 12-22.

[40] Metin Aydin and Daniel L. Akins (October 5th 2016). Geometric and Electronic Properties of Porphyrin and its Derivatives, Applications of Molecular Spectroscopy to Current Research in the Chemical and Biological Sciences, Mark T. Stauffer, IntechOpen, DOI: 10.5772/64583.

[41] Frisch, M. J.; Trucks, G. W.; Schlegel, H. B.; Scuseria, G. E.; *et al.* Gaussian 09 Revision A.02.

[42] Becke, A. D. Density-Functional Thermochemistry. III. The Role of Exact Exchange. J. Chem. Phys. **1993**, 98, 5648–5652.

[43] Lee, C.; Yang, W.; Parr, R. G. Development of the Colle-Salvetti Correlation-Energy Formula into a Functional of the Electron Density. Phys. Rev. B **1988**, 37, 785–789.

[44] Krishnan, R.; Binkley, J. S.; Seeger, R.; Pople, J. A. Self-consistent molecular orbital methods. XX. A basis set for correlated wave functions. *J. Chem. Phys.* **1980**, *72*, 650-654.

[45] Tomasi, J.; Mennucci, B.; Cammi, R. Quantum Mechanical Continuum Solvation Models. Chem. Rev. **2005**, 105, 2999–3093.

[46] Barone, V.; Cossi, M. Quantum Calculation of Molecular Energies and Energy Gradients in Solution by a Conductor Solvent Model. J. Phys. Chem. A **1998**, 102, 1995-2001.

[47] Miertus, S.; Scrocco, E.; Tomasi, J. Electrostatic Interaction of a Solute with a Continuum. A Direct Utilization of ab Initio Molecular Potentials for the Prevision of Solvent Effects. Chem. Phys. **1981**, 55, 117–129.

[48] Cossi, M.; Rega, N.; Scalmani, G.; Barone, V. Energies, structures, and electronic properties of molecules in solution with the C-PCM solvation model. J. Comput. Chem. **2003**, 24, 669–681.

[49] N. M. O'Boyle, J. G. Vos, GaussSum 0.8; Dublin City University: Dublin, Ireland, (**2004**). Available at http:// gaussum.sourceforge.net.

[50] Even, U.; Magen, J.; Jortner, J.; Friedman, J. M.; Levanon, H. Isolated ultracold porphyrins in supersonic expansions. I. Free-base tetraphenylporphyrin and Zn-tetraphenylporphyrin. *J. Chem. Phys.* **1982**, *77*, 4374-4383.

[51] Benthem, L. Tetraphenylporphyrin dimers: An optical and magnetic resonance study. Thesis Wageningen Agricultural University, Neitherlands, **1984**.

[52] Badger, G.M.; Jones, R.A.; Laslett, R.L. Porphyrins. VII. The synthesis of porphyrins by the Rothemund reaction. *Aust. J.* Chem. 1964, 17, 1028-1035.

[53] Harriman, A. Luminescence of porphyrins and metalloporphyrins. Part 1.—Zinc(II), nickel(II) and manganese(II) porphyrins. *J. Chem.*

Soc., Faraday Trans. 1, **1980, 76,** 1978-1985.

[54] Egorova, G.D.; Knyukshto, V.N.; Solovev, K.N.; Tsvirko, M.P. Intramolecular spin-orbital perturbations in ortho- and meta-halogeno-derivatives of tetraphenylporphin. Opt. Spectrosc. (USSR). **1980**, 48, 602-607.

[55] Pekkarinen, L.; Linschitz, H. Studies on Metastable States of Porphyrins. II. Spectra and Decay Kinetics of Tetraphenylporphine, Zinc Tetraphenylporphine and Bacteriochlorophyll[1]. *J. Am.* Chem. Soc. **1960**, 82, 2407-2411.

[56] Liu, Z. B.; Zhu, Y.; Zhu, Y. Z.; Tian, J. G.; Zheng, J. Y. Study on Nonlinear Spectroscopy of Tetraphenylporphyrin and Dithiaporphyrin Diacids. J. Phys. Chem. B **2007,** *111,* 14136-14142.

[57] Okada, S.; Segawa, H. Substituent-Control Exciton in J-Aggregates of Protonated Water-Insoluble Porphyrins. J. AM. CHEM. SOC. **2003**, 125, 2792-2796.

[58] Hasobe, T.; Fukuzumi, S.; Kamat, P. V. Ordered Assembly of Protonated Porphyrin Driven by Single-Wall Carbon Nanotubes. J- and H-Aggregates to Nanorods. *J. AM. CHEM. SOC.* **2005**, 127, 11884-11885.

[59] Zhang, Y.; Li, M. X.; Lu, M. Y.; Yang, R. H.; Liu, F.; and Li, K. A. Anion Chelation-Induced Porphyrin Protonation and Its Application for Chloride Anion Sensing. J. Phys. Chem. A **2005,** *109,* 7442-7448.

[60] Akins, D.L.; Özçelik, S.; Zhu, H.-R.; Guo, C. Fluorescence Decay Kinetics and Structure of Aggregated Tetrakis (p-Sulfonatophenyl)Porphyrin. J. Phys. Chem. **1996**, 100, 14390-14396.

[61] Zhang, Y.-H.; Chen, D.-M.; He, T.; Liu, F.-C. Raman and infrared spectral study of meso-sulfonatophenyl substituted porphyrins (TPPSn, n_/1, 2A, 2O, 3, 4). Spectrochim. Acta Part A. **2003**, *59*, 87-101

[62] Farajtabar, A.; Jaberi, F.; Gharib, F. Preferential solvation and solvation shell composition of free base and protonated 5, 10, 15, 20-tetrakis (4-sulfonatophenyl)porphyrin in aqueous organic mixed solvents. Spectrochimica Acta Part A. **2011**, 83, 213– 220.

[63] Roman Dedic. Antonın Svoboda, Jakub Psencık, Jan Hala, Phosphorescence of singlet oxygen and meso-tetra (4-sulfonatophenyl)porphin: Time and spectral resolved study. J. Mol. Struct. **2003**, 651-653, 301-304.

[64] Scholz, M.; Dedic, R. Thomas Breitenbach and Jan Hala. Singlet oxygen-sensitized delayed fluorescence of common water-soluble photosensitizers. Photochem. Photobiol. Sci., **2013**, 12, 1873-1884.

[65] Correa, D.S.; Boni, L. D.; Parra, G.G.; Misoguti, L.; Mendonça, C.R.; Borissevitch, I.E.; Zílio, S.C.; Neto, N.M.B.; Gonçalves, P.J. Excited-state absorption of meso-tetrasulfonato phenyl porphyrin: Effects of pH and micelles. Optical Materials. 2015, 42 516-521.

[66] Engst, P.; Kubat, P.; Jirsa, M. The influence of D20 on the photophysical properties of meso-tetra(4-sulphonatophenyl) porphine, Photosan III and tetrasulphonated aluminium and zinc phthalocyanines. J. Photo&em. Photobiol. A: Chem. *1994,* 78, 215-219.

Chapter 14

Application of Density Functional Theory in Coordination Chemistry: A Case Study of Group 13 Monohalide as a Ligand

Abstract

The chemistry of Group 13 Monohalide is of great interest due to its isoelectronic relationship with carbon monoxide and dinitrogen. In recent years, theoretical and experimental studies have been evolved on the group-13 atom-based diatomic molecules as a ligand. The synthetic, characterisation and reactivity of various metal complexes have been well discussed in recent reviews. The nature of the metal bonding of these ligands of various types has been explained in addition by the variety of theoretical studies (using DFT methods) such as FMO and EDA. This chapter has a comprehensive experimental and theoretical study of group 13 monohalides as a ligand in coordination chemistry.

Keywords: Group 13 Monohalides, Coordination Chemistry, DFT, FMO, BDE

1. Introduction

The monohalides of group 13 elements (EX) are an isoelectronic relationship with molecules like CO and N_2. The separation of metal complexes containing monohalides of group 13 elements as ligands is made possible by the recent developments in synthetic chemistry and with the investigations on electronic structural analysis and the reactivity of the coordinated diatomic group 13 monohalides. In general, the +3 oxidation states of group 13 elements have dominated the chemistry of their compounds. The applications of these compounds like catalysis, sensing, etc., are due to their inherent Lewis acidic behaviour [1–3]. Research in group 13 elements having lower oxidation states have been normally influenced by clusters of boranes. The availability of sub-valent systems reflects on the development of the ground-breaking synthetic approach in organic synthesis and the applications of some reagents specifically sub-valent indium compounds. For example, the study of Schnoeckel et al. approved to access metastable monohalides of aluminium and gallium, by utilising the entropic factor at a high temperature can be driven the equilibrium to the right and defined by Eq. (1) (E = B, Al, Ga, In and Tl) [4–14].

$$EX_3(g) + 2E(s\,or\,l) = 3EX(g) \quad (1)$$

Sub-valent aluminium and gallium compounds can be accessed by consequential entrapping and derivatization. The formation of sub-valent group 13 compounds is proposed by the theory to mimic the accession of a similar state of the elements [8–21]. For separating the discrete molecular systems, chemistry in solution has taken advantage of sterically bulkier groups (amino, guanidinate, β-diketiminate, pentamethyl cyclopentadienyl, terphenyl groups) as an approach [15–17]. Also, the competency of monovalent systems E(I) having such molecules to behave as ligands in complexes for transition metal atoms has been recognised extensively for B, Al and Ga. In organometallic chemistry, group 13 elements have been an unexplored area, even it has an isoelectronic relationship with well-known CO and N_2 molecules. Even though the thorough comparative theoretical investigations on group 13 monohalides, there is no sufficient experimental data [15–17]. Such paucities naturally mimic the coordinated group 13 monohalide fragment's high polarity and low steric loading (as projected theoretically). This book chapter explores both synthetic and theoretical advances in the topic of group 13 monohalides in great depth.

2. Free molecules of boron (I) halides

In 1935, the properties of gaseous boron monofluoride synthesised from elemental boron and CaF using techniques such as high-temperature in situs method were reported. Since then, it has been studied using several spectral and thermal analyses [22–36]. BF was first synthesised on a preparative scale in 1967, based on Peter Timms' pioneering work to develop a proportionation pathway from BF_3 and elemental boron [37–39]. The boron trifluoride gas is passed through elemental boron at very high temperature (2000°C) and very low pressure (1 mmHg) and in a specially designed reactor [37]. In addition, while BF molecule in the vapour phase maintains its diatomic nature. A variety of BF-containing molecules (such as B_8F_{12}, B_3F_5 and B_2F_4) [37, 40, 41] are then formed with the proportion based on the amount of BF_3 co-condensate. BF insertion into BF_3 produces B_2F_4, and BF insertion into B_2F_4 produces B_3F_5 [37]. Then, Timms reported how metastable boron monochloride can be prepared by cracking B_2Cl_4 (which produces BCl_3 and BCl) or by reacting elemental boron with BCl_3 at a very high temperature [39, 42]. A laser-ablated boron atom reacts with its respective X_2 molecule, resulting in combinations of BX_3, BX_2, and BX. Matrix isolation offers a feasible way to capturing and interrogating diatomics. Microwave spectroscopy [35] produced a bond length of 1.26267 Å, as well as values for Do 757 kJ mol^{-1} [36], ωe (1765 cm^{-1}) [33], and the first ionisation energy 11.115 eV [29]. A similar approach was used to determine bond lengths for the BF, BCl and BI 1.715, 1.888, 2.131 Å respectively [1, 2]. Infrared spectroscopy reveals vibrational frequencies of 564 cm^{-1} (^{11}BI), 667/666 (^{11}B^{79}Br/^{11}B^{81}Br), 815/810 (^{11}B^{35}Cl/^{11}B^{37}Cl), and 1374 (^{11}BF) for BX molecules separated in solid argon atmosphere [43]. Research with C_2F_4 and SiF_4 has shown that even though the BF insertion is simple for B-F bonds, similar procedures have yielded traces of products with other E–F couplings. When BF is combined with soft donors (such as CO, AsH_3, SMe_2, PH_3, PF_3 and PCl_3) leads to the formation of $(F_2B)_3B$-L (general formula). Based on the crystallographic description of BCl is very similar to carbonyl complexes [42, 44]. In the presence of acetylene or propene, BF produced acyclic or cyclic products containing BF or BF_2 moieties [38], whereas when BF was combined with gaseous BCl, only cyclic compounds were produced.

3. Coordination chemistry of boron (I) halides

The boron monohalides used as ligands in organometallic complexes were examined in the context of theoretical approaches were reported since 1998 [15–17] (**Figure 1**). The bonding properties of isoelectronic species of CO, N_2 and BX (X = F – I) molecules were studied in particular. Based on the various proportion of the FMOs and energies of these molecules, BX will be a better sigma donor than CO or N_2, as well as a similar π-acceptor, although the unbound molecule should have a smaller HOMO–LUMO gap due to stronger localization in the donor atom. In line with the broader focus of the synthetic endeavour, the related ligands BO and BNH_2 have undergone comparable computational research. It also has a set of non-degenerate p-type orbitals, which are similar to those found in the vinylidene (CCR_2) ligand family. The exceptionally high energy of BO^- means that it possesses no π acceptor characteristics, but it retains outstanding donor characteristics. It has been determined that BF and related metal complexes have very strong thermodynamic stability, despite heavily polar BF bonds and positive charge build-up at boron. Both methods have already been synthesised, and could be used to shield reactive boron centres: either by shielding with (NH_2 groups) BNR_2 or by incorporating haloborylene as a bridge between two metal sites [8–17]. In 2010, it was predicted that BF would adopt bridging modes of coordination (μ2 or μ3) in ruthenium bimetallic systems [45, 46]. Timms et al. synthesised and characterised a thermally unstable volatile complex $[(F_3P)_4Fe(BF)]$ using IR and ^{19}F NMR spectroscopy [39]. Aldridge and his co-workers successfully synthesised the first fluoroborylene ruthenium complex $[Cp_2Ru_2(CO)_4(\mu\text{-}BF)]$ and characterised it by the X-ray diffraction method [47]. The earlier complex was synthesised using a stable source for the BF ligand ($Et_2O\text{-}BF_3$) and $CpRu(CO)_2{}^-$ (Cp = η^5-C_5H_5). The reaction with BX_3 and $[NaMn(CO)_5]$ was yielded the haloborylene complexes $(\mu\text{-}BX)Mn_2(CO)_{10}$ [48]. It is very similar to the ruthenium complexes $(\mu\text{-}BX)Mn_2(CO)_{10}$. Several alkylborylene complexes were synthesised and characterised but studies on the haloborylene complexes were limited [49–51]. Several complexes synthesised with BX fragments attached with metal centres with additional Lewis base support structures $[Cp^*Fe(CO)_2\{(4\text{-pic})_2BBr\}]^+Br^-$ [52]. Hildendrand et al. synthesised bimetallic complex with manganese $[\{(\eta_5\text{-}C_5H_5)Mn(CO)_2\}_2(\eta{:}\eta{:}\mu\text{-}B_2Cl_2)]$ [28]. The geometrical and bonding analysis of halo and alkylborylene complexes $[(\eta^5\text{-}C_5H_5)M(BX)\ (CO)_2]$ (M = Mn and Re; R = Et, iPr, Me and tBu; X = F - I) were studied theoretically in 2011 by K. K. Pandey et al., [53]. The steric stability of terminal haloborylene complexes, as well as the pi-electron contribution of the haloborylene ligands, play a significant role in their separation. A variety of alternative synthetic techniques have been developed in response to the shortage of group 13 monohalides, such as halide abstraction/ejection [54–56], metal–metal borylene transfer [57], borane dehydrogenation [58], and salt elimination [49, 50]. The strength of metal complexes depending on the

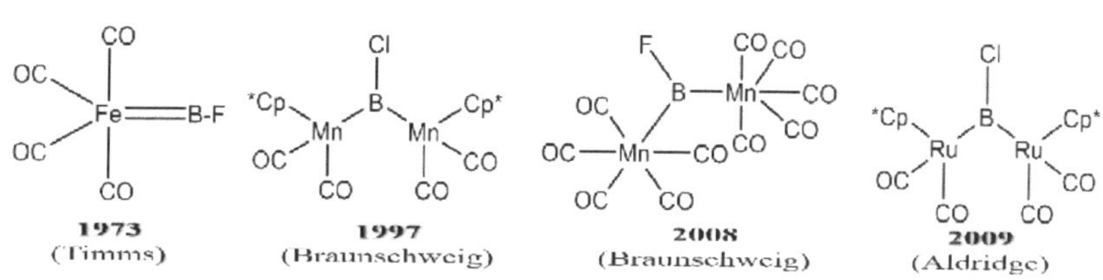

Figure 1.
Development of boron monohalides based complexes.

coordination of the metal with highly reactive ligands. The difficult separation of metal complexes with highly reactive ligands such as BF drives a strong interest in synthetic approaches to these complexes [59–64]. According to recent studies, polar diatomic BF ligands bind to transition metal centres more effectively than CO ligands, with greater σ-donor and π-acceptor properties [17, 46, 65]. Due to the electron-withdrawing halogen atoms, the haloborylene ligands of metal complexes are particularly significant because they have potential π-acceptor characteristics [66]. The theoretical analysis will be the most appropriate way for understanding the bonding character of the metal borylene complexes [67–73]. The C_{3v} point group of [$(CO)_3$M-BX] complexes have been studied using DFT, as well as the interaction between CO and BX ligands, the covalent character of M-B bonds, and bonding donation to M-B bonds [18–21].

4. Free molecules of the heavier group 13 metal monohalides

Aluminium and gallium monohalides belong to two groups: indium and thallium monohalides (excluding InF), which can be disproportionated into metal and metal trihalide under ordinary conditions, and indium and thallium trihalides, which are stable and commercially available [1, 2]. High temperature/low pressure give rise to entropically favoured Al (I) halides in the gas phase, which could be captured by inert gas matrices. In the gas phase, spectroscopic analysis of the AlX (X = F-I) molecule show bond distances of 2.537, 2.295, 2.130 and 1.654 A for Al-F to Al-I respectively [1, 2, 74–76], whereas in low-temperature matrices, symmetrical bridging halides, Al(μ-Cl)$_2$Al and Al(μ-F)$_2$Al predominate [77, 78]. The bond lengths of the gaseous GaX molecules corresponding to aluminium (1.774, 2.202, 2.352, and 2.575 A, respectively) varied slightly, which is consistent with Ga(I) having a greater covalent radius [1, 2, 75, 76]. The metastable AlX solutions can be synthesised by using mixtures of AlX molecules and donor solvents like toluene [5–7, 59, 79, 80]. In the solid-state, there are no completely stable Ga (I) halides. GaI was first reported in 1955, and it was synthesised by vacuum heating Ga and I. Green reported a new synthesis of GaI in 1990, employing an ultrasonically activated Ga with iodine in $C_6H_5CH_3$ [81], but later Raman spectroscopy investigation indicated that it was made up of a mixture of Ga subhalides and its valence salt $[Ga]_2[Ga_2I_6]$ [82]. This green's reagent acts as an accessible and versatile reagent for the synthesis of various monovalent gallium compounds, as well as a possible source of gallium iodide in addition to procedures involving different chemical bonds [81, 83, 84]. And also acts as an extreme source of the gallium iodide fragment in one newly described compound including GaI diatomic fragments [85–87]. In contrast to AlX and GaX, InX (X = Cl-I) and TlX (X = F-I) halides are stable to disproportionation in the solid-state at room temperatures, making them a potential candidate for low oxidation state In/Tl compounds and, as a result, their wider use as reagents in organic synthesis [1, 2, 4, 88–90]. Bond distances of 2.754 (I), 2.543 (Br), 2.401 (Cl) and 1.985 A (F), for the monovalent indium diatomics, have been determined in the vapour phase, with equivalent distances of 2.814, 2.618, 2.485 and 2.084 Å observed for the corresponding Tl molecules [1, 2, 4, 75]. In contrast, in an argon atmosphere, the reactivity of singlets (1S) and triplets (3P) of monovalent indium chloride towards HX (X = H, Cl, or OH) has been investigated [91–93]. Tuck previously reported that In(I) halides can be treated with Lewis bases at low temperatures to create insoluble complexes that are disproportionable. So, below 20°C, InBr (16 mM) solutions in toluene/TMEDA mixtures are stable, the crystalline complex InBr(TMEDA) was separated from the same solution, indicating long-range In/In interactions (3.7 Å) [94, 95].

5. Coordination chemistry of heavier group 13 metal(I) halides

Aluminium and gallium are the two categories of metal monohalides in Group-13 (**Figure 2**). The organometallic chemistry of these two monohalides of Al and Ga-based molecules is an expeditiously bolster object of research. The complexities in synthesising these compounds and the new situations in which they will be used as synthetic starting materials attracted a great deal of interest. From a theoretical perspective, these molecules that correspond to electronic and molecular structure and bonding have a lot of significance [5]. It has been reported that by co-condensing AlX with combinations of donor solvents and toluene, metastable solutions of aluminium monohalides can be made [6, 7, 59, 79, 80]. The AlX adducts $[(EEt_3)AlX]_4$ (X = Br, I E = N, P) were confined and geometrically characterised [74, 96–99]. The symmetrical bridging dimers, which were observable in low-temperature environments [25, 26], contradicted with the spectroscopic investigation of the vapour phase AlX (X = F – I) [74, 76]. The reactivity of monovalent Al compounds is high [100–103], and their scope exceeding that of the more reactive transition metal complexes. Fischer et al. described the chemical $[(CO)_4Fe\text{-}Al(\eta^5\text{-}C_5Me_5)]$, which was synthesised from an AlCp* unit that was bonded to a metal via a terminal non-bridging bond. When compared to dialkyl fragment ER_2 [43–45], single-source precursors of M-M bonds with ER fragments were significant in attaining the molecular force of the thin film stoichiometry [104–110]. The metal complexes $(CO)_nME[(X)L_2]$ of Cr, Mo, W, and Fe with the monohalides of aluminium and gallium ligands were explored in depth. The nature of the bonding in the donor-stabilised complexes $[(NH_3)_2(CO)_5W(ECl)]$ for group-13 atoms (E = B-Tl) was examined [110].

The structural and bonding study of monohalide of group 13 elements as a ligand in metal carbonyls was recently examined (**Figure 3**) [18–21]. Lewis bases such as $[(tmeda)(CO)_4Fe(GaCl)]$ assist in the stabilisation of GaX ligands in complexes, however, ligands such as GaR are not supported by Lewis bases [111–114]. The great reactivity of GaX complexes, as well as the scarcity of EX synthons, present difficulties in their synthesis. Aldridge et al. (2008) addressed these obstacles by employing a sterically hindered, rich-electron metalcore as well as a stable GaI_2 precursor [85, 86]. M-E complexes are more interesting because of their binding properties [115]. Bond dissociation energies for such complexes have been determined thermodynamically, in the same way, that they have been determined for boron monohalide ligands [116]. As a result, BDEs for $[(CO)_4Fe(GaX)]$ has been calculated [X = F-I(equatorial/axial): 140.6/141.8;151.5/151.0;153.6/152.7; 158.6/157.7 $kJmol^{-1}$ respectively] to be 193.7 kJ mol^{-1} (for CO equatorial/axial) lower than for $[Fe(CO)_5]$ but 91.6/88.7 kJ mol^{-1} (for N_2 equatorial/axial) less than $[Fe(CO)_4(N_2)]$. Because various terminal metal gallylene, alumylene, and borylene compounds have been computationally investigated and reported [84, 117–120], very few complexes such as dialkyl, haloaryl, and dihalogallylene complexes have been synthesised and characterised. Moreover, while the covalent to electrostatic

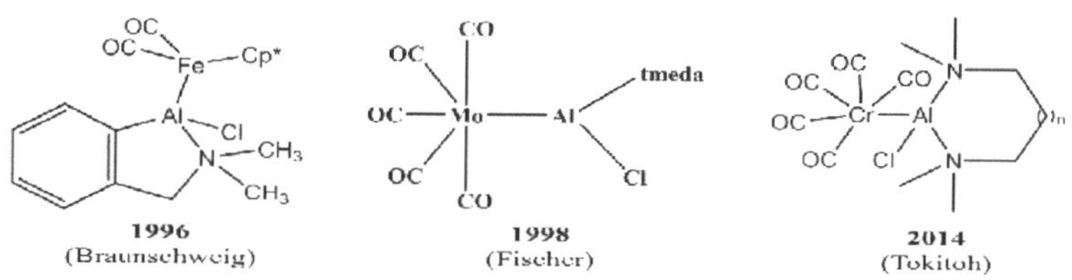

Figure 2.
Development of aluminium monohalides based complexes.

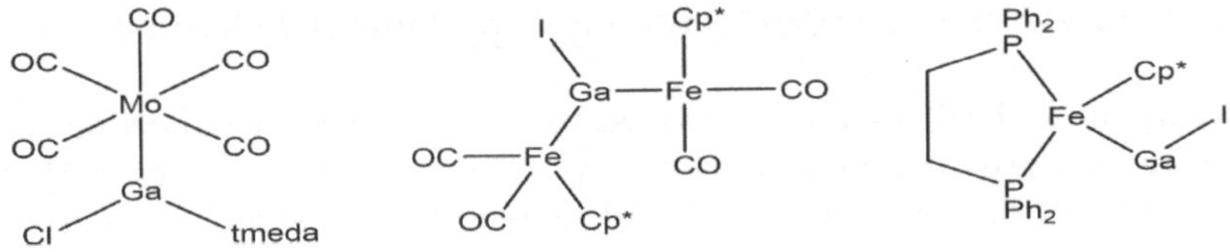

1998 (Fischer) 2005 (Aldridge) 2008 (Aldridge)

Figure 3.
Development of gallium monohalides based complexes.

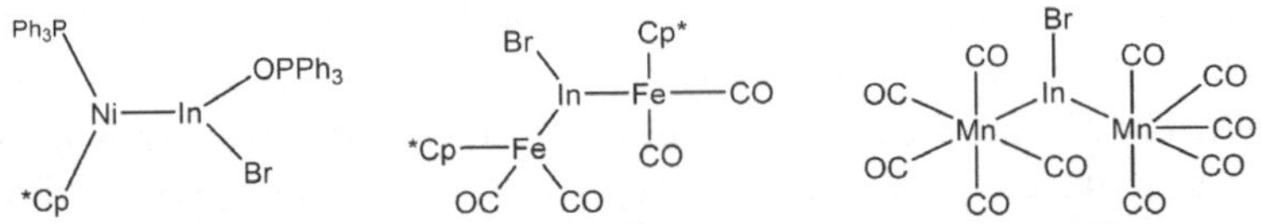

1999 (Fischer) 2005 (Aldridge) 2008 (Braunschweig)

Figure 4.
Development of indium monohalides based complexes.

interaction ratios obtained for Fe–GaX bonds are analogous to those calculated for Fe–CO and Fe–N_2, the impact of sigma donation to the covalent bonding contribution is substantially larger for GaX than for Fe–CO and Fe–N_2 bonds [115]. Although a variety of systems including these lengths EX bridging between two metal centres have previously been characterised, complexes containing the heavier EX as terminal ligands have just previously been reported experimentally [85–87]. Dimeric complexes of the kind $[(L_nM)_2E(\mu\text{-}X)_2E(ML_n)_2]$ or polymeric structure with sterically less hindered transition metal fragments [84, 121–125] are examples. The Lewis base coordinated at the group 13 centre can be used to segregate mononuclear systems in such circumstances [121, 126–128]. Pandey et al. (2010) investigated the bonding nature of group 8 and 10 metal complexes with dihalogallyl ligands. As a result, understanding the bonding behaviour of M-Ga in gallyl complexes is significant. The binding behaviour of the M-E bond in charge-neutral compounds coupled with typical metal carbonyls such as chromium hexacarbonyl, iron pentacarbonyl and nickel tetracarbonyl was systematically investigated to gain a greater understanding [69, 72, 73, 129–131].

Many researchers are still interested in synthesising metal coordination with In ligands, which has been recognised in recent decades [132–134] (**Figure 4**). The cationic derivative $[Cp^*Fe\,(GaCl)(phen)\,(CO)_2]^+$ was produced as the $[BPh_4]^-$ salt by reacting $[Cp^*Fe(CO)_2(GaCl_2)]$ with $Na[BPh_4]$ in the presence of 1,10-phenanthroline [135]. The chemistry of halide abstraction technic was initially utilised to synthesise the trimetallic indium and gallium cations in 2004 and was later extended to synthesise the iodogallylene complex $[Cp^*Fe(GaI)(dppe)]^+\,[ArBF_4]^-$ by using a more electronegative and sterically hindered bis(phosphine)iron fragment [85–87, 136, 137]. Gallium has a low coordination number, which is compatible with crystallographic parameters. For the similar bivalent ligand system in $[(OC)_4Fe(Ga\text{-}C_6H_5)(GaI)]$, the bond distance of Fe–Ga, Ga–I and Fe–Ga were 2.222, 2.444, 2.225 Å respectively and the bond angle of Fe–Ga–I was 171.4^0 [85, 115]. The weaker orbital contribution is thought to reflect the more diffuse nature of the 4s/4p orbitals derived from gallium, as well as less effective interactions with the fragment orbitals of $[CpFe(dmpe)]^+$, despite the higher energy of the HOMO for GaI (−6.08 eV vs. -9.03 eV for CO) and greater localization of the LUMO at the donor atom. The total metal–ligand bond strengths [ΔEint = −103 (GaI), −285 (BF), −213 (CO), and – 120 $kJmol^{-1}$ (N2)] after adding CO to form $[Cp^*Fe(dppe)(CO)]^+[ArBF_4]^-$ reveal very poor binding of the

GaI ligand [52]. The Rh–Ga distance [2.471 Å] becomes noticeably shorter [2.334 Å] when pyridine is coordinated at the gallium atom, this is in marked contradiction to analogous borylene systems, and it appears to indicate that the gallylene ligand is a stronger σ -donor, as well as the system's relative lack of π-back bonding representation. The availability of low valent and highly reactive [In^IR] molecules allowed for the synthesis of many complexes. [In^IR] molecules often operate as two-electron donors for metal fragments in both terminal and bridging modes of coordination [138–140]. The reason that identical complexes are coupled via numerous bonds (L_nM-EX) reflects not only the significant structural and bonding challenges brought by such complexes but also the insufficiency of strong experimental verification of possible bonding models [115, 141]. Mays *et al.* established the synthesis of the [$(\eta^5$-$C_5H_5)_2Fe(CO)_2(InCl)$] complex via the addition of InCl into the Fe-Fe bond of [$Fe_2(CO)_4(\eta C_5H_5)_2$] [142]. In most common organic solvents, commercially available monovalent indium compounds disintegrate or become insoluble [4, 71]. In this context, many studies have proposed a protonolytic approach to monovalent indium sources to improve performance and stability. The synthesis of stable metal complexes, particularly with group-13 diyl ligands ER, involved a detailed study of the bonds between the compounds. It's unusual to find a perfect computational analysis that accurately describes the bonding of the M-ER_n [68]. The ability of tri-coordinate complexes of the type [$L_nM]_2(\mu$-EX) to oligomerize through E-X-E bridges are modified by larger additional ligands of electrostatic repulsion induced by the net charge of [$(OC)_5Cr_2(\mu$-EX)$]_2$ where E = In; X = F-I [84, 136, 137, 143]. For the $Cp^*Fe(CO)_2$ systems, synthetic pathways initiating in E^I or E^{III} precursors are feasible (using addition or salt elimination techniques), with monomeric complexes generated that differ from the oligo/polymeric structures of similar [$Cp^*Fe(CO)_2$] complexes [84, 123, 144, 145]. Although the reaction of [$Cp^*Fe(CO)_2]_2$ and InI have clear mechanistic similarities to classical oxidative insertion reactions, the assignment of oxidation states in the product is rather arbitrary because iron and indium have similar electronegativities (1.83 and 1.78 on the Pauling scale, respectively).

6. Conclusions

Although monohalides of group 13 atoms have a rich history, their applications in the synthesis of limited alkyl, aryl, amide groups and related compounds have developed significantly in recent years because of the contribution of Schnockel, Power, Fischer, Jones and Aldridge. Simple diatom capture EX was just recently completed. Furthermore, at extremely high temperatures, the source of EX (E = B, Al, Ga) is widely thought to be a non-donor species (different from the equivalent electron ligand of CO or N_2), which will appear in (or close to) Ligand: general electron temperature asymmetry or aggregation problem. The complex's most recent structural characteristics, such as terminal BF and GaI bridging fragments (for example, [$\{CpRu(CO)_2\}_2(\mu BF)$] [47] and [$Cp^*Fe(dppe)(GaI)]^+[ArBF_4]^-$. The BF ligand attaches to two metal atoms in the form of a μ2CO ligand (that is, the ligand is in a singlet form) [146, 147], or the BF fragment is produced from conditional triplets if the boron centre is an effective triangle. Of course, the last case is comparable to the combination of CO units in ketones and is similar to Stalke and Braunschweig's [$CpMn(CO)_2(\mu BtBu)$] system [148]. [$CpRu(CO)_{22}(\mu BF)$] [47]. Because of the triple BF fragment and the interaction between two fragments [$CpRu(CO)_2$], the final description may have some reality in a better form. The BF singlet-triplet gap (with a singlet ground state) is determined to be around 86 kcal mol^{-1} [149], although the M-B bond value in the relevant system is around 6070 kcal mol^{-1}. The iron boron bond in [$CpFe(CO)_2(BF_2)$] complex is 66 kcal mol^{-1} [150], assuming that it is made up of three BF bonds and two $[CpRu(CO)_6]$. This corresponds to BF bond length 1.348 Å [47], which is significantly larger than the formal triple bond in BF diatoms (1.263 Å) [35], but slightly shorter than the Valence Radius co-beam (1.46 Å) [151]. The GaI distance in the terminal cationic iodoaromatic compound is 2.444 Å, which is less than the parent diatom's (2.575 Å) distance. Even with gallium's low coordination number, the Fe-Ga bond length is found to be quite short. This is owing to the strong M-Ga π-orbital interaction, rather than the s-orbital donation of a considerable amount of gallium sorbate to the M-Ga [53] bonding orbital. The orbital interaction between the model $[CpFe(PMe_3)_2]^+$ and [GaI] fragments is dominated by ΔE, indicating that GaI ligands are mostly utilised as donors. Quantum chemistry investigations of comparable neutral charge complexes containing GaI ligands, particularly the $[(CO)_4Fe(GaI)]$ axis, found significant electronic/geometric similarities to the cationic system, including the gap between Ga and GaI. The modest overall interaction energy, the short distance, and the contribution ratio of all interaction energies with similar electrostatic and covalent interactions (about 1:1). (Approximately 20%). These results suggest that GaI ligands interact similarly in each of these systems [53, 116]. Given the long history of capturing coordination chemistry and subsequent spectroscopic/structural interrogation of species with highly unstable kinetics, as well as the recent isolation of complexes containing terminally bound CF, GaI, BO^- and even Ga^+ ligands [85–87, 152–158], it appears that more progress in this field will be made soon. Because it offers precise experimental comparisons of electronic structures with CO and N_2 textbook systems, the BF terminal junction complex would be an appealing target. The key to this experiment is to establish a new preparation-scale procedure to make the most of the kinetic depressant chemical, despite quantum chemistry studies showing it to be thermodynamically stable. Comparing experimental results with theory will help to resolve the current controversy over the possible ways of binding in these molecules.

References

[1] A. J. Downs, Blackie , Glasgow, 1993

[2] S. Aldridge and A. J. Downs, Wiley, Chichester, 2011.

[3] D. W. Stephan and G. Erker, Angew. Chem., Int. Ed., 2010, 49, 56-76.

[4] J. A. J. Pardoe and A. J. Downs, Chem. Rev., 2007, 107, 2-45.

[5] C. Dohmeier, D. Loos and H. Schnockel, Angew. Chem., Int. Ed. Engl., 1996, 35, 129-149.

[6] A. Schnepf and H. Schnockel, Angew. Chem., Int. Ed., 2002, 41, 3532-3554.

[7] K. Koch, R. Burgert and H. Schnockel, Angew. Chem., Int. Ed., 2007, 46, 5795-5798.

[8] C. Jones, Coord. Chem. Rev., 2010, 254, 1273-1289

[9] H. Braunschweig, R. Dewhurst and A. Schneider, Chem. Rev., 2010, 110, 3924-3957

[10] D. Vidovic, G. A. Pierce and S. Aldridge, Chem. Commun., 2009, 1157-1171

[11] E. Rivard and P. P. Power, Inorg. Chem., 2007, 46, 10047-10064

[12] R. J. Baker and C. Jones, Coord. Chem. Rev., 2005, 249, 1857-1869

[13] C. Gemel, T. Steinke, M. Cokoja, A. Kempter and R. A. Fischer, Eur. J. Inorg. Chem., 2004, 4161-4176

[14] R. A. Fischer and J. Weiss, Angew. Chem., Int. Ed., 1999, 38, 2831-2850.

[15] F. M. Bickelhaupt, U. Radius, A. W. Ehlers, R. Hoffmann and E. J. Baerends, New J. Chem., 1998, 22, 1-3

[16] U. Radius, F. M. Bickelhaupt, A. W. Ehlers, N. Goldberg and R. Hoffmann, Inorg. Chem., 1998, 37, 1080-1090

[17] A. W. Ehlers, E. J. Baerends, F. M. Bickelhaupt and U. Radius, Chem. Eur. J., 1998, 4, 210-221.

[18] F. Paularokiadoss, A. Sekar, T. C. Jeyakumar, Comput. Theor. Chem., 2020, 1177, 112750-112758

[19] Chem. Pap., 2021, 75, 311-324

[20] Comput. Theor. Chem., 2021, 1197, 113139-113151

[21] Theor. Chem. Acc., 2021, 140, 101-114.

[22] R. L. Altman, J. Chem. Phys., 1959, 31, 1035-1038.

[23] R. F. Barrow, Trans. Faraday Soc., 1960, 56, 952-958.

[24] D. W. Robinson, J. Mol. Spectrosc., 1963, 11, 275-300.

[25] R. K. Nesbet, J. Chem. Phys., 1964, 40, 3619-3633.

[26] J. Blauer, M. A. Greenbaum and M. Farber, J. Phys. Chem., 1964, 68, 2332-2334.

[27] W. M. Huo, J. Chem. Phys., 1965, 43, 624-647.

[28] D. L. Hildenbrand and E. Murad, J. Chem. Phys., 1965, 43, 1400-1403.

[29] R. B. Caton and A. E. Douglas, Can. J. Phys., 1970, 48, 432-452.

[30] J. Singh, K. P. R. Nair and D. K. Rai, J. Mol. Struct., 1970, 6, 328-332.

[31] F. J. Lovas and D. R. Johnson, J. Chem. Phys., 1971, 55, 41-44.

[32] V. S. Kushawaha, B. P. Asthana and C. M. Pathak, Spectrosc. Lett., 1972, 5, 357-360.

[33] J. M. Dyke, C. Kirby and A. Morris, J. Chem. Soc., Faraday Trans. 2, 1983, 79, 483-490.

[34] H. Bredohl, I. Dubois, F. Melen and M. Vervloet, J. Mol. Spectrosc., 1988, 129, 145-150.

[35] G. Cazzoli, L. Cludi, C. Degli Esposti and L. Dore, J. Mol. Spectrosc., 1989, 134, 159-167.

[36] K. Q. Zhang, B. Guo, V. Braun, M. Dulick and P. F. Bernath, J. Mol. Spectrosc., 1995, 170, 82-93.

[37] P. L. Timms, J. Am. Chem. Soc., 1967, 89, 1629-1623.

[38] P. L. Timms, J. Am. Chem. Soc., 1968, 90, 4585-4589.

[39] P. L. Timms, Acc. Chem. Res., 1973, 6, 118-123.

[40] J. A. J. Pardoe, N. C. Norman, P. L. Timms, S. Parsons, I. Mackie, C. R. Pulham and D. W. H. Rankin, Angew. Chem., Int. Ed., 2003, 42, 571-573.

[41] P. L. Timms, N. C. Norman, J. A. J. Pardoe, I. D. Mackie, S. L. Hinchley, S. Parsons and D. W. H. Rankin, Dalton Trans., 2005, 607-616.

[42] J. A. J. Pardoe, N. C. Norman and P. L. Timms, Polyhedron, 2002, 21, 543-548.

[43] P. Hassanzadeh and L. Andrews, J. Phys. Chem., 1993, 97, 4910-4915.

[44] J. C. Jeffery, N. C. Norman, J. A. J. Pardoe and P. L. Timms, Chem. Commun., 2000, 2367-2368.

[45] L. Xu, Q. Li, Y. Xie, R. B. King and H. F. Schaefer III, Inorg. Chem., 2010, 49, 2996-3001.

[46] L. Xu, Q. Li, Y. Xie, R. B. King and H. F. Schaefer III, Inorg. Chem., 2010, 49, 1046-1055.

[47] D. Vidovic and S. Aldridge, Angew. Chem., Int. Ed., 2009, 48, 3669-3672.

[48] P. Bissinger, H. Braunschweig and F. Seeler, Organometallics, 2007, 26, 4700-4701.

[49] A. H. Cowley, V. Lomel_1 and A. Voigt, J. Am. Chem. Soc., 1998, 120, 6401-6402;

[50] H. Braunschweig, C. Kollann and U. Englert, Angew. Chem., Int. Ed., 1998, 37, 3179-3180.

[51] H. Braunschweig, K. Radacki, F. Seeler and G. R. Whittell, Organometallics, 2006, 25, 4605-4610.

[52] H. Braunschweig, M. Colling, C. Hu and K. Radacki, Angew. Chem., Int. Ed., 2002, 41, 1359-1361.

[53] K. K. Pandey and S. Aldridge, Inorg. Chem., 2011, 50, 1798-1807.

[54] D. L. Coombs, S. Aldridge, C. Jones and D. J. Willock, J. Am. Chem. Soc., 2003, 125, 6356-6357

[55] D. Vidovic, M. Findlater, G. Reeske and A. H. Cowley, Chem. Commun., 2006, 3786-3787

[56] D. A. Addy, G. A. Pierce, D. Vidovic, D. Mallick, E. D. Jemmis, J. M. Goicoechea and S. Aldridge, J. Am. Chem. Soc., 2010, 132, 4586-4588.

[57] H. Braunschweig, M. Colling, C. Kollann, H. G. Stammler and B. Neumann, Angew. Chem., Int. Ed., 2001, 40, 2298-2300.

[58] G. Alcaraz, U. Helmstedt, E. Clot, L. Vendier and S. Sabo-Etienne, J. Am. Chem. Soc., 2008, 130, 12878-12879.

[59] A. Kerr, T. B. Marder, N. C. Norman, A. G. Orpen, M. J. Quayle, C. R. Rice, P. L. Timms and G. R. Whittell, Chem. Commun., 1998, 319-320

[60] N. Lu, N. C. Norman, A. G. Orpen, M. J. Quayle, P. L. Timms and G. R. Whittell, J. Chem. Soc., Dalton Trans., 2000, 4032-4037.

[61] Z. Lu, C.-H. Jun, S. R. de Gala, M. Sigalas, O. Eisenstein and R. H. Crabtree, J. Chem. Soc., Chem. Commun., 1993, 1877-1880

[62] Z. Lu, C.-H. Jun, S. R. de Gala, M. P. Sigalas, O. Eisenstein and R. H. Crabtree, Organometallics, 1995, 14, 1168-1175

[63] H. Braunschweig, M. Colling, C. Kollann and U. Englert, J. Chem. Soc., Dalton Trans., 2002, 2289-2296

[64] D. L. Coombs, S. Aldridge, A. Rossin, C. Jones and D. J. Willock, Organometallics, 2004, 23, 2911-2926.

[65] H. Braunschweig, K. Kraft, T. Kupfer, K. Radacki and F. Seeler, Angew. Chem., Int. Ed., 2008, 47, 4931-4933.

[66] T. Ziegler, J. Autschbach, Chem. Rev., 2005, 105, 2695-2722.

[67] C.L.B. Macdonald, A.H. Cowley, J. Am. Chem. Soc., 1999, 121, 12113-12126.

[68] C. Boehme, G. Frenking, Chem. Eur. J., 1999, 5, 2184-2190.

[69] J. Uddin, G. Frenking, J. Am. Chem. Soc., 2001, 123, 1683-1693.

[70] C. Boehme, J. Uddin, G. Frenking, Coord. Chem. Rev., 2000, 197, 249-276.

[71] G. Frenking, N. Fröhlich, Chem. Rev., 2000, 100, 717-774.

[72] J. Uddin, C. Boehme, G. Frenking, Organometallics., 2000, 19, 571-582.

[73] S. Aldridge, A. Rossin, D.L. Coombs, D.J. Willock, Dalton Trans., 2004, 1, 2649-2654.

[74] W. Klemm and E. Voss, Z. Anorg. Allg. Chem., 1943, 251, 233-240.

[75] K. P. Huber and G. Herzberg, van Nostrand Reinhold, New York, 1979

[76] H. G. Hedderich, M. Dulick and P. F. Bernath, J. Chem. Phys., 1993, 99, 8363-8370.

[77] R. Ahlrichs, L. Zhenyan and H. Schnockel, Z. Anorg. Allg. Chem., 1984, 519, 155-164

[78] H.-J. Himmel, Eur. J. Inorg. Chem., 2005, 1886-1894.

[79] M. Tacke and H. Schnockel, Inorg. Chem., 1989, 28, 2895-2896.

[80] C. U. Doriat, M. Friesen, E. Baum, A. Ecker and H. Schnockel, Angew. Chem., Int. Ed. Engl., 1997, 36, 1969-1971.

[81] M. L. H. Green, P. Mountford, G. J. Smout and S. R. Speel, Polyhedron, 1990, 9, 2763-2765.

[82] S. Coban, Diplomarbeit, Universitat Karlsruhe, 1999.

[83] R. J. Baker and C. Jones, Dalton Trans., 2005, 1341-1348.

[84] N. R. Bunn, S. Aldridge, D. L. Kays, N. D. Coombs, J. K. Day, L.-L. Ooi, S. J. Coles and M. B. Hursthouse, Organometallics, 2005, 24, 5879-5890.

[85] N. D. Coombs, W. Clegg, A. L. Thompson, D. J. Willock and S. Aldridge, J. Am. Chem. Soc., 2008, 130, 5449-5451.

[86] N. D. Coombs, D. Vidovic, J. K. Day, A. L. Thompson, D. D. Le Pevelen, A. Stasch, W. Clegg, L. Russo, L. Male, M. B. Hursthouse, D. J. Willock and S. Aldridge, J. Am. Chem. Soc., 2008, 130, 16111-16124.

[87] H.-J. Himmel and G. Linti, Angew. Chem., Int. Ed., 2008, 47, 6326-6328.

[88] H. Schmidbaur, Angew. Chem., Int. Ed. Engl., 1985, 24, 893-904

[89] H. Schmidbaur and A. Schier, Organometallics, 2008, 27, 2361-2395.

[90] D. G. Tuck, Chem. Soc. Rev., 1993, 22, 269-276.

[91] H.-J. Himmel, Dalton Trans., 2003, 3639-3649.

[92] H.-J. Himmel, A. J. Downs and T. M. Greene, J. Am. Chem. Soc., 2000, 122, 922-930.

[93] H.-J. Himmel, J. Chem. Soc., Dalton Trans., 2002, 2678-2682.

[94] C. Peppe, D. G. Tuck and L. Victoriano, J. Chem. Soc., Dalton Trans., 1982, 2165-2168.

[95] S. P. Green, C. Jones and A. Stasch, Chem. Commun., 2008, 6285-6287.

[96] M. Mocker, C. Robl and H. Schnockel, Angew. Chem., Int. Ed. Engl., 1994, 33, 1754-1755

[97] A. Ecker and H. Schnockel, Z. Anorg. Allg. Chem., 1996, 622, 149-152

[98] A. Ecker and H. Schnockel, Z. Anorg. Allg. Chem., 1998, 624, 813-816

[99] A. Ecker, R. Koppe, C. Uffing and H. Schnockel, Z. Anorg. Allg. Chem., 1998, 624, 817-822.

[100] L. L. Liu, J. Zhou, L. L. Cao, D. W. Stephan, J. Am. Chem. Soc., 2019, 141, 16971-16982.

[101] C. Weetman, P. Bag, T. Szilvási, C. Jandl, S. Inoue, Angew. Chem., 2019, 58, 10961-10965.

[102] A. Hofmann, C. Pranckevicius, T. Tröster, H. Braunschweig, Angew. Chem., 2019, 58, 3625-3629.

[103] T. Chu, I. Korobkov, G. I. Nikonov, J. Am. Chem. Soc., 2014, 136, 9195-9202.

[104] R. A. Fischer, A. Miehr, Chem Mater., 1996, 8, 497-508.

[105] R. A. Fischer, A. Miehr, M. M. Schulte, Adv. Mater., 1995, 7, 58-61.

[106] R. A. Fischer, A. Miehr, M. M. Schulte, E. Herdtweck, J. Chem. Soc. Chem. Commun., 1995, 300, 337-338.

[107] R. A. Fischer, A. Miehr, T. Priermeier, Chem. Ber., 1995, 128, 831-843.

[108] R. A. Fischer, T. Priermeier, Organometallics, 1994, 13, 4306-4314.

[109] M. M. Schulte, E. Herdtweck, G. Raudaschl-Sieber, R. A. Fischer, Angew. Chem. Int. Ed., 1996, 35, 424-426.

[110] A. W. Ehlers, S. Dapprich, S. F. Vyboishchikov, G. Frenking, Organometallics, 1996, 15, 105-117.

[111] T. Cadenbach, C. Gemel, D. Zacher, R.A. Fischer, Angew. Chem., Int. Ed., 2008, 47, 3438-3441.

[112] R. A. Fischer, M. M. Schulte, J. Weiss, L. Z. Zolnai, A. Jacobi, G. Huttner, G. Frenking, C. Boehme, S. F. Vyboishchikov, J. Am. Chem. Soc., 1998, 120, 1237-1248.

[113] R. A. Fischer, J. Weiss, Angew. Chem., Int. Ed., 1999, 38, 2830-2850.

[114] B. Wrackmeyer, Angew. Chem., Int. Ed., 1999, 38, 771-772.

[115] J. Su, X.-W. Li, R. C. Crittendon, C. F. Campana and G. H. Robinson, Organometallics, 1997, 16, 4511-4513.

[116] J. A. Gamez, R. Tonner and G. Frenking, Organometallics, 2010, 29, 5676-5680.

[117] B. Buchin, C. Gemel, A. Kempter, T. Cadenbach, R.A. Fischer, Inorg. Chim. Acta., 2006, 359, 4833-4839.

[118] H. Braunschweig, K. Gruss, K. Radacki, Inorg. Chem., 2008, 47, 8595-8597.

[119] K. K. Pandey, P. Patidar, H. Braunschweig, Inorg. Chem., 2010, 49, 6994-7000.

[120] K. K. Pandey, P. Patidar, S. Aldridge, J. Phys. Chem. A, 2010, 114, 12099-12105.

[121] H.-J. Haupt, W. Wolfes and H. Preut, Inorg. Chem., 1976, 15, 2920-2927.

[122] H.-J. Haupt, H. Preut and W. Wolfes, Z. Anorg. Allg. Chem., 1979, 448, 93-99.

[123] L. M. Clarkson, N. C. Norman and L. J. Farrugia, Organometallics, 1991, 10, 1286-1292.

[124] J. C. Calabrese, L. M. Clarkson, T. B. Marder, N. C. Norman and N. J. Taylor, J. Chem. Soc., Dalton Trans., 1992, 3525-3529.

[125] J. J. Schneider, U. Denninger, J. Hagen, C. Kruger, D. Blaser and R. Boese, Chem. Ber., 1997, 130, 1433-1440.

[126] L. M. Clarkson, W. Clegg, D. C. R. Hockless, N. C. Norman, L. J. Farrugia, S. G. Bott and J. L. Atwood, J. Chem. Soc., Dalton Trans., 1991, 2241-2252.

[127] G. Linti, G. Li and H. Pritzkow, J. Organomet. Chem., 2001, 626, 82-91.

[128] H. Nakazawa, M. Itazaki and M. Owaribe, Acta Crystallogr., Sect. E: Struct. Rep. Online, 2005, 61, 945-946.

[129] G. Frenking, K. Wichmann, N. Fröhlich, C. Loschen, M. Lein, J. Frunzke, V.M. Rayon, Coord. Chem. Rev., 2003, 238, 55-82.

[130] K.K. Pandey, A. Lledos, F. Maseras, Organometallics, 2009, 28, 6442-6449.

[131] K.K. Pandey, D.J. Musaev, Organometallics, 2010, 29, 142-148.

[132] S. Reaktionen, C. Dohmeier, D. Loos, B. Ende, Angew. Chem., 1996, 108, 141-161.

[133] K.H. Whitmire, J. Coord. Chem., 1988, 17, 95-203.

[134] L.M. Clarkson, W. Clegg, N.C. Norman, Inorg. Chem., 1998, 27, 2653-2660.

[135] K. Ueno, T. Watanabe and H. Ogino, Appl. Organomet. Chem., 2003, 17, 403.

[136] N. R. Bunn, S. Aldridge, D. L. Coombs, A. Rossin, D. J. Willock, C. Jones and L.-L. Ooi, Chem. Commun., 2004, 1732-1733.

[137] N. R. Bunn, S. Aldridge, D. L. Kays, N. D. Coombs, A. Rossin, D. J. Willock, J. K. Day, C. Jones and L.-L. Ooi, Organometallics, 2005, 24, 5891-5900.

[138] P. Jutzi, B. Neumann, G. Reumann, Organometallics, 1999, 18, 2550-2552.

[139] W. Uhl, Z. Anorg. Allg. Chem., 2000, 626, 2043-2045.

[140] T. Steinke, C. Gemel, M. Winter, R. A. Fischer, Angew. Chem. Int. Ed., 2002, 41, 4761-4763.

[141] F. A. Cotton, X. Feng, Organometallics, 1998, 17, 128-130.

[142] A. T. T. Hsieh, M. Mays, J. Inorg. Nucl. Chem. Lett., 1971, 7, 223-225.

[143] O. J. Curnow, B. Schiemenz, G. Huttner and L. Zsolnai, J. Organomet. Chem., 1993, 459, 17-20.

[144] S. Borovik, S. G. Bott and A. R. Barron, Organometallics, 1999, 18, 2668-2676.

[145] N. D. Coombs, J. K. Day, S. Aldridge, S. J. Coles and M. B. Hursthouse, Main Group Met. Chem., 2007, 30, 195-198.

[146] E. D. Jemmis, A. R. Pinhas and R. Hoffmann, J. Am. Chem. Soc., 1980, 102, 2576-2585.

[147] J. F. Hartwig, Organotransition Metal Chemistry, Sausalito, California, USA, 2010, 29-30.

[148] U. Flierler, M. Burzler, D. Leusser, J. Henn, H. Ott, H. Braunschweig and D. Stalke, Angew. Chem., Int. Ed., 2008, 47, 4321-4325.

[149] Rozas, I. Alkorta and J. Elguero, J. Phys. Chem. A, 1999, 103, 8861-8869.

[150] Dickinson, D. J. Willock, R. J. Calder and S. Aldridge, Organometallics, 2002, 21, 1146-1157.

[151] J. Emsley, The Elements, OUP, Oxford, 1998.

[152] H. Braunschweig, K. Radacki and A. Schneider, Science, 2010, 328, 345-347

[153] S. Westcott, Angew. Chem., Int. Ed., 2010, 49, 9045-9046.

[154] H. Braunschweig, K. Radacki and A. Schneider, Angew. Chem., Int. Ed, 2010, 49, 5993-5996.

[155] H. Braunschweig, K. Radacki and A. Schneider, Chem. Commun., 2010, 46, 6473-6475.

[156] H. Huang, R. P. Hughes, C. R. Landis and A. L. Rheingold, J. Am. Chem. Soc., 2006, 128, 7454-7455.

[157] B. Buchin, C. Gemel, T. Cadenbach, R. Schmid and R. A. Fischer, Angew. Chem., Int. Ed., 2006, 45, 1674-1674;

[158] S. Aldridge, Angew. Chem., Int. Ed., 2006, 45, 8097-8099.

Chapter 15

Computational Study on Optoelectronic Properties of Donor-Acceptor Type Small π-Conjugated Molecules for Organic Light-Emitting Diodes (OLEDs) and Nonlinear Optical (NLO) Applications

Abstract

Recently, donor-acceptor type molecule that contains electron-rich (D) and electron-deficient (A) moiety has emerged as an interesting approach of molecular design strategy to develop organic light-emitting diodes (OLEDs) and non-linear optical (NLO) devices. In this work, we report a theoretical investigation based on two donor-acceptor (D-A) type small π-conjugated molecules based on dithieno [3,2-b: 2′,3′-d] pyrrole (DTP) and anthracene derivatives. All of the theoretical calculations were performed by Density Functional Theory (DFT) approach at B3LYP/6-31 g(d) level of theory. The structural, electronic, optical and charge transfer properties were investigated. The effect of acceptor blocks (DPA and DTA) on the molecular characteristics was elucidated. The obtained results clearly show that the studied compounds exhibit non-coplanar structures with low electronic band gap values. These relevant structures exhibited important optical absorption and intense emission in the green-yellow region. NLO investigation based on static polarizability (α_0), first-order hyperpolarizability (β_0) and second-order hyperpolazabilty (γ_0) demonstrated that the studied materials exhibit excellent NLO properties. Thus, the designed materials showed promising capabilities to be utilized in OLED and NLO applications.

Keywords: DTP, Anthracene, DFT, Optical properties, OLEDs, NLO properties

1. Introduction

Recently, several categories of conjugated materials have gained significant interest due to their relevant optoelectronic characteristics for the development of a wide range of organic electronic devices [1–3]. Remarkably, organic luminescent materials are increasingly attractive in the manufacture of organic light-emitting diodes (OLEDs) and non-linear optical (NLO) devices [4–6].

OLEDs are desirable electronic devices due to their valuable advantages of full color emission, high brightness, flexibility and operation stability [7–11]. OLEDs were greatly developed in electronic fields and showed their successful applications in digital displays, mobile phones and flat panels in TV screens [12]. Various functional units, such as fluorene [13], carbazole [14] and anthracene [15] were utilized to develop efficient luminescent materials for their prominent photoluminescence properties and simple modifications. Among these functional units, anthracene and its derivatives were widely studied as potential building blocks in the development of active materials for OLED devices [16].

Apart from OLEDs, recent studies have demonstrated the potential application of conjugated materials in NLO devices [17, 18]. NLO materials have reached large interest of the scientific community for prosperous use in technological areas such as telecommunications, optical information processing and data storage [19–21]. NLO studies have shown the design of promising organic materials for nonlinear effect based on the introduction of highly delocalized electron fragments and additional electron donor and acceptor groups for enhancing the molecular conjugation [22].

Previous works have shown that D-A (Donor-Acceptor) systems dispose a successful architecture for non-linear optics and OLEDs. Where, D-A systems exhibit large charge transfer (CT) in which the electrons located in the electron rich donor unit undergo an intra-molecular charge transfer (ICT) to the electron deficient acceptor unit [23, 24]. The present CT phenomenon leads to excellent optoelectronic characteristics that encourage the use of D-A systems in NLO and OLED applications.

Dithieno [3,2-b: 2′,3′-d] pyrrole (DTP) material has been recognized as one of the most efficient building blocks with high electron donation capacity [25, 26]. DTP building blocks have been widely incorporated into a variety of materials for the aim of reducing the band gap, improving the mobility of charge carriers, and reinforcing solution and solid state fluorescence.

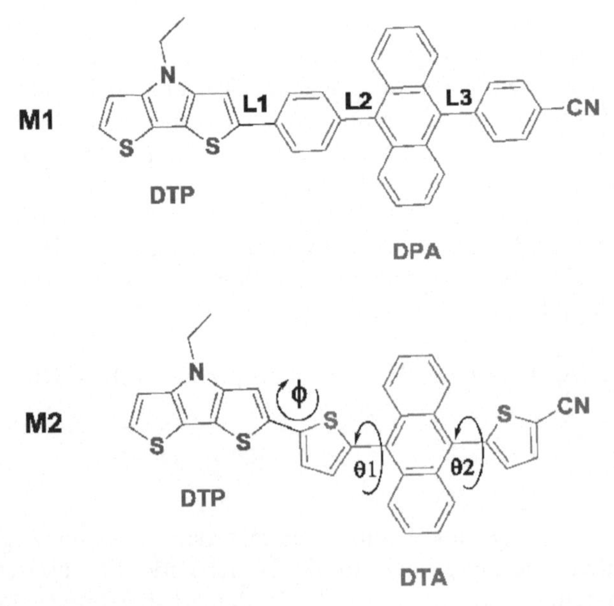

Figure 1.
Molecular structures of the investigated compounds.

Equally, anthracene has attracted significant attention to the construction of organic luminescent materials due to its unique features. Where, anthracene derivatives such as 9,10-diphenylanthracene (DPA) and 9,10-di(thiophen-2-yl)anthracene (DTA) have exhibited remarkable electronic and light-emitting properties to be applied in optical applications [27, 28].

In the present work, we have developed a theoretical investigation based on new D-A type small molecules for NLO and OLED applications. As mentioned in **Figure 1**, we have used DTP as an electron donor block and derivatives from DPA and DTA as an electron acceptor block. The addition of a strong electron withdrawing group as cyano group is for the reason of enhancing the polarization and improving the π-electrons delocalization [29].

A theoretical computational study using density functional theory (DFT) approach introduce excellent tools to predict the optoelectronic properties of molecular systems as well as the design of new materials for OLED and NLO devices [30]. Hence, the designed materials will be theoretically investigated and discussed to envisage the reliability for OLED and NLO applications.

2. Computational methods

Theoretical calculations were performed using Density Functional Theory (DFT) approach implemented in Gaussian 09 software [31]. Previous studies have shown that DFT//B3LYP/6-31 g(d) method gives better accuracy in investigating the photo-physical properties of materials based on DTP and anthracene [32, 33]. Structural properties such as dihedral angles, torsion angles and bridge bond lengths were firstly investigated based on the geometry optimization of **M1** and **M2** in their ground states. Vibrational calculations were carried out to confirm the stable conformers with no imaginary frequencies. Frontier molecular orbitals (FMOs) and electron density difference (EDD) contour plots were carried out to examine the electron delocalization within the conjugated frameworks. The absorption and emission spectra were simulated using Time Dependent DFT (TD-DFT) at B3LYP/6-31 g(d) level. Photoluminescence color coordinates of **M1** and **M2** were determined using International Commission on Illumination (usually abbreviated CIE for its French name, Commission internationale de l'éclairage) process. Bilayer OLED devices are designed based on the optoelectronic properties of the studied molecules. Hole and electron charge transfer properties (λ_h, λ_e) were carried out on neutral, anionic and cationic states. The NLO properties involving the electric dipole moment (μ), the polarizability (α), the first order hyperpolarizability (β) and the second order hyperpolarizability (γ) were investigated.

3. Results and discussion

3.1 Ground state optimized geometries

The development of high performance luminescent materials is principally based on the enhancement of π-electrons delocalization within the conjugated architectures. In fact, the geometric structure gives an idea about the delocalization of the π-electrons as well as the charge transfer (CT) within the conjugated structure [34].

Hence, the ground state geometries of the studied materials were optimized using DFT//B3LYP/6-31 g(d) method in gaseous phase. The optimized geometries of **M1** and **M2** are illustrated in **Figure 2** and the geometric parameters involving bond lengths, torsion angles and dihedral angles are listed in **Table 1**.

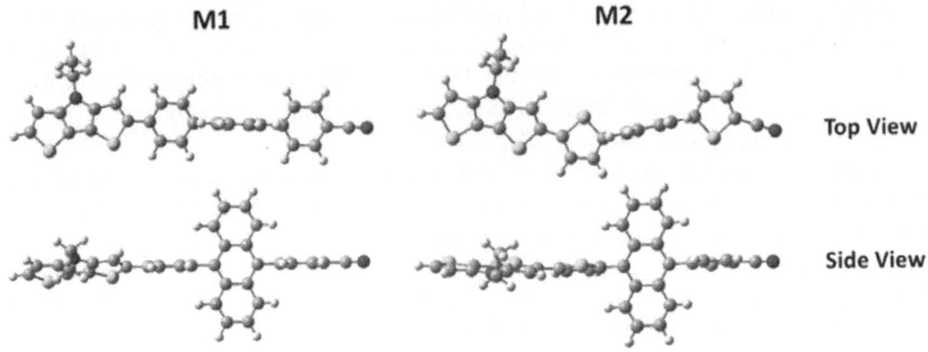

Figure 2.
Ground state optimized geometries at B3LYP/6-31 g(d) level for the studied molecules.

	M1	M2
Bond length (Å)		
L1	1.46	1.44
L2	1.49	1.48
L3	1.49	1.48
Torsion angle (deg.)		
Θ1	119.96	119.88
Θ2	119.83	119.68
Torsion angle (deg.)		
Φ	25.67	17.24

Table 1.
Optimized ground state geometry parameters of the studied molecules.

The considered materials exhibit non-coplanar structures (**Figure 2**). There are large torsion angles around 119° between anthracene and phenyl ring in **M1** and between anthracene and thiophene ring in **M2** (**Table 1**) which is a characteristic property of compounds based substituted anthracene [23, 35].

Furthermore, as mentioned in **Figure 1**, the characteristic bridge bonds between the donor and acceptor blocks (L1) and between the different groups of the acceptor units (L2 and L3) were calculated to have an idea about the charge transfer (CT) in the conjugated backbone. The calculated lengths of bond L1 were found equal to 1.46 and 1.44 Å for **M1** and **M2**, respectively. While, the bond lengths of L2 and L3 were found about 1.49 Å for **M1** and about 1.48 Å for **M2**. The calculated values are located in the interval between C-C single bond length (C-C = 1.54 Å) and C=C double-bond length (C=C = 1.33 Å) showing the high π-electron delocalization and interesting intra-molecular charge transfer (ICT) within the framework [36].

The structural analysis suggests that the studied compounds exhibit good conjugated structures with high π-electron delocalization, which is important for applications in organic electronic devices.

3.2 Frontier molecular orbitals (FMOs)

The electronic properties of the studied molecules are examined based on the frontier molecular orbital (FMOs) analysis. The highest occupied molecular orbitals

(HOMOs) and the lowest unoccupied molecular orbitals (LUMOs) contour plots were carried out using the DFT//B3LYP/6-31 g(d) method at the optimized ground state geometries. The energy values of HOMOs, LUMOs and band gap energies are listed in **Table 2**. The HOMOs and LUMOs distributions are illustrated in **Figure 3**.

As depicted in **Figure 3**, the HOMOs are mostly located over the DTP unit while the LUMOs are distributed over the DPA and DTA acceptor units that indicate the important electron charge transfer from the donor to acceptor moieties. Hence, the FMOs analysis has shown the considerable charge transfer taking place within the designed molecules.

The band gap energies are calculated from the difference between the HOMO and LUMO energy levels at DFT//B3LYP/6-31 g(d) method. The band gap energies are about 3.16 eV and 2.81 eV for **M1** and **M2**, respectively. These lower values of band gaps with the FMOs distributions demonstrate the presence of a significant intra-molecular charge transfer (ICT) that leads to enhance the electronic properties [37].

Density of states (DOS) is a helpful tool to examine the delocalization of π-electrons in the compound. The DOS plots were determined by DFT//B3LYP/6-31 g (d) method at the ground state geometry and illustrated in **Figure 4**. DOS plots of **M1** and **M2** show a large overlapping of electronic energy levels that indicates the high electron delocalization. In fact, the high electron delocalization within these materials could be explained by the mutual reactions of donor and acceptor constructive blocks.

To better understand the delocalization of π-electrons, electron density difference (EDD) was simulated between the ground state (S_0) and the first excited state (S_1). As seen in **Figure 5**, the EDD plots contain blue regions referring to electron density depletion and purple regions referring to electron density increment.

Hence, the regions of electron increment density are mainly located over the DPA and DTA blocks. While, the DTP block presents the region of depleted free carriers (depletion region). These observations decline the effective electron transfer from donor to acceptor units within the studied materials.

Compound	E_{HOMO-1} (eV)	E_{HOMO} (eV)	E_{LUMO} (eV)	E_{LUMO+1} (eV)	ΔE_{gap} (eV)
M1	−5.36	−5.04	−1.88	−1.41	3.16
M2	−5.57	−4.97	−2.16	−1.64	2.81

Table 2.
DFT//B3LYP/6-31 g(d) calculated electronic properties of ***M1*** *and* ***M2*** *molecules.*

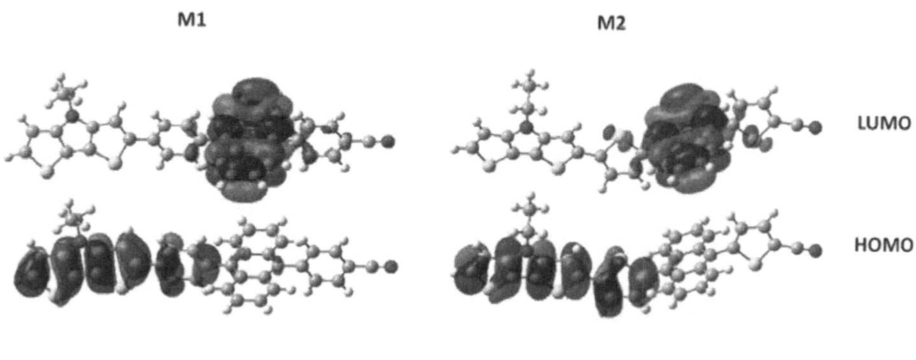

Figure 3.
Frontier molecular orbitals in the optimized ground state for the studied molecules.

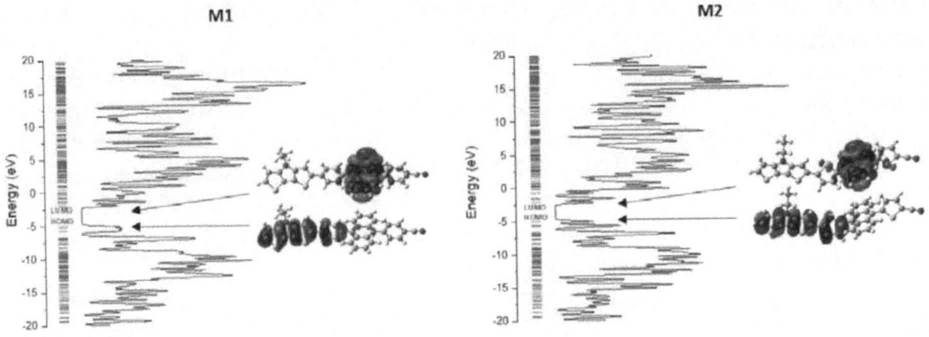

Figure 4.
Density of states (DOS) plots of the studied materials.

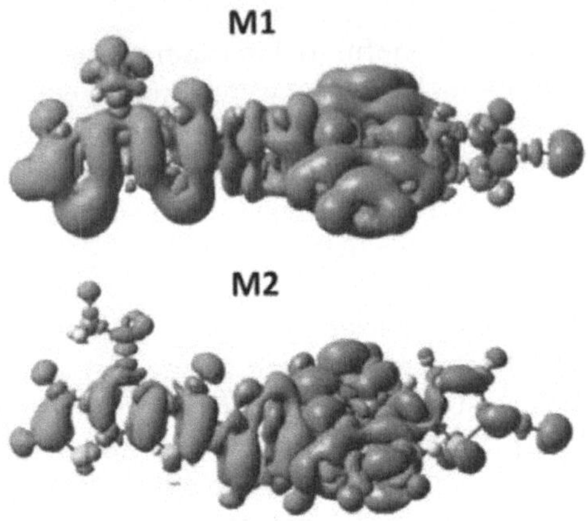

Figure 5.
Electron density difference (EDD) contour plots of the studied materials.

3.3 Optical properties

3.3.1 Optical absorption properties

The optical absorption properties of **M1** and **M2** have been investigated based on the simulated UV–Vis optical absorption spectra using TD-DFT method at 6-31 g(d) basis set. The obtained optical absorption curves with their Gaussian fitting peaks are given in **Figure 6**. The simulated corresponding maximum absorption wavelengths (λ_{max}^{abs}), the electronic transition energy (Eex), the oscillator strength (f), the full-width at half maximum (FWHM) and the main electronic transitions are listed in **Table 3**.

As it can be seen from **Figure 6**, the studied molecules exhibit large and intense absorption bands in the visible zone ranging from 300 nm to 500 nm. These electronic transitions are defined as $\pi \rightarrow \pi^*$ transitions associated with the electron migration from the HOMOs mainly located over the DTP units to the LUMOs mainly concentrated over the anthracene units [38].

From **Table 3**, the maximum absorption wavelengths (λ_{max}^{abs}) of **M1** and **M2** are found at 442 nm and 406 nm, respectively. The full-width at half maximum FWHM referring to the main absorption band for **M1** is found higher than that of **M2** (FWHM (**M1**) = 76.96 > FWHM (**M2**) = 58.70). This result shows the role played by the acceptor block in improving the optical absorption properties.

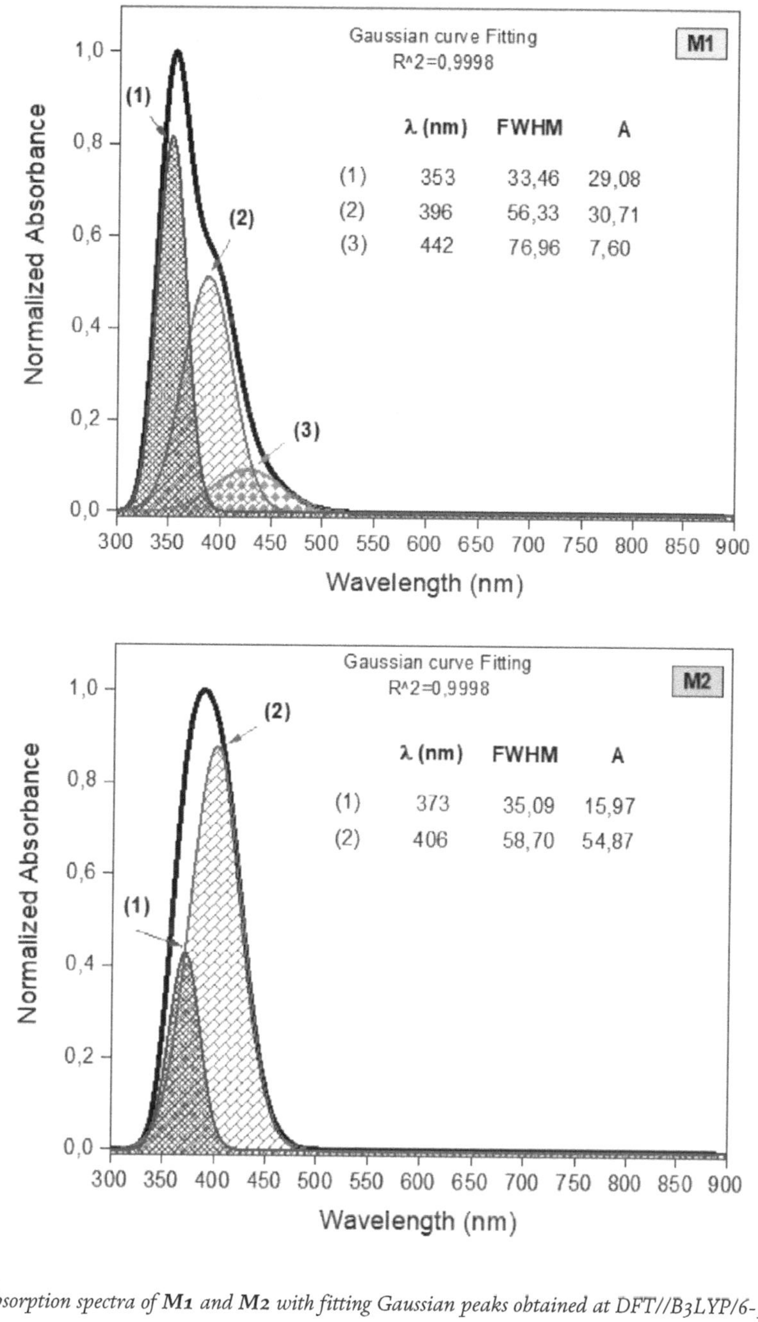

Figure 6.
Optical absorption spectra of **M1** *and* **M2** *with fitting Gaussian peaks obtained at DFT//B3LYP/6-31 g(d) level.*

From the optical absorption analysis, it is obvious that the acceptor block configuration contributes significantly in enhancing the absorption properties of conjugated materials.

3.3.2 Emission properties

The emission properties of **M1** and **M2** are simulated on their optimized geometries at the first excited state by means of TD-DFT//B3LYP/6-31 g(d) method. The

Compound	λ_{max}^{abs} (nm/eV)	f (a.u)	FWHM (nm)	Main electronic transitions
M1	442/2.80	0.05	76.96	H→L (99%)
	396/3.12	0.38	56.33	H-1→L (93%), H→L + 1 (4%)
	353/3.51	0.71	33.46	H→L + 2 (81%), H-2→L (9%)
M2	406/3.08	0.63	58.70	H-1→L (85%), H→L + 2 (12%)
	373/3.32	0.67	35.09	H→L + 2 (85%), H-1→L (12%)

Table 3.
Maximum absorption wavelengths λ_{max}^{abs} (nm), electronic transition energy, Eex (eV), oscillator strength f (a.u), full-width at half maximum FWHM (nm) and main electronic transitions calculated at B3LYP/6-31 g(d) level.

emission spectra with their fitting Gaussian peaks are depicted in **Figure** 7 and the corresponding emission characteristics are listed in **Table 4**.

As it can be seen from **Figure** 7, the studied materials exhibit large emission bands with maximum emission wavelengths of 478 nm and 554 nm for **M1** and **M2**, respectively. As mentioned in **Table 4**, the emission spectrum of **M2** exhibits larger FWHM comparing to **M2** which indicates the effect of acceptor moiety in enhancing the emission properties.

The photoluminescence chromaticity coordinates of **M1** and **M2** were carried out according to the CIE 1931 diagram. As illustrated in **Figure 8**, the CIE coordinates are found upon (x: 0.45, y: 0.45) and (x: 0.32, y: 0.43) for **M1** and **M2**, respectively. Hence, **M1** displays a pure green color while **M2** displays a pure yellow color.

From the emission investigation, it is revealed that the studied materials are promising materials for OLED applications. Indeed, OLEDs as they are promising organic electronic devices present the subject of intense research. The efficient OLED operation requires balanced charge injection, charge transport and charge recombination within the electronic device [39].

The considered materials exhibit appropriate optoelectronic characteristics allowing their use as emitting layers in two layers OLED device. Alq_3 (Tris (8-hydroxyquinoline) aluminum) is a suitable material to act as an electron transport layer [40] with respect to the LUMO energy levels of **M1** and **M2** (See **Figure 8**).

Radiative lifetime (τ) introduces the average time of the stay of a molecule at the excited state before a photon-emission. The lower value of τ leads to a relevant emission of the conjugated material. The radiative lifetimeτcould be determined according to the following expression [41]:

$$\tau = \frac{C^3}{2(E_{flu})^2 f} \tag{1}$$

Where, c, E_{flu} and f represent the velocity of light, the fluorescent energy and the oscillator strength, respectively.

The radiative lifetime values are of 10.12 ns and 5.46 ns for **M1** and **M2**, respectively. The small values of τ denote the efficient light emission of these materials. The findings of the present report corroborate a slight difference in radiative lifetime values which are explained by the effect of DPA and DTA acceptor blocks within the conjugated structures.

According to these results and compared to some previous studies reported in Refs. [28, 42], it is revealed that **M1** and **M2** exhibit promising optoelectronic properties for high performance OLED devices.

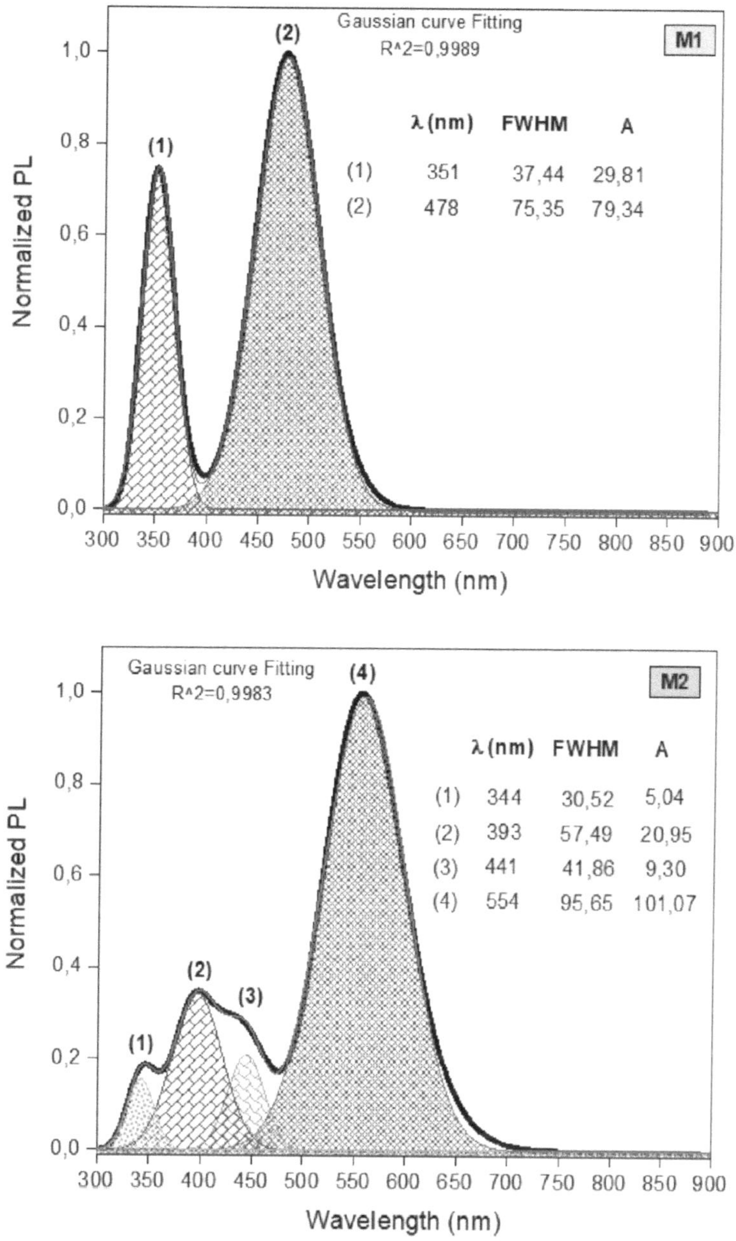

Figure 7.
*Emission spectra of **M1** and **M2** with fitting Gaussian peaks obtained at DFT//B3LYP/6-31 g(d) level.*

3.4 Charge transfer properties

Many factors are responsible for high performance organic optoelectronic devices such as hole/electron charge transfer balance. The reorganization energies of holes and electrons (λ_h and λ_e) together with the ionization potential (IP) and the electron affinity (EA) have been calculated to evaluate the hole/electron transfer abilities. The reorganization energies were carried out from neutral, cationic and anionic geometries, as detailed in **Figure 9**.

The reorganization energies of holes and electrons can be calculated following the expression above [43]:

Compound	λ_{max}^{em} (nm/eV)	f (a.u)	FWHM (nm)	Main transition	τ (ns)
M1	478/2.59	0.34	75.35	L→H (98%)	10.12
	351/3.53	0.27	37.44	L→H-2 (90%)	
M2	554/2.23	0.85	95.65	L→H (99%)	5.46
	441/2.81	0.21	41.86	L→H-1(63%)	
	393/3.15	0.18	57.49	L + 1→H (40%)	
	344/3.60	0.15	30.52	L + 1→H-1 (94%)	

Table 4.
Calculated maximum emission wavelengths λ_{max}^{em} (nm), fluorescent energy E_{flu}(eV), oscillator strength f (a.u), full-width at half maximum FWHM (nm), main electronic transitions, stokes shift (nm) and radiative lifetime (τ).

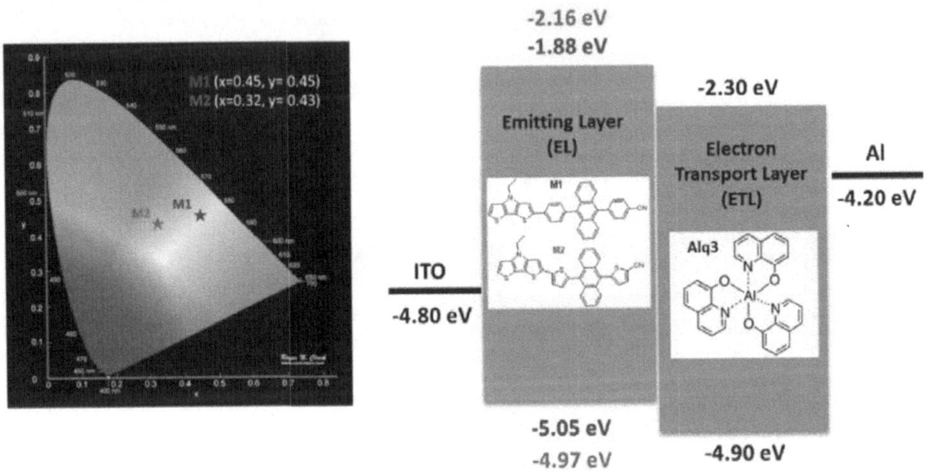

Figure 8.
CIE color coordinates for the studied materials (right), schematic structure of proposed bilayer OLED based studied materials (left).

$$\lambda_h/\lambda_e = \left(E_0^{\pm} - E_{\pm}^{\pm}\right) + \left(E_{\pm}^{0} - E_0^{0}\right) \quad (2)$$

Hence, $E_0^{\pm}$, $E_{\pm}^{0}$, $E_{\pm}^{\pm}$ and E_0^{0} represent the cation/anion energy of cation/anion at neutral geometry, the energy of neutral structure in the cation/anion state, the energy of cation/anion in the cation/anion state and the energy of the neutral structure, respectively. Charge transfer parameters have been determined using DFT//B3LYP/6-31 g(d) method and the results are presented in **Table 5**. Based on the obtained results, it is found that **M1** exhibits higher hole and electron transfer abilities regarding the lower reorganization energies of hole and electron.

To get insights into the charge transport properties, the ionization potential (IP) and the electron affinity (AE) were carried out for better evaluating the electron extraction and attraction abilities, respectively [44]. In comparison with **M2, M1** exhibits relatively lower EA (0.94 eV versus 1.17 eV) which demonstrates that our materials exhibit low performance of grasping electrons. However, the low values of IP (6.02 eV for **M1** and 6.00 eV for **M2**) indicated the high ability of these materials to grasp hole (See **Table 5**). It is possible to better improve the mobility of charge carriers through the modification of the conjugate structure [45].

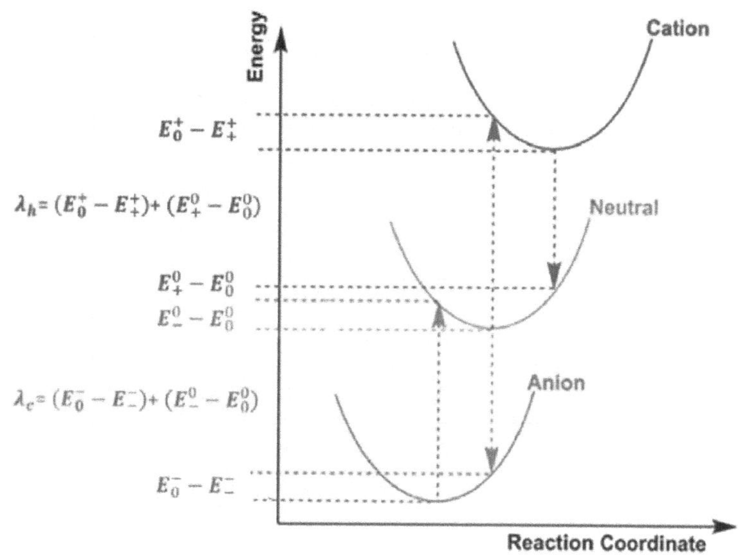

Figure 9.
Calculation details of reorganization energies from neutral cation and anion states.

Compound	IP	EA	$(E_0^+ - E_+^+)$	$(E_+^0 - E_0^0)$	$(E_0^- - E_-^-)$	$(E_-^0 - E_0^0)$	λ_h	λ_e
M1	6.02	0.94	0.171	0.129	0.283	0.144	0.300	0.427
M2	6.00	1.17	0.248	0.158	0.298	0.224	0.406	0.522

Table 5.
Calculated charge transfer parameters (expressed in eV) of ***M1*** *and* ***M2*** *at DFT//B3LYP/6-31 g(d) level of theory.*

Overall, the charge properties analysis of **M1** and **M2** shows that they are considered as promising materials for organic optoelectronic applications.

3.5 Nonlinear optical (NLO) properties

The principle of nonlinear optics represents the interaction between an incident electromagnetic field with a particular material leading to the generation of an electromagnetic field modified in wave number, phase or frequency [17]. NLO materials are increasingly applied in emerging technological fields such as telecommunications, optical memory, optical information processing, etc. [46].

NLO properties produced from the high delocalization of electrons within the molecule increase while increasing the molecular conjugation [47]. Further, the presence of electron donor blocks (D) and electron acceptor blocks (A) contributes to the improvement of the NLO properties [48–50]. Studies have shown that nonlinear organic optical materials possess higher optical nonlinearity compared to inorganic materials [51].

The electric dipole moment μ, the polarizabilities α, the first (β) and the second-order hyperpolarizability (γ) describe the nonlinear optical response of an isolated molecule within an electric field [52]. The total dipole moment μ_{tot}, the static polarizability α_0, the static first hyperpolarizability β_0 and the static second order hyperpolarizability $\gamma_0$are calculated using the expressions above [22, 53]:

	M1	M2		M1	M2		M1	M2		M1	M2
μ_x	0	0	α_{xx}	63.382	74.513	β_{xxx}	5.006	5.457	γ_{xxxx}	69.828	95.548
μ_y	0	0	α_{xy}	5.051	−6.979	β_{xxy}	1.042	−1.122	γ_{yyyy}	42.211	40.665
μ_z	7.383	7.362	α_{yy}	55.263	55.263	β_{xyy}	0.220	0.139	γ_{zzzz}	831.140	624.809
μ_{tot}	7.383	7.362	α_{xz}	−33.103	−36.709	β_{yyy}	0.056	−0.034	γ_{xxyy}	3.191	1.728
			α_{yz}	−5.787	5.383	β_{xxz}	−8.816	−7.756	γ_{xxzz}	247.105	241.440
			α_{zz}	109.672	97.457	β_{xyz}	−1.979	1.656	γ_{yyzz}	11.081	10.138
			α_0	76.106	75.744	β_{yyz}	−0.475	−0.347	γ_0	293.186	253.527
						β_{xzz}	18.993	14.347			
						β_{yzz}	4.340	−3.168			
						β_{zzz}	−36.233	−24.125			
						β_0	51.851	38.145			

Table 6.
Calculated electric dipole moment μ_{tot} (D), polarizability α_o ($\times 10^{-24}$ esu), first-order hyperpolarizability $\boldsymbol{\beta_o}$ ($\times 10^{-30}$ esu) and second-order hyperpolarizability γ_o ($\times 10^{-36}$ esu) of the studied materials at DFT//B3LYP/6-31 g(d) level.

$$\mu_{\text{tot}} = \left(\mu_x^{\,2} + \mu_y^{\,2} + \mu_z^{\,2}\right)^{\frac{1}{2}} \quad (3)$$

$$\alpha_0 = \frac{1}{3}\left(\alpha_{xx} + \alpha_{yy} + \alpha_{zz}\right) \quad (4)$$

$$\beta_0 = \left[\left(\beta_{xxx} + \beta_{xyy} + \beta_{xzz}\right)^2 + \left(\beta_{yyy} + \beta_{yzz} + \beta_{yxx}\right)^2 + \left(\beta_{zzz} + \beta_{zxx} + \beta_{zyy}\right)^2\right]^{\frac{1}{2}} \quad (5)$$

$$\gamma_0 = (1/5)\left[\gamma_{xxxx} + \gamma_{yyyy} + \gamma_{zzzz} + 2\,\gamma_{xxyy} + 2\,\gamma_{yyzz} + 2\,\gamma_{zzxx}\right] \quad (6)$$

DFT approach is used as a reliable method for the determination of NLO properties of organic materials [54]. To get insights into the NLO properties, theoretical calculations were performed on the ground state optimized geometries of **M1** and **M2** at DFT//B3LYP/6-31 g(d) level of theory and the results are listed in **Table 6**.

From **Table 6**, the first- and second-order hyperpolarizabilities of **M2** are found lower than those of **M1** explained by the distinct electron delocalization in the conjugated structures. Thus, as compared to **M2**, it is important to note that **M1** presents the best NLO properties.

Urea is a prototypical organic molecule used as a threshold comparison value in the study of the NLO properties of molecular materials [55]. The NLO parameters of urea calculated at DFT//B3LYP/6-31 g(d) level of theory are found of: μ_{tot} = 4.259 D, α_0 =3.749 × 10–24 esu and β_0 =0.557× 10–30 esu and γ_0 = 0.746× 10–36 esu.

Compared to urea, the high values of NLO parameters of **M1** and **M2** confirm the design of high performance nonlinear optical materials.

4. Conclusion

In this study, we reported a DFT study based on structural, optoelectronic and nonlinear optical (NLO) properties of D-A small π-conjugated molecules based on

DTP and anthracene. The optimized structures have shown the non-planarity of the investigated molecules **M1** and **M2** arising from the anthracene derivatives (DPA and DTA) conjugated configuration. FMOs analysis shows the appropriate HOMO/LUMO energy levels with the low band gap energies. To support the FMOs analysis, EDD contour plots have been computed to identify the donor and acceptor moiety within **M1** and **M2** structures. The TD-DFT study demonstrated the role played by the acceptor block in improving the absorption properties of the studied materials. The emission properties revealed an intense emission in the pure green and pure yellow for **M1** and **M2**, respectively. These molecules have shown their promising abilities to be used in OLED devices. The charge transfer properties analysis attested the relevant hole/electron transport abilities of these materials. Computed static polarizability (α_0), first-order hyperpolarizability (β_0) and second-order hyperpolazabilty (γ_0) indicated the excellent NLO properties of **M1** and **M2**. This NLO response suggested these compounds to be used as potential candidates for NLO applications. Overall, this study provided an insight into a promising D-A conjugated architecture with the role of the acceptor block on enhancing the optoelectronic performances of organic materials.

References

[1] A. Kivrak, H. Çalış, Y. Topal, H. Kivrak, M. Kuş, Synthesis of thiophenyl-substituted unsymmetrical anthracene derivatives and investigation of their electrochemical and electrooptical properties, Solar Energy Materials Solar Cells. 161 (2017) p. 31. https://doi.org/10.1016/j.solmat.2016.11.006

[2] S.R. Forrest, M.E. Thompson, Introduction: Organic Electronics and Optoelectronics, Chem. Rev. 107 (2007) p. 923. DOI: 10.1021/cr0501590

[3] R. Zaier, S. Ayachi, DFT molecular modeling studies of D-π-A-π-D type cyclopentadithiophene-diketopyrrolopyrrole based small molecules donor materials for organic photovoltaic cells, Optik. 239 (2021) p. 166787.https://doi.org/10.1016/j.ijleo.2021.166787

[4] U.R. Felscia, B.J. Rajkumar, M.B. Mary, Theoretical investigations on nonlinear fused 4-ring systems: Application to OLED and NLO devices, Synt. Met. 246 (2018) p. 31.https://doi.org/10.1016/j.synthmet.2018.09.008.

[5] S. Vijayalakshmi, S. Kalyanaraman, DFT and TD-DFT approach for the analysis of NLO and OLED applications of 9-anthraldehyde, Optik. 125 (2014) p. 2429. DOI: 10.1016/j.ijleo.2013.10.104

[6] J. George, D. Sajan, J. Alex, A. Aravind, G. Vinitha, R.J.O. Chitra, L. An experimental and computational approach to electronic and optical properties of Diglycine barium chloride monohydrate crystal: Applications to NLO and OLED, Technology, Opt. Laser Technol. 105 (2018) p. 207.https://doi.org/10.1016/j.optlastec.2018.02.056

[7] S. Yuvaraja, A. Nawaz, Q. Liu, D. Dubal, S.G. Surya, K.N. Salama, P. Sonar,Organic field-effect transistor-based flexible sensors, Chem. Soc. Rev., 49 (2020) p. 3423. https://doi.org/10.1039/C9CS00811J.

[8] L. Dou, Y. Liu, Z. Hong, G. Li, Y. Yang, Low-Bandgap Near-IR Conjugated Polymers/Molecules for Organic Electronics, Chem. Rev. 115 (2015) p. 12633. https://doi.org/10.1021/acs.chemrev.5b00165.

[9] J. Song, H. Lee, E.G. Jeong, K.C. Choi, S. Yoo, Organic Light-Emitting Diodes: Pushing Toward the Limits and Beyond, Adv. Mat. 32 (2020) p. 1907539. https://doi.org/10.1002/adma.201907539

[10] O. Ostroverkhova, Organic Optoelectronic Materials: Mechanisms and Applications, Chemical reviews. 116 (2016) p. 13279. https://doi.org/10.1021/acs.chemrev.6b00127

[11] X. Guo, M. Baumgarten, K. Müllen, Designing π-conjugated polymers for organic electronics, Prog. Poly. Sci. 38 (2013) p. 1832. https://doi.org/10.1016/j.progpolymsci.2013.09.005.

[12] B. Yang, S.K. Kim, H. Xu, Y.I. Park, H. Zhang, C. Gu, F. Shen, C. Wang, D. Liu, X. Liu, The Origin of the Improved Efficiency and Stability of Triphenylamine-Substituted Anthracene Derivatives for OLEDs: A Theoretical Investigation, ChemPhysChem. 9 (2008) p. 2601. https://doi.org/10.1002/cphc.200800513.

[13] S. Tao, Z. Peng, X. Zhang, P. Wang, C.S. Lee, S.T. Lee, Highly Efficient Non-Doped Blue Organic Light-Emitting Diodes Based on Fluorene Derivatives with High Thermal Stability, Adv. Funct. Mater. 15 (2005) p. 1716. https://doi.org/10.1002/adfm.200500067.

[14] C. Yin, D. Zhang, L. Duan, A perspective on blue TADF materials based on carbazole-benzonitrile derivatives for efficient and stable

OLEDs, Appl. Phys. Lett. 116 (2020) p. 120503. https://doi.org/10.1063/1.5143501.

[15] R. Malatong, C. Kaiyasuan, P. Nalaoh, S. Jungsuttiwong, T. Sudyoadsuk, V. Promarak, Rational design of anthracene-based deep-blue emissive materials for highly efficient deep-blue organic light-emitting diodes with CIEy ≤ 0.05, Deyes Pigm. 184 (2021) p. 108874. https://doi.org/10.1016/j.dyepig.2020.108874.

[16] X. Zhu, Y. Li, Z. Wu, C. Lin, D. Ma, Z. Zhao, B.Z. Tang, Anthracene-based bipolar deep-blue emitters for efficient white OLEDs with ultra-high stabilities of emission color and efficiency, J. Mat. Chem. C. 9 (2021) p. 5198.https://doi.org/10.1039/D1TC00432H.

[17] Y.S. Mary, H.T. Varghese, C.Y. Panicker, T. Thiemann, A.A. Al-Saadi, S.A. Popoola, C. Van Alsenoy, Y. Al Jasem, Molecular conformational analysis, vibrational spectra, NBO, NLO, HOMO-LUMO and molecular docking studies of ethyl 3-(E)-(anthracen-9-yl)prop-2-enoate based on density functional theory calculations, Spectrochim. Act. A. 150 (2015) p. 533. https://doi.org/10.1016/j.saa.2015.05.092.

[18] A. Ekbote, P. Patil, S.R. Maidur, T.S. Chia, C.K. Quah, Structural, third-order optical nonlinearities and figures of merit of (E)-1-(3-substituted phenyl)-3-(4-fluorophenyl) prop-2-en-1-one under CW regime: New chalcone derivatives for optical limiting applications, Deyes Pigm. 139 (2017) p. 720. https://doi.org/10.1016/j.dyepig.2017.01.002

[19] P.S. Patil, S.R. Maidur, J.R. Jahagirdar, T.S. Chia, C.K. Quah, M. Shkir, Crystal structure, spectroscopic analyses, linear and third-order nonlinear optical properties of anthracene-based chalcone derivative for visible laser protection, Applied Physics B. 125 (2019) p. 1. https://doi.org/10.1007/s00340-019-7275-z

[20] N.N. Ayare, V.K. Shukla, N. Sekar, Charge transfer and nonlinear optical properties of anthraquinone D-π-A dyes in relation with the DFT based molecular descriptors and perturbational potential, Comput. Thoer. Chem. 1174 (2020) p. 112712. https://doi.org/10.1016/j.comptc.2020.112712

[21] R. Zaier, F. Mahdhaoui, S. Ayachi, T. Boubaker, Prediction of structural, vibrational and nonlinear optical properties of small organic conjugated molecules derived from pyridine, J. Mol. Struct. 1182 (2019) p. 131. https://doi.org/10.1016/j.molstruc.2019.01.043

[22] M. Rajeshirke, N. Sekar, NLO properties of ester containing fluorescent carbazole based styryl dyes – Consolidated spectroscopic and DFT approach Opt. Mater.76 (2018) p. 191. DOI:10.1016/j.optmat.2017.12.035.

[23] J. Zhang, Y. Zhao, H. Xu, D. Zhang, Y. Miao, R. Shinar, J. Shinar, H. Wang, B. Xu, Y. Wu, Novel blue fluorescent emitters structured by linking triphenylamine and anthracene derivatives for organic light-emitting devices with EQE exceeding 5%, J. Mat. Chem. C. 7 (2019) p. 10810. https://doi.org/10.1039/C9TC02773D.

[24] F. Liu, X. Man, H. Liu, J. Min, S. Zhao, W. Min, L. Gao, H. Jin, P. Lu, Highly efficient nondoped blue organic light-emitting diodes with high brightness and negligible efficiency roll-off based on anthracene-triazine derivatives, J. Mat. Chem. C. 7 (2019) p. 14881. https://doi.org/10.1039/C9TC05040J.

[25] S.P. Mishra, A.K. Palai, R. Srivastava, M.N. Kamalasanan, M. Patri, Dithieno[3,2-b:2′,3′-d]pyrrole–alkylthiophene–benzo[c][1,2,5]thiadiazole-based highly stable and low

band gap polymers for polymer light-emitting diodes, J. Poly. Sci. Pol. Chem. 47 (2009) p. 6514. https://doi.org/10.1002/pola.23694.

[26] S.C. Rasmussen, S.J. Evenson, Dithieno[3,2-b,2′,3′-d]pyrrole-based materials: Synthesis and application to organic electronics, Prog. polym. Sci. 38 (2013) p. 1773. https://doi.org/10.1016/j.progpolymsci.2013.04.004.

[27] L. Diao, J. Zhang, R. Wang, G. Liu, S.Pu, Synthesis and properties of asymmetric 9, 10-dithienylanthracene derivatives with AIEE properties and their applications in cell imaging, J. Photochem. Photobio. A. Chem.. 400 (2020) p. 112663. https://doi.org/10.1016/j.jphotochem.2020.112663.

[28] R. Wang, Y. Liang, G. Liu, S. Pu, Aggregation-induced emission compounds based on 9,10-diheteroarylanthracene and their applications in cell imaging, RSC Adv. 10 (2020) p. 2170. DOI: 10.1039/C9RA09290K.

[29] J.K. Fang, D.L. An, K. Wakamatsu, T. Ishikawa, T. Iwanaga, S. Toyota, S. I. Akita, D. Matsuo, A. Orita, J. Otera, Synthesis and spectroscopic study of phenylene–(poly)ethynylenes substituted by amino or amino/cyano groups at terminal(s): electronic effect of cyano group on charge-transfer excitation of acetylenic π-systems, Tetrahedron 66 (2010) p. 5479. https://doi.org/10.1016/j.tet.2010.05.016.

[30] J. George, A. K. Thomas, D. Sajan, S. Sathiyamoorthi, P. Srinivasan, N. Joy, R. Philip, Experimental and DFT/TD-DFT approach on photo-physical and NLO properties of 2, 6-bis (4-Chlorobenzylidene) cyclohexanone, Opt. Mater. 100 (2020) p. 109620. https://doi.org/10.1016/j.optmat.2019.109620

[31] G.W.T.] M.J. Frisch, H.B. Schlegel, G.E. Scuseria, M.A. Robb,, G.S. J.R. Cheeseman, V. Barone, B. Mennucci, G. A. Petersson,, M.C. H. Nakatsuji, X. Li, H.P. Hratchian, A.F. Izmaylov, J. Bloino,, J.L.S. G. Zheng, M. Hada, M. Ehara, K. Toyota, R. Fukuda,, M.I. J. Hasegawa, T. Nakajima, Y. Honda, O. Kitao, H. Nakai, T. Vreven,, J.E.P. J.A. Montgomery Jr., F. Ogliaro, M. Bearpark, J.J. Heyd, E. Brothers,, V.N.S. K.N. Kudin, R. Kobayashi, J. Normand, K. Raghavachari,, J.C.B. A. Rendell, S.S. Iyengar, J. Tomasi, M. Cossi, N. Rega, J. M. Millam,, J.E.K. M. Klene, J.B. Cross, V. Bakken, C. Adamo, J. Jaramillo, R. Gomperts,, O.Y. R.E. Stratmann, A.J. Austin, R. Cammi, C. Pomelli, J.W. Ochterski,, K.M. R.L. Martin, V.G. Zakrzewski, G.A. Voth, P. Salvador,, S. D. J.J. Dannenberg, A.D. Daniels, O. Farkas, J.B. Foresman, J.V. Ortiz,, D.J.F. J. Cioslowski, Gaussian 09 (Revision A.01), Gaussian, Inc., Wallingford, CT.

[32] Z. Lu, P. Dai, C. Wang, M. Liang, X. Zong, Z. Sun, S. Xue, Synthesis of new dithieno[3,2-b:2′,3′-d]pyrrole (DTP) dyes for dye-sensitized solar cells: effect of substituent on photovoltaic properties, Tetrahedron. 72 (2016) p. 3204. https://doi.org/10.1016/j.tet.2016.04.044.

[33] D.C. Santra, M.K. Bera, P.K. Sukul, S. Malik, Charge-Transfer-Induced Fluorescence Quenching of Anthracene Derivatives and Selective Detection of Picric Acid, CHEM-EUR J. 22 (2016) p. 2012. https://doi.org/10.1002/chem.201504126.

[34] R. Zaier, M.P. De La Cruz, F.L. De La Puente, S. Ayachi, Optoelectronic properties of cyclopentadithiophene-based donor–acceptor copolymers as donors in bulk heterojunction organic solar cells: A theoretical study, J. Phys. Chem . Sol. 145 (2020) p. 109532. DOI: 10.1016/j.jpcs.2020.109532.

[35] C. Chitpakdee, S. Namuangruk, P. Khongpracha, S. Jungsuttiwong, R. Tarsang, T. Sudyoadsuk, V. Promarak,

Theoretical studies on electronic structures and photophysical properties of anthracene derivatives as hole-transporting materials for OLEDs, Spectrochim. Act. A. 125 (2014) p. 36. https://doi.org/10.1016/j.saa.2013.12.111.

[36] L. Zhang, W. Shen, R. He, X. Liu, X. Tang, Y. Yang, M. Li, Fine structural tuning of diketopyrrolopyrrole-cored donor materials for small molecule-fullerene organic solar cells: A theoretical study, Org. Elect. 32 (2016) p. 134. https://doi.org/10.1016/j.orgel.2016.01.023.

[37] L. Bhattacharya, S.R. Sahoo, S. Sharma, S. Sahu, Inter. Effect of electron-withdrawing groups on photovoltaic performance of thiophene-vinyl-thiophene derivative and benzochalcogenadiazole based copolymers: A computational study, J. Quant. Chem. 119 (2019) p. e25982. https://doi.org/10.1002/qua.25982.

[38] D. Zhang, X. Song, H. Li, M. Cai, Z. Bin, T. Huang, L. Duan, High-Performance Fluorescent Organic Light-Emitting Diodes Utilizing an Asymmetric Anthracene Derivative as an Electron-Transporting Material, Adv. Mat. 30 (2018) p. 1707590. https://doi.org/10.1002/adma.201707590.

[39] C. Cao, G.X. Yang, J.H. Tan, D. Shen, W.C. Chen, J.X. Chen, J.L. Liang, Z.L. Zhu, S.H. Liu, Q.X. Tong, C.S. Lee, Deep-blue high-efficiency triplet-triplet annihilation organic light-emitting diodes using donor- and acceptor-modified anthracene fluorescent emitters Materials Today Energy. 21 (2021) p. 100727. https://doi.org/10.1016/j.mtener.2021.100727

[40] M.M. Duvenhage, M. Ntwaeaborwa, H.G. Visser, P.J. Swarts, J.C. Swarts, H.C. Swart, Determination of the optical band gap of Alq3 and its derivatives for the use in two-layer OLEDs, Opt. Mater. 42 (2015) p. 193. https://doi.org/10.1016/j.optmat.2015.01.008.

[41] R. Jin, X. Zhang, W. Xiao, Theoretical Studies of Photophysical Properties of D−π−A−π−D-Type Diketopyrrolopyrrole-Based Molecules for Organic Light-Emitting Diodes and Organic Solar Cells, Molecules. 25 (2020) p. 667. https://doi.org/10.3390/molecules25030667.

[42] S.C. Rasmussen, S.J. Evenson, C.B. McCausland, Fluorescent thiophene-based materials and their outlook for emissive applications, Chem. Comm. 51 (2015) p. 4528. https://doi.org/10.1039/C4CC09206F.

[43] R. Oshi, S. Abdalla, M. Springborg, Study of the influence of functionalization on the reorganization energy of naphthalene using DFT, Comput. Thoer. Chem. 1099 (2017) p. 209. https://doi.org/10.1016/j.comptc.2016.12.002.

[44] R. Zaier, S. Hajaji, M. Kozaki, S. Ayachi, DFT and TD-DFT studies on the electronic and optical properties of linear π-conjugated cyclopentadithiophene (CPDT) dimer for efficient blue OLED, Opt. Mater. 91 (2019) p. 108. https://doi.org/10.1016/j.optmat.2019.03.013.

[45] R. Zaier, S. Ayachi, Toward designing new cyclopentadithiophene-naphthalene derivatives based small molecules for organic electronic applications: A theoretical investigation, Mater Today Commun. (2021) p. 102370. https://doi.org/10.1016/j.mtcomm.2021.102370.

[46] K. Singh, I. Bala, R. Kataria, Crystal structure, Hirshfeld surface and DFT based NBO, NLO, ECT and MEP of benzothiazole based hydrazone, Chem. Phys. 538 (2020) p. 110873. DOI: 10.1016/j.chemphys.2020.110873.

[47] S. Zongo, K. Sanusi, J. Britton, P. Mthunzi, T. Nyokong, M. Maaza, B.

Sahraoui, Nonlinear optical properties of natural laccaic acid dye studied using Z-scan technique, Opt. Mater. 46 (2015) p. 270. https://doi.org/10.1016/j.optmat.2015.04.031.

[48] I. Kucuk, Y. Kaya, A.A. Kaya, Structural, spectroscopic (FT-IR, NMR, UV–visible), nonlinear optical (NLO), cytotoxic and molecular docking studies of 4-nitro-isonitrosoacetophenone (ninapH) by DFT method, J. Mol. Struct. 1139 (2017) p. 308. https://doi.org/10.1016/j.molstruc.2017.03.032.

[49] A. Hussain, M.U. Khan, M. Ibrahim, M. Khalid, A. Ali, S. Hussain, M. Saleem, N. Ahmad, S. Muhammad, A. Al-Sehemi, Structural parameters, electronic, linear and nonlinear optical exploration of thiopyrimidine derivatives: A comparison between DFT/TDDFT and experimental study, J. Mol. Struct. 1201 (2020) p. 127183. https://doi.org/10.1016/j.molstruc.2019.127183

[50] S. Selvakumar, M.S. Boobalan, S.A. Babu, S. Ramalingam, A.L. Rajesh, Crystal growth and DFT insight on sodium *para*-nitrophenolate *para*-nitrophenol dihydrate single crystal for NLO applications, J. Mol. Struct. 1125 (2016) p. 1. https://doi.org/10.1016/j.molstruc.2016.05.104.

[51] S. Muthu, E.I. Paulraj, Molecular structure and spectroscopic characterization of ethyl 4-aminobenzoate with experimental techniques and DFT quantum chemical calculations, Spectrochim. Act. A. 112 (2013) p. 169. https://doi.org/10.1016/j.saa.2013.04.024.

[52] N. Islam, S.S. Chimni, DFT investigation on nonlinear optical (NLO) properties of novel borazine derivatives, Comput. Thoer. Chem. 1086 (2016) p. 58. https://doi.org/10.1016/j.comptc.2016.04.016.

[53] S. Wild, N. Tice, DFT study of structural and electronic properties of 1,4-diarylcyclopenta[d] pyridazines and oxazines for non-linear optical applications, J. Mol. Model. 27 (2021) p. 1. DOI: 10.1007/s00894-021-04676-6.

[54] C.W. Ghanavatkar, V.R. Mishra, N. Sekar, E. Mathew, S.S. Thomas, I.H. Joe, Benzothiazole pyrazole containing emissive azo dyes decorated with ESIPT core: Linear and non linear optical properties, Z scan, optical limiting, laser damage threshold with comparative DFT studies, J. Mol. Str. 1203 (2020) p. 127401. https://doi.org/10.1016/j.molstruc.2019.127401.

[55] R. Jayarajan, R. Satheeshkumar, T. Kottha, S. Subbaramanian, K. Sayin, G. Vasuki, Water mediated synthesis of 6-amino-5-cyano-2-oxo-N-(pyridin-2-yl)-4-(p-tolyl)-2H-[1,2′-bipyridine]-3-carboxamide and 6-amino-5-cyano-4-(4-fluorophenyl)-2-oxo-N-(pyridin-2-yl)-2H-[1,2′-bipyridine]-3-carboxamide – An experimental and computational studies with non-linear optical (NLO) and molecular docking analyses, Spectrochim. Act. A. 229 (2020) p. 117861. https://doi.org/10.1016/j.saa.2019.117861.

Chapter 16

DFT Study of Structure and Radical Scavenging Activity of Natural Pigment Delphinidin and Derivatives

Abstract

A theoretical evaluation of the antioxidant activity of natural pigment delphinidin (**1a**) and derivatives **1b**, **1c**, **1d** & **1e** was performed using the DFT-B3LYP/6–311 + G (d, p) level of theory. Three potential working mechanisms, hydrogen atom transfer (HAT), stepwise electron transfer proton transfer (SET-PT), and sequential proton loss electron transfer (SPLET), have been investigated. The physiochemical parameters, including O–H bond dissociation enthalpy (BDE), ionization potential (IP), proton dissociation enthalpy (PDE), proton affinity (PA), and electron transfer enthalpy (ETE), have been calculated in the gas phase and aqueous phase. The study found that the most suitable mechanism for explaining antioxidant activity is HAT in the gas phase and SPLET in the aqueous medium in this level of theory. Spin density calculation and delocalization index of studied molecules also support the radical scavenging activity. When incorporated into natural pigment delphinidin, the gallate moiety can enhance the activity and stability of the compounds.

Keywords: DFT, Multiwfn, Delphinidin, Radical scavenger, Gallic acid

1. Introduction

Dietary polyphenols have interesting spectrum biological properties, including radical scavenging activity [1]. Anthocyanins are one of essential classes in the polyphenol family [2]. These are the plant pigments responsible for the bright red-orange to blue-violet colors of many fruits and vegetables [3]. Since these are natural colored compounds, many pieces of literature come with their application as coloring or coloring material, especially in the food industry [4–9]. Naturally, these compounds found in glycoside form can collectively be called anthocyanins. At the same time, their aglycon forms are commonly called anthocyanidins [10]. The common aglycon forms are cyanidin, delphinidin, peonidin, petunidin, malvidin, and pelargonidin. The color pigments are most abundant in berries like black currants, elderberries, blueberries, strawberries, etc., red and purple grapes, red wine, sweet cherries, eggplants, black plums, and red cabbage, etc.

The basic structure of anthocyanins is the flavylium cationic ring in which oxygen carries a positive charge. The positively charged species exhibit several equilibrium structures due to different transformations like proton transfer, isomerization, and tautomerization under various pH conditions [11]. All molecules of a group of anthocyanins have their absorption range in the visible spectrum due to effective π conjugation within the molecule. These are the largest group of water-soluble pigments in the plant kingdom. Experimental and theoretical reports show an exponential increase in findings related to its properties, color, co-pigmentation, pH effect, antiradical properties, etc. [12–17].

The delphinidin (**1a**) and its four modifications (**1b, 1c, 1d & 1e**) are selected for the study. The structure of **1a** having three rings A, B, and C, where A & C are fused rings and B connected with A-C through a single bond. The structures of compounds under consideration are shown in **Figure 1**. The colored pigments with other colorless natural products are now proven to be very impressive due to their improved activity in the sense of color and property. Hence to enhance the property, the colorless, most important, small, widely studied polyphenol gallic acid is taken and coupled with **1a** through their 3 and 4' OH bonds and respectively formed delphinidin-3-O-gallate (**1c**) and delphinidin-4'-O-gallate (**1e**). Also, the results are compared with its glucose forms delphinidin-3-O-glucoside (**1b**) and delphinidin-4'-O-glucoside (**1d**).

Gallic acid (3,4,5- Trihydroxy benzoic acid) and its derivatives have been found in various natural sources like nuts, tea, grapes, gallnut, oak bark, etc. Biological studies show that gallic acid has variable effects, including antiviral, antioxidant, and anticancer activities [18–20]. Due to its potent antioxidant activity against free radicals is used in food, cosmetics, and pharmaceutical industries and as a source material for ink and paints [21]. GA possesses the most robust antiradical property than Trolox, and hence, in many cases, this molecule is widely used as a reference compound for antioxidant studies. Green tea contains the highest concentration of GA-based compounds responsible for the plant's antioxidant capacity [22, 23].

Once the compound possesses potent antioxidant activity and a specific range of absorbance in the visible spectrum can be effectively used for various purposes, especially in the food industry, they seek less hazardous, highly protective materials

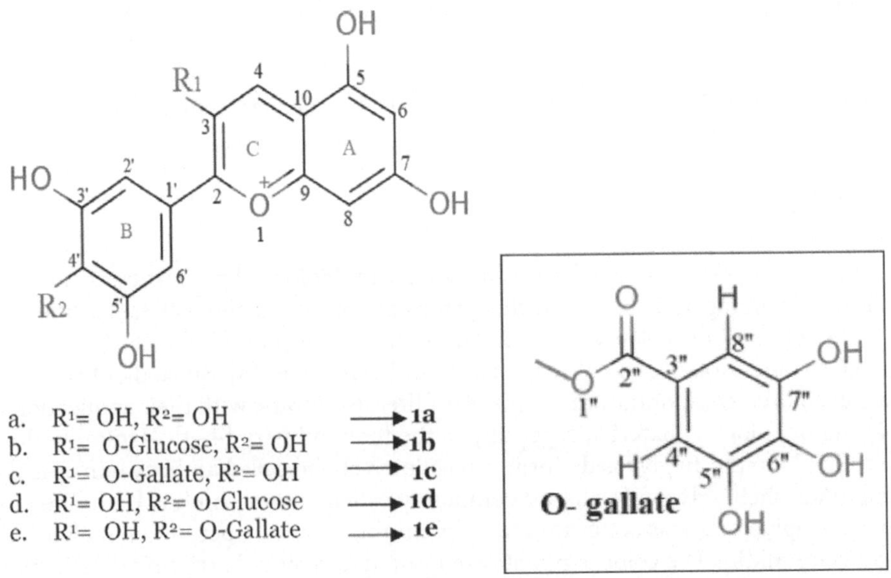

Figure 1.
Structures of compound considered for the study.

for coloring purposes. Hence these are colored compounds, and this work concludes the antiradical property of delphinidin and its derivatives with the help of computational methods.

2. Materials and method

2.1 Materials

The present work has utilized a theoretical approach to study the structure and properties of all compounds under consideration. The molecular forms of **1a** and **1b** were downloaded from the PubChem database in SDF file format and converted to GJF file format by Open Babel [24]. The structures of **1c**, **1d**, and **1e** were drawn using the Gaussview 5.0.8 graphical user interface [25]. All the computational works have been carried out through Gaussian 09 software package to get the output [26]. The spin density (SD) and electron delocalization (DI) analysis are explained with the help of the software Multiwfn 3.6 [27].

2.2 Computational methodology

The present work was carried out using density functional theory (DFT) because it is based on electron density, and antioxidant activity is mainly influenced by electron density [28–32]. In DFT calculation, 6–311+ G (d, p) basis set with B3LYP (Becke's exchange functional in conjunction [33] with Lee-Yang–Parr [34]) correlational functional has been used for geometry optimization, computation of harmonic vibrational frequencies, BDE, IP, PDE, PA and ETE calculations. To obtain antioxidant parameters BDE, IP, PDE, PA, and ETE, the geometry optimization of neutral, radicals, anions, and radical cation structures of all the studied molecules are conducted in the ground state both in the gas phase and aqueous phases. Solvent effects on the calculated systems were investigated with the self-consistent reaction field (SCRF) method via the integral equation formalism polarized continuum model (IEF-PCM).

2.2.1 Frontier molecular orbital (FMO) analysis

Frontier molecular orbital theory is an application of MO theory that explains HOMO/LUMO interactions. HOMO is the highest occupied molecular orbital, and LUMO is the lowest unoccupied molecular orbital. Frontier molecular orbital analysis is fundamental because HOMO or LUMO energies and bandgap energies are the key factors that drive the antiradical property of molecules. Since HOMO is in the highest energy state, so easier to remove an electron from this orbital. So in a chemical reaction or bond formation, HOMO is donating electrons, or it acts as a Lewis base or undergoes oxidation. LUMO is the lower-lying orbital; it is empty, so it is easier for LUMO to accept electrons into its orbital or acts as a Lewis acid or undergoes reduction. Reactivity becomes lower when the molecule has a higher bandgap. The distribution of HOMO orbitals and energies is calculated using the DFT-B3LYP/6–311 + G (d, p) level of theory from the optimized structures of **1a** and its derivatives.

2.2.2 Radical scavenging activity

Several mechanisms have theoretically explained the radical scavenging mechanism of phenolic compounds. The widely used mechanisms are hydrogen atom transfer (HAT) mechanism (Eq. (1)), single-electron transfer followed by proton transfer (SET-PT) mechanism (Eqs. (2)–(4)), and sequential proton loss electron transfer (SPLET) mechanism (Eqs. (5) and (6)) [35–40]. These mechanisms have been briefly addressed here.

1. HAT (hydrogen atom transfer) mechanism

$$ArOH + X^{\bullet} \rightarrow ArO^{\bullet} + XH \tag{1}$$

By transferring the hydrogen atom of the –OH group to the radical species, the antioxidant (ArOH) scavenges the free radical ($X^{\bullet}$) and transforms it into phenoxide radical. The descriptor associated with the HAT mechanism is bond dissociation enthalpy (BDE), and the lower value indicates good radical scavenging activity. The ArO• radical species' stabilizing features, like the resonance delocalization of the electron within the aromatic ring, are the basis of lowest energy and increased antiradical activity.

2. SET (single electron transfer) mechanism

$$ArOH + X^{\bullet} \rightarrow ArOH^{\bullet +} + X^{-} \tag{2}$$

Here, the reactive free radicals are neutralized by transferring electrons to them, resulting in anions. The most reactive hydroxyl group in antioxidant compounds provides these electrons and finally becomes a radical cation. The descriptor associated with this mechanism is AIP (adiabatic ionization potential).

3. SET-PT (single electron transfer followed by proton transfer) mechanism

$$ArOH + X^{\cdot} \rightarrow ArOH^{\bullet +} + X^{-} \tag{3}$$

$$ArOH^{\bullet +} \rightarrow ArO^{\bullet} + H^{+} \tag{4}$$

The first process involves an electron transfer from the antioxidant (Eq. (2)), and the second step consists of a proton transfer from the radical cation (Eq. (4)) generated in the first step. Proton dissociation enthalpy is the descriptor connected with the second phase (PDE).

4. SPLET (sequential proton loss electron transfer) mechanism

$$ArOH \rightarrow ArO^{-} + H^{+} \tag{5}$$

$$ArO^{-} + X^{\bullet} + H^{+} \rightarrow ArO^{\bullet} + XH \tag{6}$$

This is also a two-stage mechanism, with the dissociation of the antioxidant into phenoxide anion and proton as the first step (Eq. (5)). The first-step phenoxide anion then interacts with free radicals at a certain pH (Eq. (6)); the compounds generated are similar to those developed in the HAT mechanism. Proton affinity (PA) is the regulating descriptor for the first stage, and electron transfer enthalpy is the driving descriptor for the second stage (ETE).

The Eqs. (7)–(11) are used for analyzing the type of mechanism involved by the compound

$$BDE = H(ArO^{\bullet}) + H(H^{\bullet}) - H(ArOH) \tag{7}$$

$$AIP = H(ArOH^{\bullet+}) + H(e^{-}) - H(ArOH) \tag{8}$$

$$PDE = H(ArO^{\bullet}) + H(H^{+}) - H(ArOH^{\bullet+}) \tag{9}$$

$$PA = H(ArO^{-}) + H(H^{+}) - H(ArOH) \tag{10}$$

$$ETE = H(ArO^{\bullet}) + H(e^{-}) - H(ArO^{-}) \tag{11}$$

Thus, in the present study, BDE, IP, PDE, PA, and ETE values were used as the primary molecular descriptors to elucidate the radical scavenging activity of the investigated compounds. The enthalpies of hydrogen radicals in the water and gas phase are calculated by G09 software using the DFT/B3LYP/6–311 + G (d, p) level of theory. The enthalpies of the electron (e^{-}) and proton in the gas phase are taken from the commonly accepted values 0.00236 Hartree for proton and 0.00120 Hartree for electron. In contrast, for water, these corresponding values are calculated with the help of the DFT/B3LYP/6–311 + G(d, p) level of theory. The enthalpies of electron and proton in solvent water were computed using the same level of theory with methodology suggested by Markovic et al. (Eqs. (12) and (13)) [41, 42].

$$H^{+}_{gas} + S_{sol} \rightarrow (S - H)^{+}_{sol} \tag{12}$$

$$e^{-}_{gas} + s_{sol} \rightarrow (S - e)^{-}_{sol} \tag{13}$$

Where S_{sol} is the solvent molecule solvated by the same kind of molecule, $(S - H)^{+}_{sol}$ and $(S - e)^{-}_{sol}$ are the charged particles formed. The solvation enthalpies of proton and electron are calculated using the Eqs. (14) and (15), respectively, and that of $H^{\bullet}$ by optimizing hydrogen atom using the same level of theory. The enthalpy of hydrogen radical taken for gas is −0.49764 Hartree and − 0.497466 Hartree for water. The enthalpies of proton and electron in water, respectively, are −0.37725431 Hartree and 0.093551 Hartree.

$$\Delta H_{sol}(H^{+}) = H(S - H)^{+}_{sol} - H(S_{sol}) - H\left(H^{+}_{gas}\right) \tag{14}$$

$$\Delta H_{sol}(e^{-}) = H(S - e)^{-}_{sol} - H(S_{sol}) - H\left(e^{-}_{gas}\right) \tag{15}$$

3. Results and discussion

3.1 Conformation analysis and geometry optimization

To elucidate the reactivity of the compounds towards free radicals, the conformational and geometrical features of compounds are very significant. The structures of **1a** and **1b** are downloaded from the PubChem database and **1c**, **1d**, and **1e** are constructed from the optimized structure of **1a**. The potential energy surface of all five molecules is scanned by using the B3LYP/3-21G level of theory by varying the dihedral angles values in 12 steps of 30° from 0 to 360°. The dihedral angle between aromatic ring B and AC bicyclic in **1a** is performed on the dihedral Φ1 (C3, C2, C1', C2'), and the five OH groups are also scanned as above mentioned procedure. All phenolic OH is in a position that forms hydrogen bonding with the nearest OH. The dihedral Φ1 about (C3, C2, C1', C2') of **1a** completely reveals the planar geometry of the molecule, whereas other molecules are slightly becoming nonplanar because of the steric cloud around flavylium cationic ring. The dihedral (C3, C2, C1', C2') of **1a, 1b, 1c, 1d,** and **1e** respectively are −179.95499, −158.12940, −167.05569, −169.70713, and 171.29496 degrees.

The lowest energy conformer obtained after the last scan was then subjected to geometry optimization, and B3LYP/6–311 + G(d, p) level of theory with a correlation coefficient of 0.89281 are selected for the study. The crystal structure data of cyanidin are considered for experimental validation of results obtained from the computational analysis since these are not available for delphinidin molecules and tabulated in **Table 1** [43]. But the basic structural unit of **1a** is similar, except in cyanidin, one OH is missing from the 5′ position. The optimized structures of the five compounds are shown in **Figure 2**. These optimized structures are considered for further calculations.

3.2 Molecular orbital analysis

Frontier molecular orbital analysis gives a detailed account of HOMO/LUMO interactions in the molecule. It is very useful in describing optical and electronic properties as well as the reactivity of a molecule. The HOMO value of the molecule is strongly influenced by radical scavenging activity. The higher the HOMO energy, the easier the electron is being excited and acts substantial donor of the electron. The chemical reactivity of the molecule can be described by knowing the HOMO–LUMO gap. Like that, the distribution of orbitals among the molecule reveals probable sites for the attack of free radicals. The HOMO distribution among molecule **1a** is completely distributed on each atom, whereas the LUMO orbitals have distributed all atoms except 3- OH, 3'- OH, and 5'-OH. For the molecules studied here, the distribution of the LUMO distributed among each atom other than 3- OH, 3′- OH, and 5′-OH. So these three bonds are involved in the HOMO-LUMO transition. In **1c** and **1e**, HOMO is present only on the gallate moiety, whereas the LUMO is distributed only in the flavylium ring. The HOMO of **1b** and **1d** occur throughout the flavylium ring though only negligible HOMO/LUMO contributions are present on glucoside moiety.

Bond	Experimental bond length (A^O)	Theoretical			
		B3LYP 6–31 G (d)	B3LYP 6–31+ G (d, p)	B3LYP 6–311+ G (d, p)	B3LYP 6–311++ G (d, p)
C 2-C1'	1.453	1.44380	1.44380	1.44173	1.44176
C10-C5	1.432	1.43007	1.43002	1.42770	1.42769
C1'-C2'	1.409	1.41731	1.41735	1.41422	1.41423
C6 - C7	1.413	1.41392	1.41390	1.41107	1.41106
C2 –C3	1.396	1.42173	1.42177	1.41918	1.41920
C3'-C4'	1.400	1.40858	1.40863	1.40599	1.40602
C4–C10	1.382	1.40615	1.40614	1.40553	1.40351
C6'-C1'	1.404	1.41542	1.41539	1.41216	1.41218
C9–C10	1.408	1.41223	1.41222	1.40886	1.40888
C7 – C8	1.387	1.39860	1.39860	1.39517	1.39517

Table 1.
Comparison of bond length with experimental values in different basis sets.

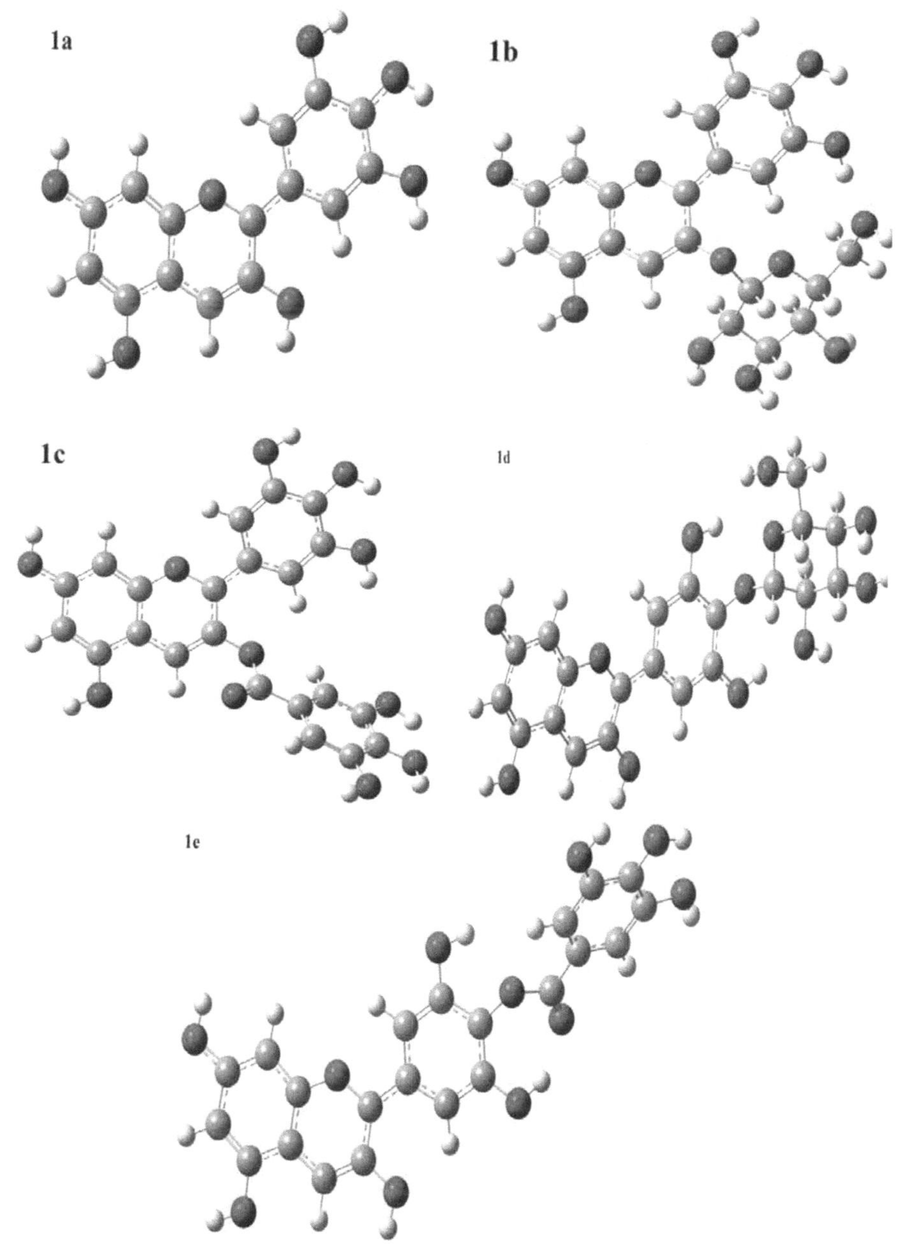

Figure 2.
The obtained optimized structures of the compounds using B3LYP /6–311 + G(d, p) level of theory (delphinidin (1a), delphinidin-3-O-glucoside (1b), delphinidin-3-O-gallate (1c), delphinidin-4'-O-glucoside (1d), and delphinidin-4'-O-gallate (1e).)

The increasing order of bandgap energies in the gas phase at **1e** < **1c** < **1d** < **1a** < **1b** reveals that the gallate-based compounds have the lowest bandgap energies and higher in activity. In the aqueous phase, the band gaps are higher compared to the gas phase, and the difference between the bandgap of water and gas is also represented in **Table 2**. One of the difficulties associated with anthocyanins as color pigments is their stability in polar solvents. When the medium changes from the gas phase to water, the only difference in the bandgap is 0.17 for **1a** and 0.20 for **1b**, but it is more significant for its gallates. So gallates are stable than others; hence expect higher reactivity in water (**Figure 3**).

	Gas (eV)			Water(eV)			
	E_{HOMO}	E_{LUMO}	Band gap (ΔE_{gas})	E_{HOMO}	E_{LUMO}	Band gap (ΔE_{water})	ΔE_{water} - ΔE_{gas}
1a	−9.27	−6.63	2.64	−6.46	−3.65	2.81	0.17
1b	−8.80	−6.10	2.70	−6.47	−3.57	2.9	0.2
1c	−8.91	−6.36	2.55	−6.62	−3.69	2.93	0.38
1d	−9.10	−6.56	2.54	−6.71	−3.74	2.97	0.43
1e	−8.60	−6.55	2.05	−6.64	−3.78	2.86	0.81

Table 2.
FMO analysis of studied compounds in gas and water media at B3LYP/6–311 + G(d, p) level of theory.

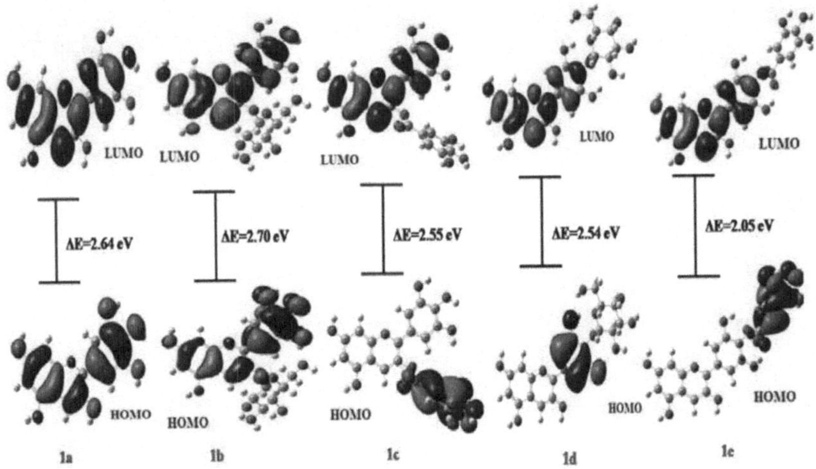

Figure 3.
*FMO of **1a, 1b, 1c, 1d,** & **1e** in the gas phase at B3LYP/6–311 + G(d, p) level of theory.*

3.3 Radical scavenging activity

The antioxidant activities of the compounds are studied using the antioxidant mechanism described in Section 2.2.2. The BDE, IP, PA, PDE, and ETE values obtained from the corresponding mechanism are used to analyze the activity of compounds. Among the five parameters, one with the lowest value and the corresponding mechanism is followed by the compound. The parameters of antioxidant activities are represented in the gas phase and aqueous phase, respectively, in **Tables 3** and **4**.

3.3.1 Analysis of HAT mechanism

In the gas phase, all studied molecules follow the HAT mechanism because of its lower BDE values. Hence, all compounds tending to form radicals by donating hydrogen atoms in respective positions are higher in the gas phase. In the case of **1a**, the positions 3-OH (81.56 kcal/mol), 4'-OH (83.19 kcal/mol), and 5'-OH (84.51 kcal/mol) have the lowest values of BDE, and hence these positions are involved in radical scavenging activity in the gas phase. For a compound possessing more than one phenolic hydroxyl group, its radical scavenging activity is determined by the one with the lowest value of BDE. Hence, in this case, 3-OH is the most active site, followed by 4'-OH and 5'-OH.

	BDE	IP	PDE	PA	ETE
1a					
3-OH	81.56	242.73	153.33	245.94	150.12
5-OH	88.99		160.77	246.47	156.03
7-OH	90.99		162.76	246.89	158.61
3'-OH	92.17		163.94	269.82	136.85
4'-OH	83.19		154.96	246.50	151.19
5'-OH	84.51		156.28	258.11	140.90
1b					
5-OH	87.97	228.97	173.50	227.47	175.01
7-OH	90.53		176.05	253.17	151.86
3'-OH	90.25		175.78	278.86	125.89
4'-OH	82.15		167.69	256.71	139.96
5'-OH	89.41		174.94	267.13	136.79
1c					
5-OH	89.80	174.95	229.35	249.97	154.33
7-OH	91.99		231.55	248.88	157.61
3'-OH	91.49		231.04	273.24	132.76
4'-OH	83.12		222.67	249.81	147.81
5'-OH	84.18		223.73	260.56	138.12
5"-OH	89.16		228.71	256.98	166.80
6"-OH	83.09		222.64	266.51	131.08
7"-OH	83.60		223.15	270.08	128.02
1d					
5-OH	88.91	232.11	171.30	248.31	155.10
7-OH	91.05		173.45	248.11	157.45
3'-OH	83.84		166.23	249.38	148.97
3'-OH	92.94		175.34	267.53	139.92
5'-OH	91.55		173.95	269.50	136.56
1e					
3-OH	88.81	227.74	175.57	248.48	154.83
5-OH	90.88		177.65	248.24	157.15
7-OH	90.30		177.06	270.75	134.06
3'-OH	83.90		170.66	249.64	148.76
5'-OH	91.04		177.80	272.33	133.22
5"-OH	89.60		176.36	282.61	121.50
6"OH	82.80		169.56	267.24	130.06
7"-OH	83.22		169.98	271.13	126.59

Table 3.
The antioxidant mechanism study of compounds ***1a, 1b, 1c, 1d,*** *and* ***1e*** *in gas using B3LYP/6–311 + G(d, p) level of theory. All values are represented in kcal/Mol.*

	BDE	IP	PDE	PA	ETE
1a					
3-OH	80.87	201.36	13.64	37.88	177.12
5-OH	85.05		17.83	37.43	181.76
7-OH	86.13		18.91	36.98	183.28
3’-OH	85.33		18.10	49.18	170.29
4’-OH	77.52		10.30	35.27	176.39
5’-OH	81.79		14.57	43.17	172.76
1b					
5-OH	86.30	200.34	20.10	37.08	183.36
7-OH	87.77		21.57	36.41	185.49
3’-OH	84.02		17.82	50.14	168.02
4’-OH	77.42		11.22	37.11	174.45
5’-OH	86.08		19.87	46.66	173.55
1c					
5-OH	87.22	206.11	13.83	35.85	185.50
7-OH	89.19		15.88	35.10	188.22
3’-OH	85.39		15.16	42.42	177.10
4’-OH	78.65		11.53	34.39	178.38
5’-OH	82.09		14.16	42.42	173.80
5”-OH	84.08		13.04	51.41	166.81
6”-OH	78.66		7.47	42.22	170.58
7”-OH	81.34		10.13	45.07	170.40
1d					
5-OH	85.72	204.22	15.64	36.36	155.10
7-OH	87.71		17.62	35.80	157.45
3’-OH	81.34		11.26	35.21	148.97
3’-OH	88.12		18.04	49.73	139.92
5’-OH	88.99		18.91	42.69	136.56
1e					
3-OH	85.82	206.11	13.83	36.73	183.22
5-OH	87.86		15.88	36.07	185.93
7-OH	87.14		15.16	47.37	173.90
3’-OH	83.51		11.53	36.78	180.86
5’-OH	86.15		14.16	46.43	173.85
5”-OH	85.02		13.04	41.67	167.55
6”-OH	79.45		7.47	45.24	171.92
7”-OH	82.11		10.13	51.60	171.00

Table 4.
The antioxidant mechanism study of compounds ***1a, 1b, 1c, 1d,*** *and* ***1e*** *in water using B3LYP/6–311 + G(d, p) level of theory. All values are represented in kcal/Mol.*

The glucose substituted derivatives of **1a, 1b,** and **1d** in the gas phase also follow the HAT mechanism due to its lowest value of BDEs comparing with other parameters. In **1b**, 4'-OH has a higher tendency to participate in radical scavenging mechanism due to its lowest BDE value (82.15 kcal/mol) compared to other phenolic OH groups in the compound. When 4'-OH is substituted with glucose moiety or **1d**, 3-OH is contributing to the radical mechanism. So in **1b** and **1d** radical scavenging activity is mainly through B-ring (4'-OH) and C-ring (3-OH).

As from **Table 3** the antioxidant activity of compound **1c** are through its phenolic hydroxyl group at 4' OH (83.12 kcal/mol), 5' OH (84.18 kcal/mol), 6"OH (83.09 kcal/mol), and 7" OH (83.60 kcal/mol) positions with an increasing order of 6"OH < 4' OH < 7" OH < 5' OH. The phenolic hydroxyl groups at 4' OH and 5' OH situates on the B-ring of **1a**, whereas 6"OH and 7" OH at gallic acid moiety. Here, both rings, the gallate ring, and delphinidin ring moieties, contribute to the radical scavenging activity through the hydroxyl groups. Since an ester relation connects gallate and delphinidin moieties, the radicals produced in one ring do not delocalize much to another ring and acts as separate contributors to radical scavenging activity. Like that, **1e** also shows antioxidant activity through 3' OH (83.90 kcal/mol), 6" OH (82.80 kcal/mol) and 7" OH (83.22 kcal/mol). The 6" OH has a slightly lower value of BDE than the other two OH groups because the radical formed at para OH is highly stabilized through the aromatic system of gallate moiety. Similar to 1c, compound **1e** also possesses radical scavenging activity through B-ring and gallic acid moiety. When comparing **1c** and **1e** with other molecules, when gallic acid is substituted, the number of hydroxyl groups under lower BDE values increases. Hence, the chance of enhancing activity is clear.

To explain the differences in BDE and consequently the differences in the reactivity of OH sites, the spin density distribution of radicals was calculated and presented in **Table 5**. The stability of radicals formed can be explained with the help

		Bond								
		5 OH	7 OH	3 OH	3' OH	4' OH	5' OH	5" OH	6"OH	7"OH
1a										
	SD	0.027	0.022	0.025	0.061	0.018	0.032	—	—	—
	DI	1.889	1.911	1.855	1.831	1.937	1.825	—	—	—
1b										
	SD	0.027	0.022	—	0.033	0.019	0.032	—	—	—
	DI	1.885	1.903	—	1.843	1.923	1.819	—	—	—
1c										
	SD	0.027	0.023	—	0.032	0.020	0.032	0.0317	0.025	0.032
	DI	1.888	1.901	—	1.845	1.923	1.825	1.847	1.884	1.832
1d										
	SD	0.026	0.022	0.026	0.032	—	0.033	—	—	—
	DI	1.892	1.910	1.860	1.829	—	1.813	—	—	—
1e										
	SD	0.026	0.022	0.026	0.031	—	0.032	0.031	0.025	0.032
	DI	1.891	1.910	1.861	1.839	—	1.824	1.850	1.888	1.830

Table 5.
*The SD distribution for the O-radical and delocalization index (DI) of C-O bond computed for **1a, 1b, 1c, 1d,** & **1e** in the gas-phase at B3LYP/6–311 + G (d, p) level of theory.*

of SD values; more delocalized SD means to be more stable is the radical formed. Moreover, the delocalization index is also a supporting parameter for explaining the stability of radicals created. The more stable the radical formed from an -OH bond, the more the corresponding C-O bond will be the delocalization index. The DI of

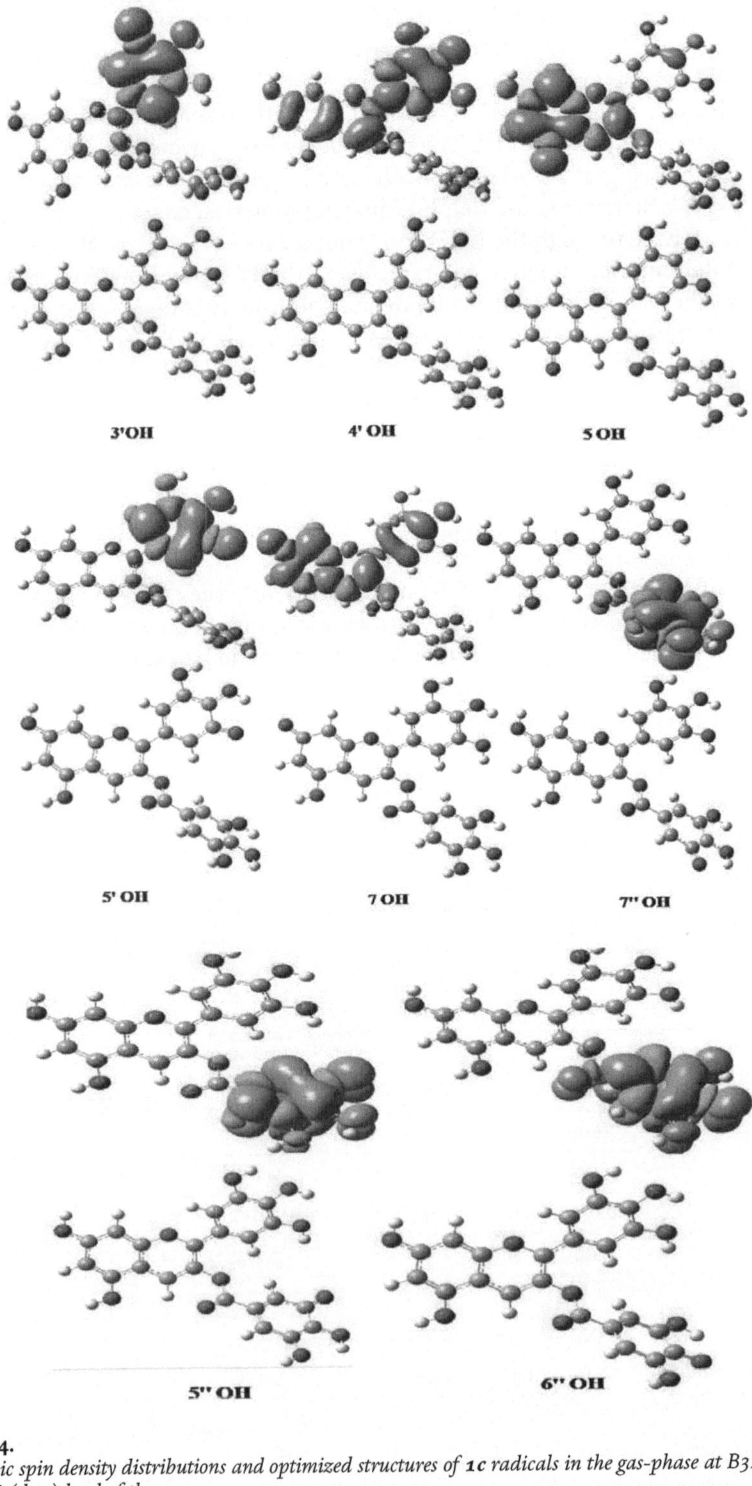

Figure 4.
Electronic spin density distributions and optimized structures of **1c** *radicals in the gas-phase at B3LYP/6–311 + G (d, p) level of theory.*

C-O bonds in each radical site are calculated using Fuzzy atomic space analysis [44–46]. The SD contours of **1c** are represented in **Figure 4.**

3.3.2 Analysis of SET-PT mechanism

In the SET-PT mechanism, the parameters involved are IP and PDE. In all derivative IP is found to be higher in both media. The PDE values are higher in the gas phase but lower in the water medium. But the lower value in the water medium cannot be used for final judgment about contribution in antioxidant activity. In the SET-PT mechanism, the PDE comes from the second step of this mechanism, whereas the first step is the IP. Since all cases have the highest values for IP in the water medium, they have to overcome this large energy barrier of IP to reach the second step.

3.3.3 Analysis of SPLET mechanism

In water, the parameters BDE, IP, and ETE having higher enthalpy than PA and are represented in **Table 4**. Hence SPLET is the mechanism followed by each molecule in the water medium. In the SPLET mechanism, the heterolytic bond cleavage of the phenolic hydroxyl group is considered, and the neutral molecule is split into an anion and a proton. The numerical parameter associated with this step is PA. The anion produced donates one electron to free radical species, and the free radical receives one electron and forms an anion. The anion of free radical react with proton forms a neutral compound by leaving the anion starting compound as radical. The numerical parameter associated with this step is ETE. The values of PA are found to be lower than BDE and other parameters in all studied cases when the solvent was aqueous. In the case of **1a**, 4' OH possesses the lowest value of PA (35.27 kcal/mol), contributing to the radical scavenging mechanism.

In the case of **1b**, the PA value of 7 OH (36.41 kcal/mol) has a lower value, and 5 OH (37.08 kcal/mol) and 4' OH (37.11 kcal/mol) have valued at the nearest. When glucose at position 3 enhances the electron-donating capacity of A ring. When glucose at 4′ position called **1d**, the lowest PA value at 3-OH (35.21 kcal/mol) and followed nearest at 7 OH (35.80 kcal/mol) and 5-OH (36.36 kcal/mol) also enhances the radical scavenging activity of A- ring. The lowest PA value of **1c** in water is 4' OH (34.39 kcal/mol), which is the lowest of all studied compounds in the water medium. The PA values 35.10 kcal/mol of 7 OH and 5 OH are near the 4' OH. For **1e**, the lowest value of PA at 5-OH (36.07 kcal/mol). In the water medium, the gallate moiety containing hydroxyl group does not affect radical scavenging activity due to its higher PA values. In the gas phase, a considerable contribution is provided by this group.

4. Conclusion

A theoretical study on the antioxidant activity of natural pigment delphinidin and its derivative has been evaluated. The antiradical properties of all studied molecules are finalized through an antioxidant mechanism. All compounds follow the HAT mechanism in the gas phase and the SPLET mechanism in the aqueous medium. In gas-phase gallic acid, substituted compounds possess considerable enhancement in activities by providing more hydroxyl groups of near BDEs. In the water medium, **1c** posses a lower value of PA than other compounds. Frontier molecular orbital analysis also supports the radical scavenging activity of compounds. The HOMO-LUMO gap of each molecule increases when the medium changes from the gas phase to the water medium. Hence all these are more stable in water especially gallic acid-based pigments, so the stability issue of bare pigments in the solvent can be solved by its gallates.

References

[1] M. Abbas, F. Saeed, F.M. Anjum, M. Afzaal, T. Tufail, M.S. Bashir, A. Ishtiaq, S. Hussain, H.A.R. Suleria, Natural polyphenols: An overview, Int. J. Food Prop. 20 (2017) 1689–1699. doi:10.1080/10942912.2016.1220393.

[2] K.E. Heim, A.R. Tagliaferro, D.J. Bobilya, Flavonoid antioxidants: chemistry, metabolism and structure-activity relationships, J. Nutr. Biochem. 13 (2002) 572–584. doi:doi:10.1016/S0955-2863(02)00208-5.

[3] A. Castañeda-ovando, M.D.L. Pacheco-hernández, M.E. Páez-hernández, J.A. Rodríguez, C.A. Galán-vidal, Chemical studies of anthocyanins : A review, Food Chem. 113 (2009) 859–871. doi:10.1016/j.foodchem.2008.09.001.

[4] M. Moloney, R.J. Robbins, T.M. Collins, T. Kondo, K. Yoshida, O. Dangles, Red cabbage anthocyanins: The influence of d-glucose acylation by hydroxycinnamic acids on their structural transformations in acidic to mildly alkaline conditions and on the resulting color, Dye. Pigment. 158 (2018) 342–352. doi:doi:10.1016/j.dyepig.2018.05.057.

[5] P. Araújo, N. Basílio, J. Azevedo, A. Fernandes, N. Mateus, F. Pina, V. de Freitas, J. Oliveira, Colour modulation of blue anthocyanin-derivatives. Lignosulfonates as a tool to improve the water solubility of natural blue dyes, Dye. Pigment. 153 (2018) 150–159. doi: doi:10.1016/j.dyepig.2018.02.019.

[6] M. Norman, P. Bartczak, J. Zdarta, H. Ehrlich, T. Jesionowski, Anthocyanin dye conjugated with Hippospongia communis marine demosponge skeleton and its antiradical activity, Dye. Pigment. 134 (2016) 541–552. doi:doi:10.1016/j.dyepig.2016.08.019.

[7] M. Ogawa, R. Takee, Y. Okabe, Y. Seki, Bio-geo hybrid pigment; clay-anthocyanin complex which changes color depending on the atmosphere, Dye. Pigment. 139 (2017) 561–565. doi: doi:10.1016/j.dyepig.2016.12.054.

[8] K. Phan, E. Van Den Broeck, V. Van Speybroeck, K. De Clerck, K. Raes, S. De Meester, The potential of anthocyanins from blueberries as a natural dye for cotton: A combined experimental and theoretical study, Dye. Pigment. 176 (2020) 108180. doi:doi:10.1016/j.dyepig.2019.108180.

[9] B. Tang, Y. He, J. Liu, J. Zhang, J. Li, J. Zhou, Y. Ye, J. Wang, X. Wang, Kinetic investigation into pH-dependent color of anthocyanin and its sensing performance, Dye. Pigment. 170 (2019) 107643. doi:doi:10.1016/j.dyepig.2019.107643.

[10] H.E. Khoo, Anthocyanidins and anthocyanins : colored pigments as food, pharmaceutical ingredients, and the potential health benefits, Food Nutr. Res. 61 (2017). doi:10.1080/16546628.2017.1361779.

[11] F. Pina, M.J. Melo, C.A.T. Laia, A.J. Parola, J.C. Lima, Chemistry and applications of flavylium compounds: a handful of colours, Chem. Soc. Rev. 41 (2012) 869–908. doi:10.1039/C1CS15126F.

[12] A. Bimpilas, M. Panagopoulou, D. Tsimogiannis, V. Oreopoulou, Anthocyanin copigmentation and color of wine: The effect of naturally obtained hydroxycinnamic acids as cofactors, Food Chem. 197 (2016) 39–46. doi:doi:10.1016/j.foodchem.2015.10.095.

[13] R. Boulton, The Copigmentation of Anthocyanins and Its Role in the Color of Red Wine: A Critical Review, Am. J. Enol. Vitic. 52 (2001) 67 LP – 87. http://www.ajevonline.org/content/52/2/67.abstract.

[14] J. Martín, E.M. Kuskoski, M.J. Navas, A.G. Asuero, Antioxidant Capacity of Anthocyanin Pigments, in: 2019. doi:10.5772/67718.

[15] T. Mu, H. Sun, M. Zhang, C. Wang, Chapter 6- Sweet Potato Anthocyanins, in: T. Mu, H. Sun, M. Zhang, C.B.T.-S.P. P.T. Wang (Eds.), Academic Press, 2017: pp. 279–355. doi:doi:10.1016/B978-0-12-812871-8.00006-4.

[16] V.K. Rajan, S.A. TK, H. CK, K. Muraleedharan, A non toxic natural food colorant and antioxidant 'Peonidin' as a pH indicator: A TDDFT analysis, Comput. Biol. Chem. 76 (2018) 202–209. doi:doi:10.1016/j.compbiolchem.2018.07.015.

[17] P. Trouillas, J.C. Sancho-García, V. De Freitas, J. Gierschner, M. Otyepka, O. Dangles, Stabilizing and Modulating Color by Copigmentation: Insights from Theory and Experiment, Chem. Rev. 116 (2016) 4937–4982. doi:10.1021/acs.chemrev.5b00507.

[18] A.A. Zanwar, S.L. Badole, P.S. Shende, M. V Hegde, S.L. Bodhankar, Chapter 80- Role of Gallic Acid in Cardiovascular Disorders, in: R.R. Watson, V.R. Preedy, S.B.T.-P. in HH and D. Zibadi (Eds.), Academic Press, San Diego, 2014: pp. 1045–1047. doi:doi:10.1016/B978-0-12-398456-2.00080-3.

[19] S. Choubey, L.R. Varughese, V. Kumar, V. Beniwal, Medicinal importance of gallic acid and its ester derivatives: a patent review, Pharm. Pat. Anal. 4 (2015) 305–315. doi:10.4155/ppa.15.14.

[20] V.K. Rajan, K. Muraleedharan, A computational investigation on the structure, global parameters and antioxidant capacity of a polyphenol, Gallic acid, Food Chem. 220 (2017) 93–99. doi:doi:10.1016/j.foodchem.2016.09.178.

[21] B. Badhani, N. Sharma, R. Kakkar, Gallic acid: a versatile antioxidant with promising therapeutic and industrial applications, RSC Adv. 5 (2015) 27540–27557. doi:10.1039/C5RA01911G.

[22] J. Wang, H. Tang, B. Hou, P. Zhang, Q. Wang, B.L. Zhang, Y.W. Huang, Y. Wang, Z.M. Xiang, C.T. Zi, X.J. Wang, J. Sheng, Synthesis, antioxidant activity, and density functional theory study of catechin derivatives, RSC Adv. 7 (2017) 54136–54141. doi:10.1039/c7ra11496f.

[23] D. Botten, G. Fugallo, F. Fraternali, C. Molteni, Structural Properties of Green Tea Catechins, J. Phys. Chem. B. 119 (2015) 12860–12867. doi:10.1021/acs.jpcb.5b08737.

[24] NM. O'Boyle, M. Banck, C.A. James, C. Morley, T. Vandermeersch, G.R. Hutchison, Open Babel: An open chemical toolbox., J. Cheminform. 3 (2011) 33. doi:10.1186/1758-2946-3-33.

[25] R. Dennington, T.A. Keith, J.M. Millam, GaussView {V}ersion {5.0.8}, (2008).

[26] M.J. Frisch, G.W. Trucks, H.B. Schlegel, G.E. Scuseria, M.A. Robb, J.R. Cheeseman, G. Scalmani, V. Barone, G. A. Petersson, H. Nakatsuji, X. Li, M. Caricato, A. V Marenich, J. Bloino, B.G. Janesko, R. Gomperts, B. Mennucci, H. P. Hratchian, J. V Ortiz, A.F. Izmaylov, J.L. Sonnenberg, D. Williams-Young, F. Ding, F. Lipparini, F. Egidi, J. Goings, B. Peng, A. Petrone, T. Henderson, D. Ranasinghe, V.G. Zakrzewski, J. Gao, N. Rega, G. Zheng, W. Liang, M. Hada, M. Ehara, K. Toyota, R. Fukuda, J. Hasegawa, M. Ishida, T. Nakajima, Y. Honda, O. Kitao, H. Nakai, T. Vreven, K. Throssell, J.A. Montgomery Jr., J.E. Peralta, F. Ogliaro, M.J. Bearpark, J.J. Heyd, E.N. Brothers, K.N. Kudin, V.N. Staroverov, T.A. Keith, R. Kobayashi, J. Normand, K. Raghavachari, A.P. Rendell, J.C. Burant, S.S. Iyengar, J. Tomasi, M. Cossi, J.M. Millam, M. Klene, C. Adamo, R. Cammi, J.W. Ochterski, R.L. Martin, K. Morokuma,

O. Farkas, J.B. Foresman, D.J. Fox, Gaussian 09, Revision A.02, (2016).

[27] T. Lu, F. Chen, Multiwfn: A multifunctional wavefunction analyzer, J. Comput. Chem. 33 (2012) 580–592. doi:10.1002/jcc.22885.

[28] G. Wang, Y. Xue, L. An, Y. Zheng, Y. Dou, L. Zhang, Y. Liu, Theoretical study on the structural and antioxidant properties of some recently synthesised 2,4,5-trimethoxy chalcones, Food Chem. 171 (2015) 89–97. doi:h ttps://doi.org/10.1016/j.foodche m.2014.08.106.

[29] V.K. Rajan, C.K. Hasna, K. Muraleedharan, The natural food colorant Peonidin from cranberries as a potential radical scavenger – A DFT based mechanistic analysis, Food Chem. 262 (2018) 184–190. doi:doi:10.1016/j. foodchem.2018.04.074.

[30] V.K. Rajan, T.K. Shameera Ahamed, K. Muraleedharan, Studies on the UV filtering and radical scavenging capacity of the bitter masking flavanone Eriodictyol, J. Photochem. Photobiol. B Biol. 185 (2018) 254–261. doi:doi:10.1016/j. jphotobiol.2018.06.017.

[31] P. Ajmala Shireen, V.M. Abdul Mujeeb, K. Muraleedharan, Theoretical insights on flavanones as antioxidants and UV filters: A TDDFT and NLMO study, J. Photochem. Photobiol. B Biol. 170 (2017) 286–294. doi:doi:10.1016/j. jphotobiol.2017.04.021.

[32] P. Ajmala Shireen, K. Muraleedharan, V.M. Abdul Mujeeb, Theoretical studies on antioxidant potential of alpinetin, Mater. Today Proc. 5 (2018) 8908–8915. doi:doi: 10.1016/j.matpr.2017.12.325.

[33] AD Becke, Density-functional thermochemistry. III. The role of exact exchange, J. Chem. Phys. 98 (1993) 5648–5652. doi:10.1063/1.464913.

[34] Lee, Yang, Parr, Development of the Colle-Salvetti correlation-energy formula into a functional of the electron density., Phys. Rev. B. Condens. Matter. 37 (1988) 785–789. doi:10.1103/ physrevb.37.785.

[35] M. Leopoldini, T. Marino, N. Russo, M. Toscano, Antioxidant Properties of Phenolic Compounds: H-Atom versus Electron Transfer Mechanism, J. Phys. Chem. A. 108 (2004) 4916–4922. doi: 10.1021/jp037247d.

[36] M. Leopoldini, N. Russo, M. Toscano, The molecular basis of working mechanism of natural polyphenolic antioxidants, Food Chem. 125 (2011) 288–306. doi:10.1016/j. foodchem.2010.08.012.

[37] G. Litwinienko, K.U. Ingold, Solvent Effects on the Rates and Mechanisms of Reaction of Phenols with Free Radicals, Acc. Chem. Res. 40 (2007) 222–230. doi: 10.1021/ar0682029.

[38] G. Litwinienko, K.U. Ingold, Abnormal Solvent Effects on Hydrogen Atom Abstraction. 2. Resolution of the Curcumin Antioxidant Controversy. The Role of Sequential Proton Loss Electron Transfer, J. Org. Chem. 69 (2004) 5888–5896. doi:10.1021/ jo049254j.

[39] M. Musialik, G. Litwinienko, Scavenging of dpph• Radicals by Vitamin E Is Accelerated by Its Partial Ionization: the Role of Sequential Proton Loss Electron Transfer, Org. Lett. 7 (2005) 4951–4954. doi:10.1021/ ol051962j.

[40] I. Nakanishi, T. Kawashima, K. Ohkubo, H. Kanazawa, K. Inami, *M. Mochizuki*, K. Fukuhara, H. Okuda, T. Ozawa, S. Itoh, S. Fukuzumi, N. Ikota, Electron-transfer mechanism in radical-scavenging reactions by a vitamin E model in a protic medium, Org. Biomol. Chem. 3 (2005) 626–629. doi:10.1039/ b416572a.

[41] J. Tošović, S. Marković, D. Milenković, Z. Marković, Solvation enthalpies and Gibbs energies of the proton and electron: Influence of solvation models, J. Serbian Soc. Comput. Mech. 10 (2016) 66–76. doi: 10.5937/jsscm1602066T.

[42] Z. Marković, J. Tošović, D. Milenković, S. Marković, Revisiting the solvation enthalpies and free energies of the proton and electron in various solvents, Comput. Theor. Chem. 1077 (2016) 11–17. doi:10.1016/j.comptc.2015.09.007.

[43] J.N. Woodford, A DFT investigation of anthocyanidins, 410 (2005) 182–187. doi:10.1016/j.cplett.2005.05.067.

[44] E. Matito, J. Poater, M. Solà, M. Duran, P. Salvador, Comparison of the AIM Delocalization Index and the Mayer and Fuzzy Atom Bond Orders, J. Phys. Chem. A. 109 (2005) 9904–9910. doi:10.1021/jp0538464.

[45] I. Mayer, P. Salvador, Overlap populations, bond orders and valences for 'fuzzy' atoms, Chem. Phys. Lett. 383 (2004) 368–375. doi:doi:10.1016/j.cplett.2003.11.048.

[46] TK. Shameera Ahamed, V.K. Rajan, K. Sabira, K. Muraleedharan, DFT and QTAIM based investigation on the structure and antioxidant behavior of lichen substances Atranorin, Evernic acid and Diffractaic acid, Comput. Biol. Chem. 80 (2019) 66–78. doi:doi: 10.1016/j.compbiolchem.2019.03.009.

Chapter 17

Structure-Property Relationships in Benzofurazan Derivatives: A Combined Experimental and DFT/TD-DFT Investigation

Abstract

In this work we seek to understand and to quantify the reactivity of benzofurazan derivatives toward secondary cyclic amines, like pyrrolidine, piperidine and morpholine, acting as nucleophile groups in S_NAr reactions. For this aim, physico-chemical and structural descriptors were determined experimentally and theoretically using the DFT/B3LYP/6-31+ g (d,p) methodology. Thus, different 4-X-7-nitrobenzofurazans (X = OCH_3, OC_6H_5 and Cl) and products corresponding to the electrophilic aromatic substitution by pyrrolidine, piperidine and morpholine, were investigated. Particularly, the HOMO and LUMO energy levels of the studied compounds, determined by Cyclic Voltammetry (CV) and DFT calculations, were used to evaluate the electrophilicity index (ω). The latter was exploited, according to Parr's approach, to develop a relationship which rationalizes the kinetic data previously reported for the reactions of the 4-X-7-nitrobenzofurazans with nucleophiles cited above. Moreover, the Parr's electrophilicity index (ω) of these benzofurazans determined in this work were combined with their electrophilicity parameters (E), reported in preceding papers, was found to predict the unknown electrophilicity parameters E of 4-piperidino, 4-morpholino and 4-pyrrolidino-7-nitrobenzofurazan. In addition, the relationship between the Parr's electrophilicity index (ω) and Hammett constants σ, has been used as a good model to predict the electronic effect of the nucleophile groups. Finally, we will subsequently compare the electrophilicity index (ω) and the electrophilicity parameters (E) of these series of 7-X-4-nitrobenzofurazans with the calculated dipole moment (μ) in order to elucidate general relationships between E, ω and μ.

Keywords: Benzofurazan, Structure–property correlations, Optical properties, Mayr's approach, Electrophilicity, Parr's approach, Voltammetry Cyclic, DFT, ICT, NLO properties

1. Introduction

4-Nitrobenzofurazan (NBD) is neutral 10π electron-deficient heteroaromatic substrates. Organic materials containing the NBD moiety have been tested in potential biomedical and bio-analytical applications [1–7]. They have been also used as an effective electron-withdrawing building block for organic solar cell materials [8], and have been discovered to be nonlinear optical (NLO) materials [9]. In addition, the strong electron withdrawing properties of nitrobenzofurazan (NBD) derivatives make it an exceptional electrophilic product of the nucleophilic aromatic substitution (S_NAr).

In addition, the activated chlorine atom in 4-chloro-7-nitrobenzofurazan (NBD chloride, namely NBD-Cl) can undergo electrophilic aromatic substitution by phenoxide, primary or secondary cyclic amines [10–13] and some of 4,7 di-substituted benzofurazan compounds to generate weakly or non-fluorescents products [14–18].

A previous kinetic study of amino-substituted NBD showed that methoxy and aryloxy substituent's are better leaving groups than chloro and that a red-colored Meisenheimer complex immediate was formed during the reactions [19].

On the other hand, the frontier molecular orbital descriptors including HOMO and LUMO energy levels play major roles in governing many chemical reactions as function of reactivity and stability of related molecules [20, 21]. Alternatively, contacts from intra-molecular interactions have been identified as an important driving force on the stabilization of various planar molecular structures [22, 23]. However, these interactions lead to intra-molecular charge transfer of formed compounds and thus provides a non-radiative decay from the excited state [24–26].

Recently, we reported a kinetic study for S_NAr reactions of NBD with secondary amines [27–30]. The present chapter is focused on the elucidation of the combined experimental–theoretical investigation relationships between the optical (UV–Vis, PL and time resolved photoluminescence (TR-PL), electrochemical (CV) properties and the chemical structures of 4.7-di-substituted benzofurazans (NBDs), based on the kinetic reactivity process of compounds. Parr's approach was employed to develop a relationship which rationalizes the kinetic data previously reported for the reactions of 4-X-7-nitrobenzofurazans with various nucleophiles [27–29].

Then, the substituent effects were particularly limited only to morpholine, piperidine and pyrrolidine groups. Thus, we seek to study the structure–property relationships for molecules containing 4-nitrobenzofurazan moiety for a better understanding of their optical behavior and their chemical reactivity. For this aim, theoretical calculations using quantum chemical calculations within density functional theory (DFT) and its extension TD-DFT in the 6-31 + g(d,p) basis set have been developed to facilitate the in-depth understanding of structure–property relationships.

The Molecular structures of the investigated compounds are illustrated in **Figures 1** and 2.

To the best of our knowledge, no similar study has been published on electrochemical and photo-physical properties of the studied compounds. Cyclic Voltammetry (CV) is used to estimate the band gaps, electron affinities (i.e. the

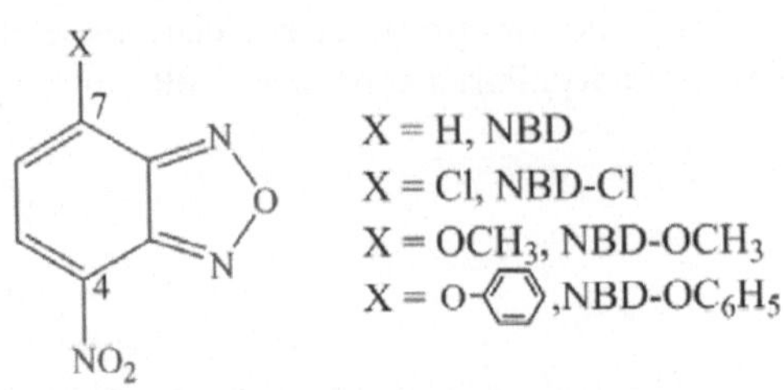

Figure 1.
Chemical structures of the studied electrophile compounds.

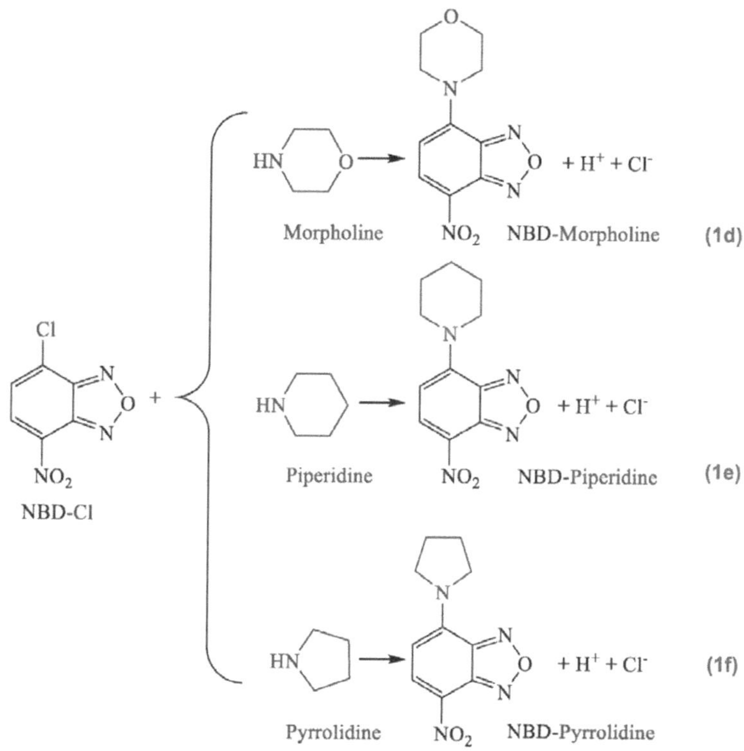

Figure 2.
Reaction mechanism between NBD-Cl and three examples of nucleophiles (morpholine, piperidine and pyrrolidine) groups.

energy of LUMO, ε_{LUMO}) and ionization potentials (i.e., the energy of HOMO, ε_{HOMO}) [31]. Besides, the computed results from DFT and TD-DFT calculations enable the access to the structure geometry, the Mullikan charge distribution and the dipole moment of the 4,7 di-substituted benzofurazan at the ground- (S_0) and excited (S_1)-states [32, 33].

2. Experimental details

2.1 Materials

The 7-chloro-4-nitrobenzofurazan (NBD-Cl) and the tetraethyl ammonium tetrafluoroborate (TEAF) were commercial products (Aldrich, Darmstadt-Germany). Samples of 7-methoxy-4-nitrobenzofurazan (NBD-OCH_3) and 7-phenoxy-4-nitrobenzofurazan (NBD-OC_6H_5) were prepared according to the standard methods described in the literature [27–29]. The secondary cyclic amines (morpholine, piperidine and pyrrolidine) were commercial products (Aldrich, Darmstadt-Germany) and were redistilled before use whenever necessary.

2.2 Apparatus

Cyclic Voltammetry (CV) measurements were carried out under nitrogen gas with a Voltalab 10 apparatus from Radiometer driven by the Volta Master software at a potential scan rate of 0.1 V/s, at room temperature, in 25 mL of CH_3CN solution containing TEAF (0.1 M) as a supporting electrolyte. The cell used three electrodes. The working electrode was a 2 mm diameter platinum disk (Tacussel type EDI) and a platinum (Pt) wire as a counter electrode. The concentrations of the substrates were 10^{-3} M. The potentials E_{Ox} and E_{Red} give us the information on HOMO and LUMO energy levels of studied materials.

The UV–visible absorption spectra of compounds have carried out on a Shimadzu UV–visible model 1650 spectrometer. The samples were dissolved in acetonitrile medium. The optical properties of the sample are secondly investigated by photoluminescence spectroscopy. The 514.5 nm line of the continuous-wave argon Ar + laser is used as an excitation source. Spectral analysis of the PL was performed using a Jobin Yvon iHR320 monochromator through the lock-in amplifier technique.

For the Time-Resolved PL (TRPL) study, the sample was illuminated by a pulsed laser diode with photon energy of 1.58 eV with a repetition rate of 80 MHz, using the time correlated photo counting technique. The details experiment was reported elsewhere [34].

2.3 Computational methods

Ground-state (S_0) optimized geometries of the studied compounds were performed by means of DFT at B3LYP level of theory [35, 36] in conjunction with the 6-31+ g(d,p) basis set. No constraints were used and all structures were free to optimize in an acetonitrile solution by applying the conductor polarizable continuum model (CPCM) with the Gaussian 09 program [37] and the output files were visualized with the Gauss View (5.0.8) molecular visualization program [38]. To predict the optical absorption properties of the studied compounds, UV–Vis absorption spectra were simulated at the time-dependent TD-DFT level with the same basis set. The electronic properties such as highest occupied molecular orbital (ε_{HOMO}) and lowest unoccupied molecular orbital (ε_{LUMO}) energy levels and dipole moment have been extracted from calculations.

The semi-empirical quantum-chemical ZINDO level was used to predict the optical emission spectra on S_1 optimized structures. Electronic transitions assignment and oscillator strengths were also calculated using the same method of calculations.

3. Results and discussion

3.1 Electrochemical analysis

Electrochemical comparative studies were undertaken using Cyclic Voltammetry (CV) for the six compounds 4.7 di-substituted benzofurazan (See **Figures 3** and **4**). The CVs profiles were performed in the potential range of −2.0 to +3.0 V.

The potentials (E_{OX} and E_{Red}) peaks can be used to determine the HOMO and LUMO energy levels of the studied compounds. Indeed, the value of their energy gaps can be estimated from the potentials difference.

The highest occupied molecular orbital (HOMO) energy and the lowest unoccupied molecular orbital (LUMO) energy were estimated from the first oxidation and reduction potentials, respectively. The energy levels were calculated according to an empirical method [39] and by assuming that the energy level of the ferrocene/ferrocenium is 4.8 V below the vacuum level.

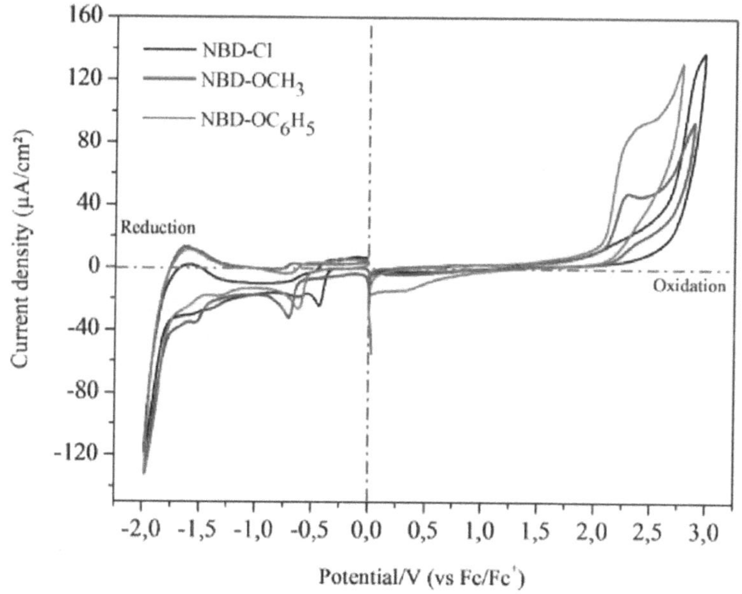

Figure 3.
Oxidation and reduction voltammograms of NBD-Cl, NBD-OCH$_3$ and NBD-OC$_6$H$_5$ products (10^{-3} M) recorded in acetonitrile.

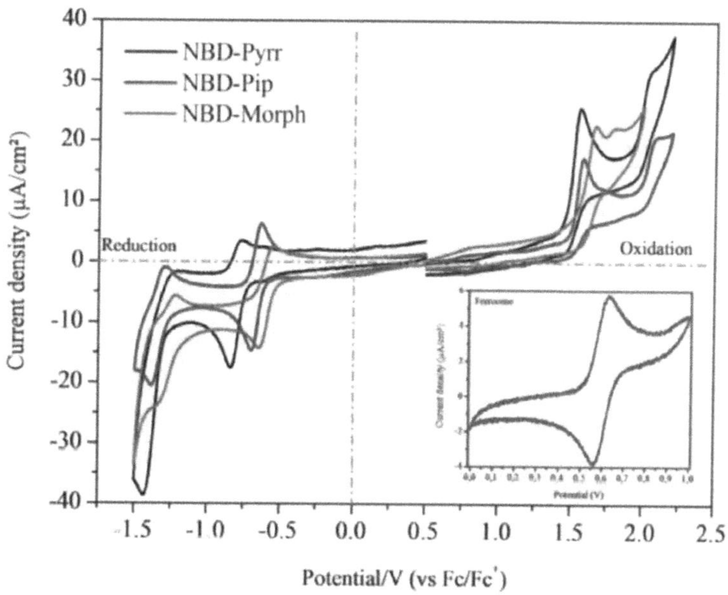

Figure 4.
Oxidation and reduction voltammograms of NBD-Pyrr, NBD-pip and NBD-Morph products (10^{-3} M) recorded in acetonitrile.

$$\varepsilon_{HOMO} = -e\left[E_{OX} - E^0_{Fc/Fc^+}\right] - 4.8 \quad (1)$$

$$\varepsilon_{LUMO} = -e\left[E_{Red} - E^0_{Fc/Fc^+}\right] - 4.8 \quad (2)$$

Here, E_{Ox} and E_{Red} are the potential of the oxidation peak and reduction peak, respectively. These values were obtained from the CV curve for each compound;

E^0_{Fc/Fc^+} is 0.59 eV, the average of oxidation and reduction peak potentials of ferrocene measured under the same experimental condition; 4.8 eV is the energy difference between the energy level of ferrocene and the vacuum. Note that when the oxidation/reduction peak of the studied compound was reversible, the Eox/red was replaced in the equation by E°ox/red determined experimentally.

For the electrophile compounds (**Figure 3** and **Table 1**), the peaks appeared at the reduction potentials of about: -0.440 V (NBD-Cl), −0.716 (NBD-OCH_3) and – 0.610 V (NBD-OC_6H_5). They correspond to LUMO energy levels at −3.77 eV, −3.49 eV and – 3.60 eV, respectively. In the same way, the oxidation potentials located at: 2.90 V, 2.28 V and 2.40 V led to HOMO energy level values of −7.11 eV, −6.49 eV and – 6.61 eV for NBD-Cl, NBD-OCH_3 and NBD-OC6H_5, respectively. The electrochemical energy gaps deduced from these measurements were: 3.34 eV (NBD-Cl), 3.00 eV (NBD-OCH_3) and 3.01 eV (NBD-OC_6H_5).

For the three first compounds used as electrophile groups, they were oxidized and reduced at relatively distinct potentials. Indeed, the NBD-Cl is oxidized and reduced at higher potentials than methoxy and phenoxy substituted NBD.

For the amino-substitued NBD (**Figure 4** and **Table 1**), the oxidation potential peak values were detected nearly 1.66 V, 1.58 V and 1.55 V for morpholine, piperidine and pyrrolidine substituted NBD, respectively. They correspond to HOMO energy levels at −5.87 eV, −5.79 eV and – 5.76 eV, respectively. However, the reduction peaks of the studied compounds were reversible and thus the E_{Red} was replaced in the equation by E^0_{Red} determined experimentally. Accordingly, the estimated LUMO energy levels are of about: −3.62 eV (NBD-Morph), −3.55 eV (NBD-Pip) and – 3.41 eV (NBD-Pyrr).

Compounds	Cyclic voltammetry (CV)							
	E_{OX} (V vs. FC/FC⁺)	E_{Red} (V vs. FC/FC⁺)	$E^0_{Red}(V)$	Energy levels		Chemical reactivity descriptors		
				ε^{CV}_{HOMO} (eV)	ε^{CV}_{LUMO} (eV)	μ (eV)	η (eV)	W^{CV} (eV)
NBD-Cl	2.90	-0.44	—	−7.11	−3.77	−5.44	1.67	8.86
NBD-OCH3	2.28	−0.71	—	−6.49	−3.49	−4.99	1.50	8.30
NBD-OC6H5	2.40	−0.61	—	−6.61	−3.60	−5.10	1.51	8.65
NBD-Morph	1.66	−0.64	−0.59	−5.87	−3.62	−4.74	1.13	9.94
NBD-Pip	1.58	−0.70	−0.66	−5.79	−3.55	−4.66	1.12	9.67
NBD-Pyrr	1.55	−0.84	−0.80	−5.76	−3.41	−4.58	1.18	8.88

Table 1.
Calculated HOMO and LUMO energy levels and extracted chemical reactivity descriptors for studied compounds.

3.2 Photo-physical properties

3.2.1 UV-Vis optical absorption and emission (PL) analysis

The optical absorption spectra of the four substituted benzofurazans (NBDs) compounds are illustrated in **Figure 5**. The NBD-Cl, as starting material, has two distinct absorption bands (262 nm and 337 nm). The spectral characteristics of NBD-Cl were maintained and appeared as the same general features for

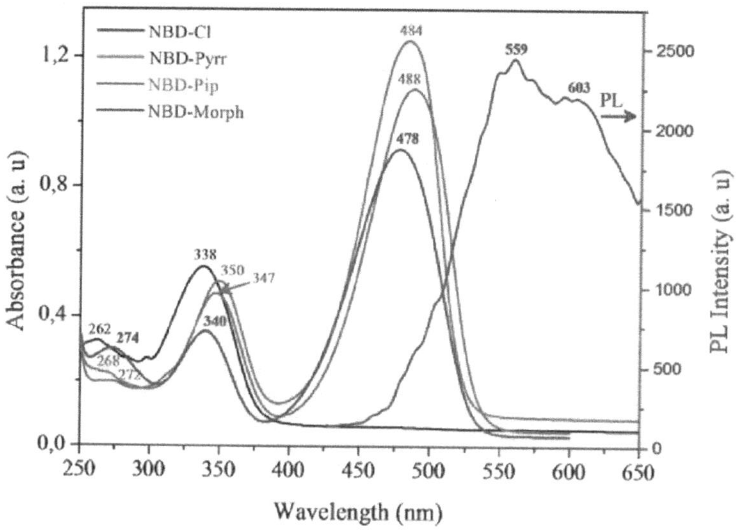

Figure 5.
Experimental UV–vis optical absorption spectra of studied compounds as well as experimental PL spectrum of NBD-Morph.

amino-NBD derivatives. For the later systems, a new optical band in the range 478-488 nm was assigned to the formation of amino-NBD derivatives by S_NAr reactions. In fact, typical three distinct absorption bands are presented for NBD substituted with pyrrolidine, piperidine or morpholine, similar to that previously reported NBD-amino derivatives [1–3]. Besides, the maximum absorption wavelengths of amino-NBD derivatives shifted more bathochromically than that of NBD-Cl.

According to assignments reported previously by Heberer and coworkers for NBD-amino derivatives [40] the short-wavelength band is associated to the aromatic benzofurazan compounds. The middle band at around 340 nm was attributed to a $\pi \rightarrow \pi^*$ electronic transition. However, the band with lower energy in the range of 478-487 nm has ascribed to ICT arranged between the electron-donor and the nitro (NO_2) electron-withdrawing groups within the molecule) (NBD-Cl, free band). Importantly, the maxima for the charge transfer band in these systems are detected in the visible region of the spectra.

In addition, similar profiles of optical absorption spectra for amino-NBD were obtained. Thus, the excitation corresponding to the lowest energy band of title compounds was used in order to investigate the excitation process from the ground (S_0) to the first excited state (S_1).

Experimentally, the 4,7-disubstitued benzofurazan compound, NBD-Morph was excited at the maximum absorption wavelength and show two distinct fluorescence bands (See **Figure 5**). The PL spectrum shows of two dominant features peaking at approximately 2.21 eV (559 nm) and 2.05 eV (603 nm), thus, a broad green to orange emission band has been identified for the tested NBD-Morph compound.

3.2.2 Time-resolved photoluminescence (TR-PL) decay kinetics

The performed time-resolved photoluminescence (TR-PL) decay kinetics is analyzed and illustrated by experiment (**Figure 6**). A particular attention is devoted to the thorough analysis of non-exponential decay kinetics. However, the TR-PL decay recorded at longer wavelength emission was relatively affected. It should be mentioned that lifetimes are inaccessible from the decay process by the use

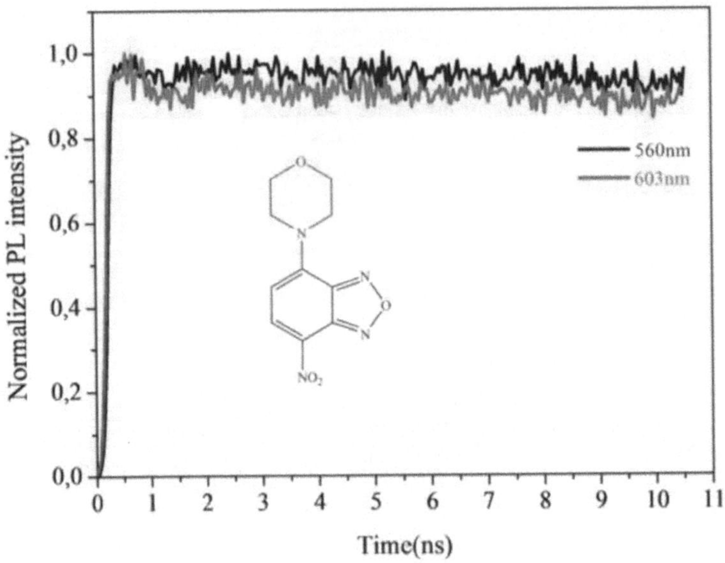

Figure 6.
Time-resolved decay curves recorded at emission energy of 2.21 eV (560 nm) and 2.05 eV (603 nm) for NBD-Morph film.

experimental equipment and could be related to an efficient intra-molecular charge transfer occurred in NBD-Morph molecule. The result was supported by the ICT characteristic optical band for a compound that had a lower intensity compared to those of analogue compounds (See **Figure 5**). The flat TR-PL spectrum means that the photo-carriers have a long lifetime (more than the nanosecond regime), leading to an important diffusion length.

3.3 Quantum chemistry computation results

Complete geometry optimizations of studied compounds were firstly carried out at DFT using B3LYP functional and the polarized 6-31 g+(d,p) basis set. The optimized molecular structures are illustrated in **Figure** 7. Particularly, the title amino-NBD compounds contain mainly three types of chemical bonds, namely O—H, N—H and N—O which are associated to an intra-molecular non covalent interactions.

The computed geometrical parameters related to the short contacts at ground (S_0) and excited (S_1) states are given in **Figure** 7 and tabulated in **Table** 2. The S_0 calculated short contact O—H was 2.39 Å (versus 2.38 Å, S_1) in all systems,

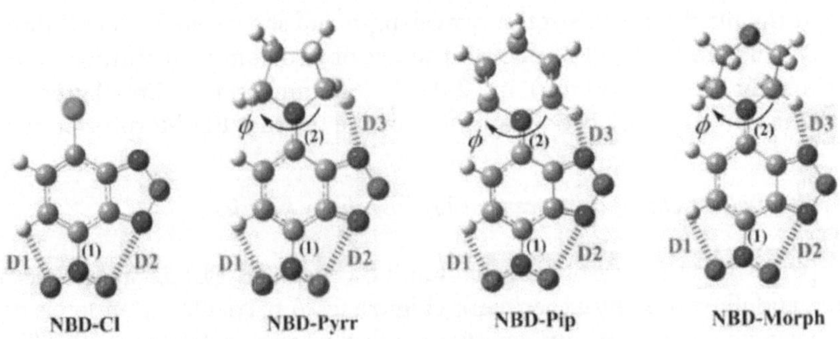

Figure 7.
Molecular optimized structures with intra-molecular short contacts.

	Intra-molecular Short Contacts			Covalent contacts		Dihedral Angle	Dipole Moment (D)
	D1 O—H (Å)	D2 N—O (Å)	D3 N—H (Å)	dC-N (1) (Å)	dC-N (2) (Å)	ϕ (°)	
NBD-Pyrr	2.39 (2.38)	2.78 (2.74)	2.47 (2.53)	1.41 (1.37)	1.33 (1.32)	3.47 (5.75)	17.08 (17.45)
NBD-Pip	2.39 (2.38)	2.78 (2.74)	2.11 (2.25)	1.41 (1.37)	1.34 (1.33)	9.66 (16.76)	16.84 (16.53)
NBD-Morph	2.39 (2.38)	2.78 (2.74)	2.11 (2.22)	1.41 (1.37)	1.35 (1.33)	5.16 (12.34)	13.57 (13.95)

Table 2.
Calculated non covalent interactions for O—H N—O and N—H participating in short intra-molecular contacts within molecular at ground (S_0) and excited (S_1) states. The values presented in parentheses are obtained at S_1.

considerably smaller than the sum of the van der Waals radii (2.72 Å). The N—O short contact is fond 2.78 Å versus 2.74 Å, S1) significantly too smaller than the sum of the van der Waals radii (3.07 Å). In addition, the third short contact of N—H (smaller than the sum of the van der Waals radii (2.75 Å)) is allowed to vary from 2.11 Å -2.47 Å (S_0) to 2.22 Å-2.53 Å (S_1). The later (N—H) contacts are sensitive to amino substituted groups. As a result, these contacts have been identified as an important driving force on the stabilization of the studied co-planar molecular structures [41]. The major difference in the ground and excited state is the change in torsional angle ϕ and in the dipole moment. Calculations reveal that the existence of the amino groups significantly modifies the properties of their excited states. While the ground state is predicted to be planar, the excited-state geometry is twisted by about 2-7°. Due to conformational changes, the singlet excited-state dipole moment was found to be greater than ground-state dipole moment, unexpected for NBD-Pip (16.84 D (S_1) versus 16.53 D (S_0)). Importantly, the C-NO_2 bond lengths indexed as (1) decreases by 0.4 Å upon excitation, due the nitro strong electron-withdrawing group effect.

Note that the band gap values estimated using the electrochemical method may be different from those calculated by optical or theoretical methods (see **Table 3**). This result could be explained by the fact that the oxidation/reduction process at the electrode are reactions generating species in ground states, whereas the optical electron transition leads to the formation of excited states. Moreover, the electrogeneration of a radical cation or radical anion in solution involves other thermodynamic (salvation ...) and kinetic effects. Consequently, it is expected that the peak

Compounds	Energy levels		Chemical reactivity descriptors			IP(eV)	EA(eV)
	ε_{HOMO}(eV)	ε_{LUMO}(eV)	μ (eV)	η (eV)	W (eV)		
X = Cl	−7.648	−3.929	−5.788	1.859	9.009	7.481	4.195
X = OCH_3	−6.988	−3.561	−5.274	1.713	8.117	6.806	3.830
X = OC_6H_5	−6.985	−3.684	−5.334	1.650	8.620	6.629	4.041
NBD-MORPH	−6.349	−3.296	−4.822	1.526	7.617	6.081	3.672
NBD-Pip	−6.238	−3.190	−4.714	1.524	7.290	6.013	3.607
NBD-Pyrr	−6.262	−3.151	−4.706	1.555	7.120	6.044	3.420

Table 3.
Calculated HOMO and LUMO energy levels and extracted chemical reactivity descriptors for studied compounds using DFT//B3LYP/6-31+ g(d,p) in acetonitrile.

potential will be lowered due to fast chemical reactions following the primarily electron transfer generating chemically reactive radical ions.

To better understand the optical responses, we used the time-dependent density functional theory (TD-DFT) to simulate the optical absorption spectra (See **Figure 8**). Their corresponding electronic transition assignments are given in **Table 4**. The maximum absorption wavelengths and the molecular extinction coefficients (ε) of the chloro- and amino-substituted NBD, in acetonitrile, are also given.

It is worthwhile mentioning that the extensive DFT and TD-DFT calculations show a good correlation between observed absorption and theoretical vertical excitation.

For the emission properties, it should be noted that the ZINDO semi-empirical quantum chemistry [42] was used to predict the emission spectra (See **Figure 9**). **Table 4** lists the emission optical bands of the amino substituted benzofurazan and their assignments. The emission properties are mainly from LUMO and LUMO+1 to HOMO.

It is found that compounds emit at appreciably higher wavelengths in the range 620-640 nm. It should be noted also that bathochromic effect is due to the NBD moiety and the amino substituted groups. It should be noted that NBD-Morph present absorption at wavelength maximum of 478 nm and emit orange light at 559-603 nm. This provides a large apparent Stokes shift of 81 nm attributed to the ICT process (**Table 5**). The combined experimental and computed results suggest that the new NBD-Morph compound is still considered fluorescent. Then, considerable efforts have been undertaken to test the other analogue compounds.

4-amino substituted NBD is recognized among to the broad family of intramolecular charge transfer (ICT) complexes. The ICT characteristics are identified by the presence of both electron donor and withdrawing acceptor substituent moieties within the same molecule. Thus, the photo-initiated electron transfer from the donor to the acceptor sites yields two kinds positive and negative charges within separated functional parts of the molecule. Herein, we have extracted the atomic

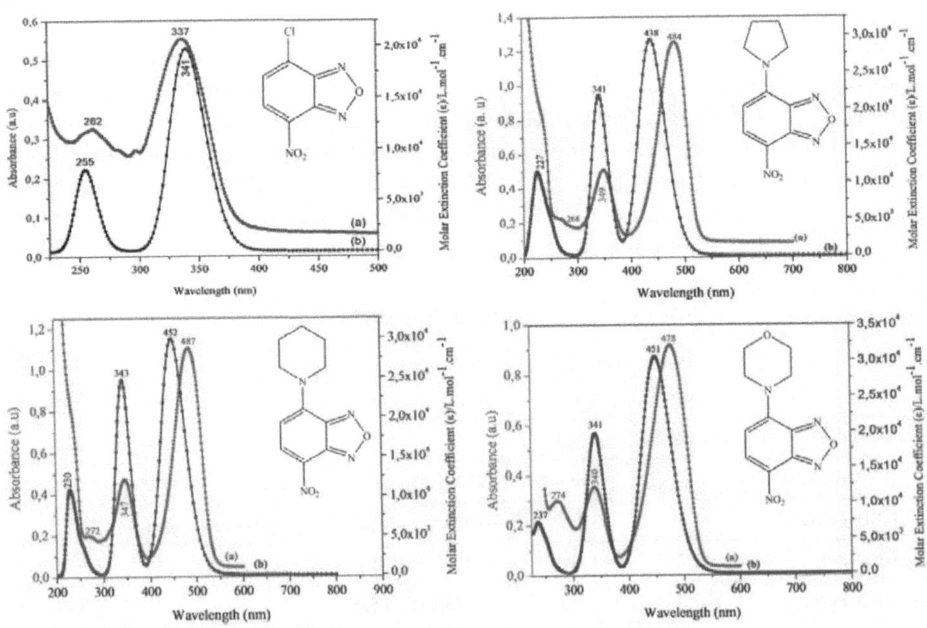

Figure 8.
Simulated optical absorption spectra of Cl-NBD and its related amio-substituted NBD by means of DFT// B3LYP/6-31+ g(d,p) level of theory, in acetonitrile.

Compounds	λ_{max}^{abs} (nm)	Osc. strength	Major contribs	Minor contribs
NBD-Cl	341	0.2729	H → L (99%)	—
	255	0.0709	H-4 → L (79%)	H → L + 2 (10%) H → L + 1 (6%)
	λ_{max}^{Exp}: 262 nm, 337 nm			
NBD-Pyrr	227	0.0181	H → L + 3 (94%)	H → L + 5 (3%)
	341	0.2781	H → L + 1 (87%)	H-1 → L (6%) H → L (5%)
	438	0.4065	H → L (94%)	H → L + 1 (5%)
	λ_{max}^{Exp}: 268 nm, 349 nm, 484 nm			
NBD-Pip	230	0.0012	H-7 → L (78%) H-4 → L + 1 (13%)	H-9 → L (3%) H-4 → L(2%)
	343	0.2083	H-1 → L (30%) H1 → L + 1 (59%)	H-1 → L + 1 (7%) H → L (2%)
	452	0.4129	H → L (94%)	H → L + 1 (5%)
	λ_{max}^{Exp}: 272 nm, 347 nm, 487 nm			
NBD-Morph	237	0.0859	H → L + 2 (64%) H-3 → L + 1 (15%) H-6 → L (12%)	H-6 → L + 1 (2%)
	341	0.2608	H → L + 1 (89%)	H-1 → L (5%) H → L (4%)
	451	0.4209	H → L (95%)	H → L + 1 (4%)
	λ_{max}^{Exp}: 274 nm, 340 nm, 478 nm			

Table 4.
The vertical excited energies (nm) and their oscillator strengths (f) for the ground ($S_o \rightarrow S_1$) states of studied compounds by TD//B3LYP/6-31+ g(d,p) level of theory.

Mulliken charges from both ground- and excited-geometry structures. The results are illustrated in **Figure 10**. In fact the 4-amino substituted NBD belongs to the broad family of ICT molecules, with the amino group acting as an electron donor upon photo-excitation, and the nitro (NO_2) group as an electron acceptor.

As illustrated in **Figure 11**, we have presented the frontier molecular orbitals (FMOs) to obtain insight into the molecular structure and optoelectronic changes from the ground (S_0) to excited (S_1) States. We can observe that for all systems, the nitro and the 4-amino substituent groups significantly contribute to the optoelectronic properties. Particularly, the NBD-Morph differs from the other two compounds, where the oxygen atom did not contribute exclusively in excited state. This could explain the electron transfer from the nitrogen group of the electron donor moiety.

3.4 Relationships of chemical structure and reactivity properties

It is relevant to note that good accuracy of computational predictions of optical properties is already attainable. Accordingly, we utilize this approach for accurate prediction of the global reactivity descriptors including chemical potential (μ), Chemical hardness (η) and electrophilicity index (ω) (see **Table 5**).

Parr's electrophilicity ω values are calculated according to Eq. (3) [43, 44] based on μ and η which can be evaluated using Eqs. (4) and (5).

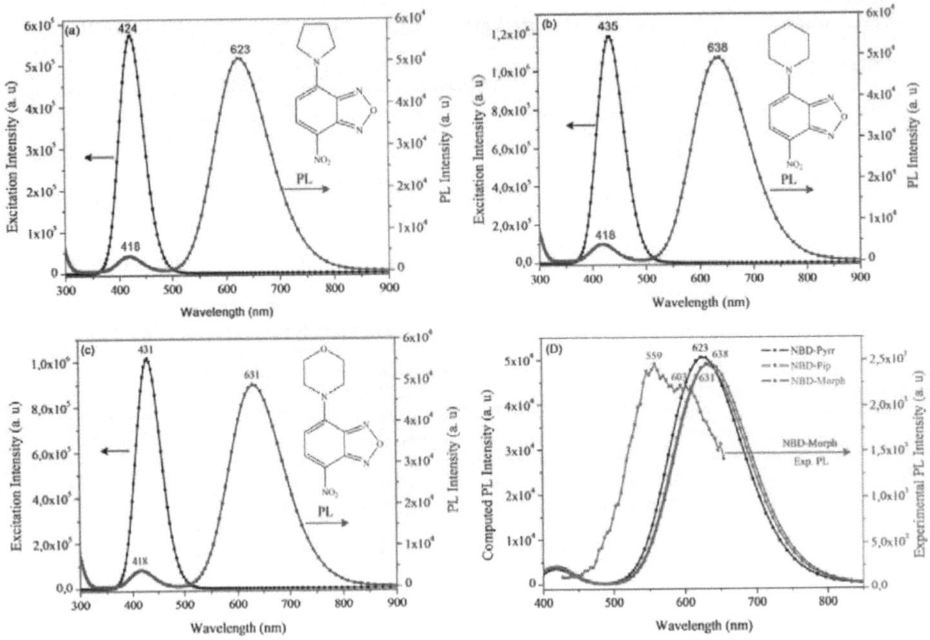

Figure 9.
*Simulated emission spectra of amino-substituted NBD compounds (a,b,c). Superposition of emission spectra with experimental data for NBD-Morph is given (**Figure 9D**).*

Compounds	λ_{max}^{Em} (nm)	Osc. strength	Major contribs	Minor contribs
NBD-Pyrr	418	0.0498	L + 1 → H (95%)	—
	623	0.6989	L → H (97%)	—
NBD-Pip	418	0.0564	L + 1 → H (95%)	—
	638	0.6806	L → H (97%)	—
NBD-Morph	418	0.0552	L + 1 → H (95%)	—
	631	0.6747	L → H (97%)	—
	λ_{max}^{Exp}: 559 nm, 603 nm			

Table 5.
The vertical excited energies (nm) and their oscillator strengths (f) for the $S_1 \rightarrow S_0$ states of studied compounds obtained by ZINDO method. The wavelengths excitation was ranging from 424 to 435 nm.

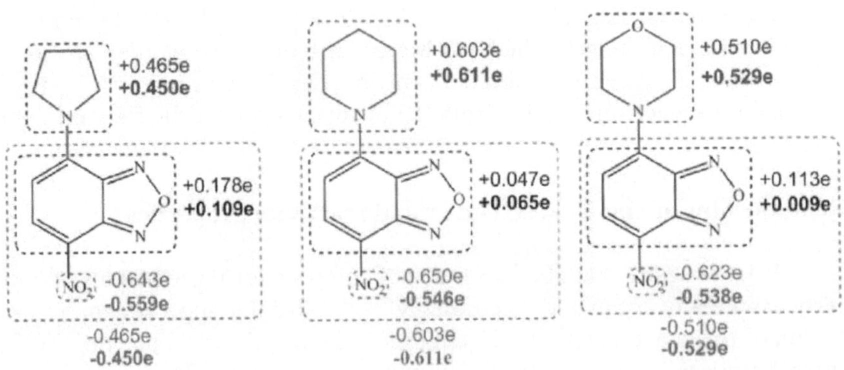

Figure 10.
Mulliken charge distribution of ground- and excited (bold values) states optimized structures of studied compounds.

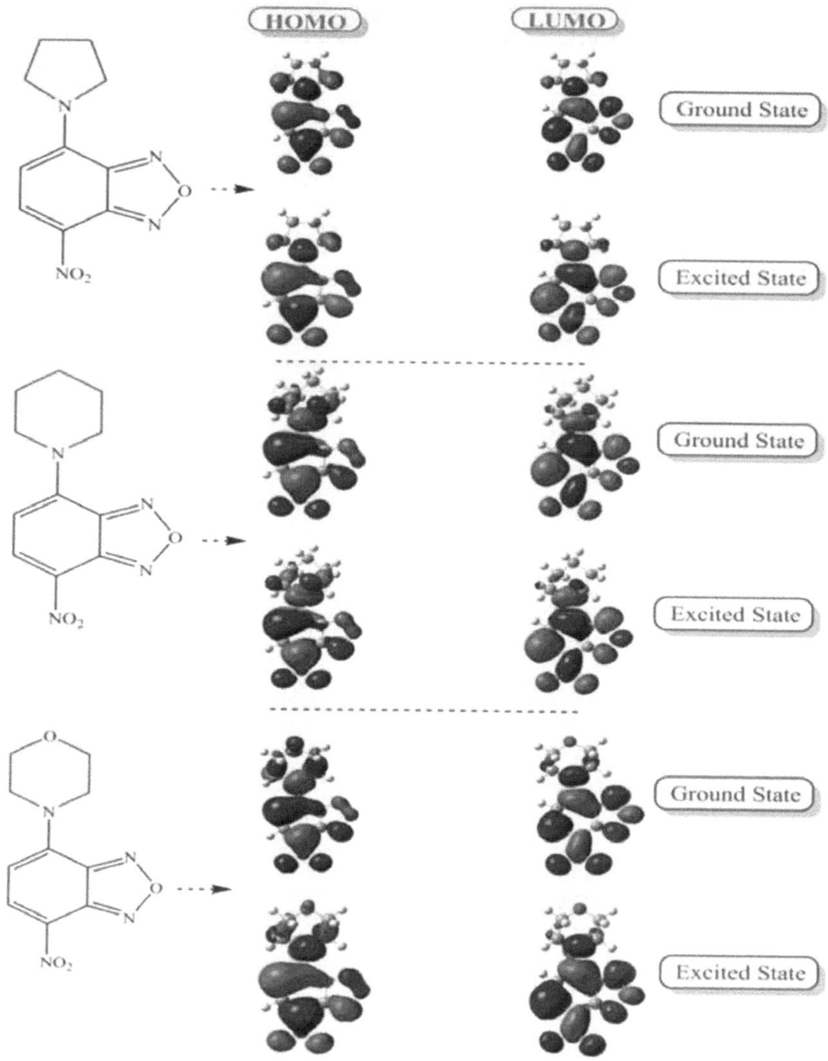

Figure 11.
Frontier molecular orbitals (FMOS) obtained at the ground- (S_0) and excited- (S_1) states of the optimized geometries of studied compounds.

$$w = \frac{\mu^2}{2\eta} \tag{3}$$

$$\mu \approx \frac{1}{2}(\varepsilon_{HOMO} + \varepsilon_{LUMO}) \tag{4}$$

$$\eta \approx \frac{1}{2}(\varepsilon_{LUMO} - \varepsilon_{HOMO}) \tag{5}$$

3.4.1 Correlation analysis

In a recent study, we showed that the second-order rate constants for the reactions of 7-X-4-nitrobenzofurazans **1** (**1a**, X = Cl, **1b**, X = OC_6H_5 and **1c**: X = OCH_3) with secondary cyclic amines **2** (**2a**, morpholine, **2b**, piperidine and **2c**, pyrrolidine) in acetonitrile at 20°C [27, 29] can be described by the linear free energy relationship *log k* = $s_N(N+E)$ [45–48], and we determined the *E* parameters of these electrophiles **1a-c** which are collected in **Figure 12**. We have now used these parameters *E* in order to elucidate the relationship between Mayr's electrophilicities *E* [45] and Parr's global electrophilicity index *w* values [49, 50]. As will be seen, satisfactory correlations between experimentally electrophilic reactivities *E* and other properties of these series of para-substituted nitrobenzofurazans **1a-c** are found and discussed. On the other hand, we discuss how the structure–property relationships can be used to evaluate electrophilicity parameter *E* and Hammett constant *σ* values which are not directly accessible.

3.4.2 Correlation between electrophilicity parameter (E) and global electrophilicity index (w)

Based on the data listed in **Table 6**, the plot of electrophilicity parameters *E* for these series of benzofurazans **1a-c** against their global electrophilicity index ***w***, determined in this work, was constructed. As can be seen in **Figure 13**, the relationship between *E* and *w* values is linear ($r^2 > 0.9799$) and results in the following equation:

$$E = -29.785 + 1.545\,w \quad (6)$$

It is interesting to note that the relationships parameters *E* versus *w* have also been reported by many authors [51–56]. The data in **Table 7** are illustrative in this regard. Interestingly, the plot of Eq. (6) can be used to estimate the unknown electrophilicity parameter *E* values of other para-substituted benzofurazans. Using the *w* values calculated in the present work by DFT//B3LYP/6-31+ g(d,p), the *E* of **1d** (X = morpholine, *E* = −18.01), **1e** (X = piperidine, *E* = −18.52), and **1f** (X = pyrrolidine, *E* = -18.78) have been obtained (The structures of title compounds are given in **Figure 2**). The detailed results are already listed in **Table 6**.

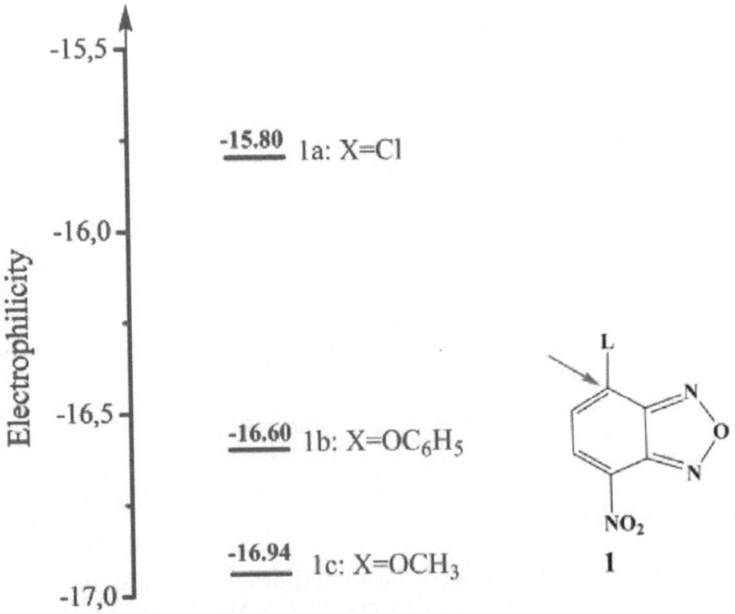

Figure 12.
*Electrophilicity parameters E of 7-X-4-nitrobenzofurazans **1a-c**.* [27, 29].

Benzofurazan	1a	1b	1c	1d	1e	1f
E	−15.80[a]	−16.60[b]	−16.94[a]	−18.01[c] −16.80[d]	−18.52[c] −17.04[d]	−18.78[c] −17.20[d]
w[e]	9.0090	8.6200	8.2670	7.6199	7.2908	7.1228

[a]*The E values were taken from Ref. [29].*
[b]*The E value was taken from Ref. [27].*
[c]*The E values estimated in this work.*
[d]*The E values were taken from Ref. [27].*
[e]*The w values calculated in this work.*

Table 6.
Electrophilicity parameter (E) and global electrophilicity index (w).

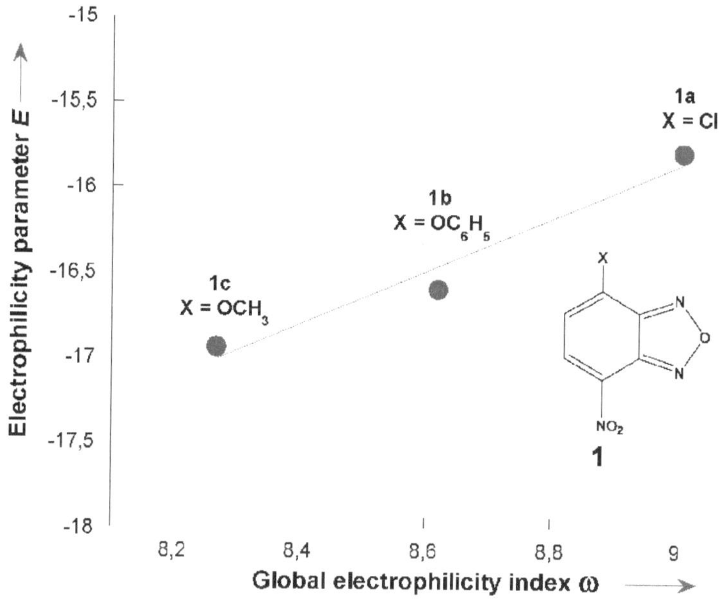

Figure 13.
*Correlation between experimentally determined electrophilicities (E) and global electrophilicity index (ω) calculated at DFT//B3LYP/6-31+ g(d,p) in acetonitrile for the benzofurazans **1a-c**.*

Electrophile	Eq.	Ref.	Electrophile	Eq.	Ref.
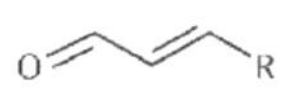	$E = -0.24 + 0.153\,w$	[51]	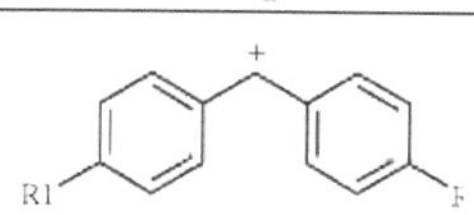	$E = -39.26 + 3.36\,w$	[54]
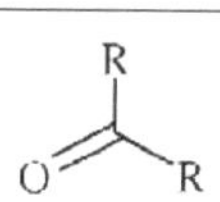	$E = 1.47 + 0.025\,w$	[52]	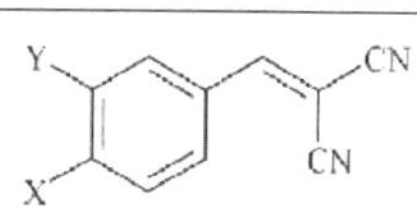	$E = -27.71 + 6.18\,w$	[55]
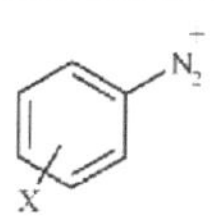	$E = -12.30 + 0.45\,w$ $E = -25.60 + 1.56\,w$	[53]	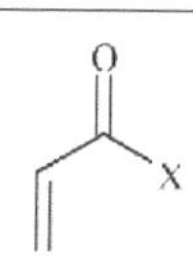	$E = -34.73 + 10.37w$	[56]

Table 7.
Correlations of the electrophilicity parameters E of some representative electrophiles with their global electrophilicity w values.

It should be noted that these predicted values of electrophilicity parameters *E* are smaller by 1.5 units than those reported values by Raissi and co-workers [27]. In all cases, the agreement between the *E* values of benzofurazans **1d-f** estimated using DFT//B3LYP/6-31+ g(d,p) and previously reported with the same method of calculation, but recorded in water medium [27] is remarkably good [57–61]. It is noteworthy to mention that the observed deviations between results give lower accuracy owing to their dependence on solvent polarity and to the added diffuse basis functions, that have also been observed in numerous systems. Chamorro and co-workers, have also observed a satisfactory correlations between the electrophilicity parameter E of various benzhydrylium cations and their global electrophilicity index ω calculated at B3LYP/6- 31 g(d) and HF/6-31 g(d) levels of theory Model I (E = −41.60+ 3.57 ω+), Model II (E = −40.42+ 2.08 ω+), Model III (E = −38.66 + 2.44 ω+) and Model IV (E = −47.2+ 20.0 ω+) [62].

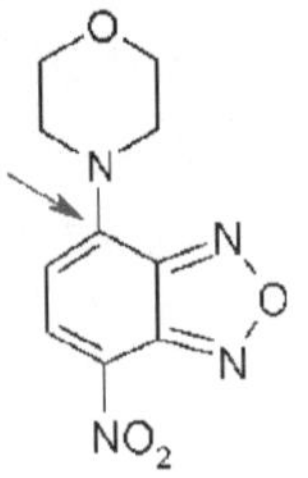
1d

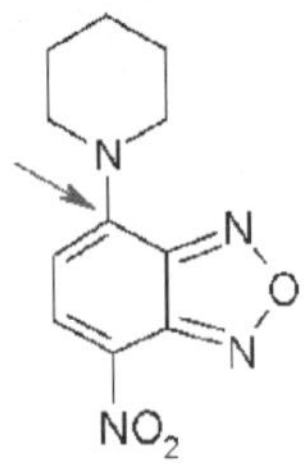
1e

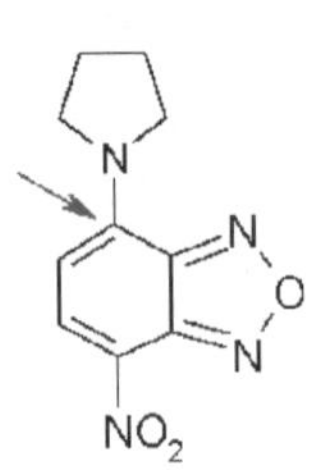
1f

3.5 Effect of substituents X on electrophilicity: *E* vs. *σ* and *w* vs. *σ* correlations

Examination of the data in **Tables 6** and **8** show that the electrophilicity parameter (*E*) or global electrophilicity index (*w*) of benzofurazans **1a-c** appears to be significantly dependent on the electronic nature of the substituents X, i.e. the *E* or *w* decrease regularly from X = Cl to X = OCH_3 leading to a nice Hammett-type plots [63, 64] (**Figure 14**), which are defined by the following Equations

$$E = -16.475 + 2.878\,\sigma\,\left(r^2 = 0.9984\right) \tag{7}$$

$$w = 8.614 + 1.810\,\sigma\,\left(r^2 = 0.9897\right) \tag{8}$$

Correlations between *E* and *σ* parameters or *E* and *w* of various system have also been reported by many authors [65–72]. Zenz and Mayr, have established that the *E* parameters for a series of trans-β-nitrostyrenes correlate well with σ_p values of their substituents (E = −13.95+ 2.08 σ_p (r^2 = 0.9847)) [69]. Recently, Rammah and

Substituted	Cl	OC_6H_5	OCH_3	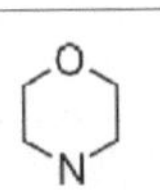	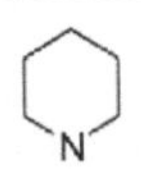	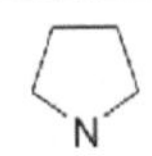
σ	0.23^a	-0.03^a	-0.17^a	-0.52^b -0.54^c	-0.70^b -0.72^c	-0.79^b -0.81^c
μ^d	7.23	10.29	9.26	13.57	16.84	17.08

[a]*The σ values were taken from Ref. [63, 64].*
[b]*The σ values calculated by using Eq. (7) with E from* ***Table 6****.*
[c]*Calculated by using Eq. (8) with w from* ***Table 6****.*
[d]*The dipole moment μ values calculated in this work.*

Table 8.
Hammett constants (σ) values and dipole moment (μ, D) values.

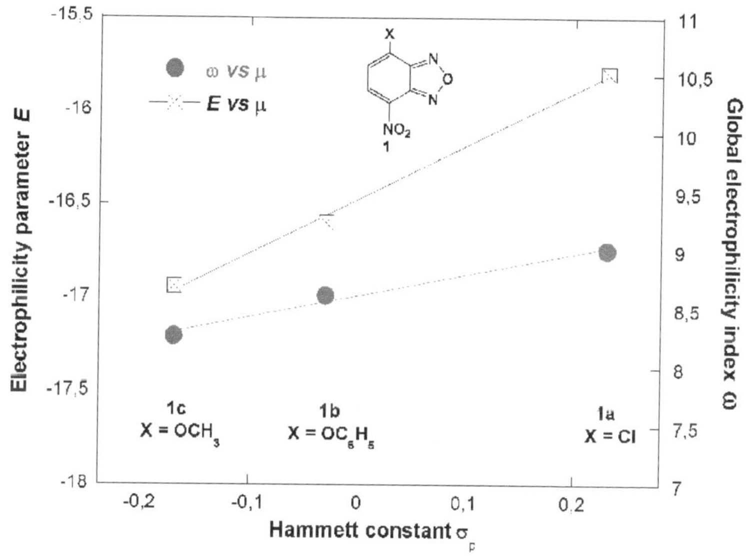

Figure 14.
*Correlations of the electrophilicity parameter E and global electrophilicity index w versus the Hammett's substituted σ_p values. The values of E and w are given in **Table 6**. The σ_p values were taken from Ref. [63, 64].*

co-workers, have also observed a linear correlation between the electrophilicity parameters E of a series of 2-N-(4′-X-pheny1)-4,6-dinitrobenzotriazole 1-oxides and the Hammett's substituent constants σ ($E = -11.15 + 2.28\,\sigma$ ($r^2 = 0.9878$)). [65]

In order to test the applicability of Eq. (7) for the prediction of the Hammett's constants σ_p of other groups, we have calculated the global electrophilicity index (w) of benzofurazans **1 g-j** (**1 g**: X = HNOH, **1 h:** X = $HNNH_2$, **1i:** X = HNH and **1j:** X = $HNCH_3$). The results are listed in **Table 9** which also includes the σ_p values

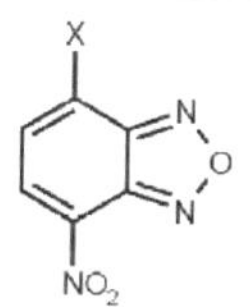 1	**Substituent X**	w[a]	μ[b]	$\sigma_p^{Cal,\,c}$	$\sigma_p^{Exp,\,d}$
1 g	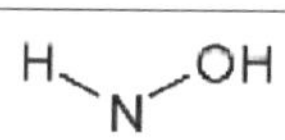	7.893	12.88	−0.34	−0.34
1 h	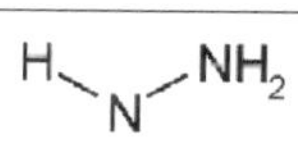	7.630	12.26	−0.53	−0.55
1i	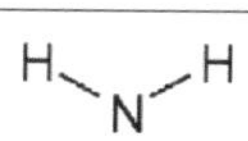	7.501	13.69	−0.60	−0.66
1j	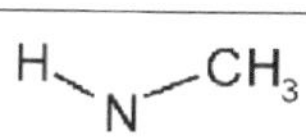	7.356	15.30	−0.68	−0.70

[a]*The w values calculated in this work.*
[b]*The dipole moment μ values calculated in this work.*
[c]*The σ_p^{Cal} values calculated by using Eq. (8) with w from **Table 5**.*
[d]*The σ_p^{Exp} values were taken from Ref. [63, 64].*

Table 9.
Comparison between calculated and experimental reported Hammett's constants σ_p for the HNOH, $HNNH_2$, HNH and $HNCH_3$ groups.

previously published [63, 64]. As well be seen, **Table 2** clearly shows that the calculated σ_p values are in good agreement with the experimental data.

Most importantly, the satisfactory correlations shown in **Figure 14** can be employed to obtain more information about the unknown Hammett's substituent constants (σ) of morpholine, piperidine and pyrrolidine groups.

Substitution of the *E* and *w* values for benzofurazans **1d-f** into the correlation Eqs. (7) and (8), the Hammett's σ_p values are thus obtained for the three leaving-groups. As can be seen in **Table 8,** the σ_p values estimated in the present work using work using Eqs. (7) or (8) are similar but relatively large compared with those previously reported ($-0.47 < \sigma_p < -0.24$) [27]. The major reason for the overestimation can, at least in part, be understood in terms of solvent polarity and also the used on the basis set with diffuse functions.

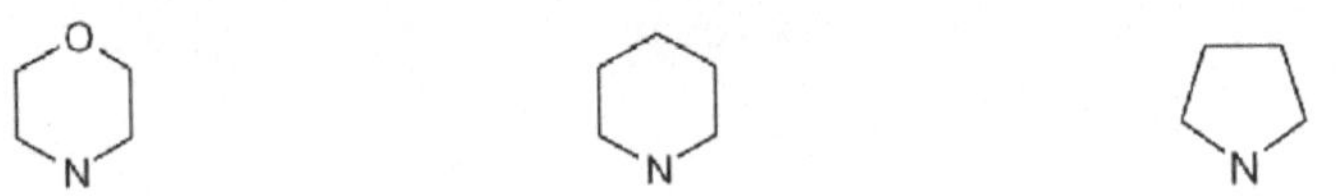

$\sigma_p = 0.51 \pm 0.01$ $\sigma_p = 0.71 \pm 0.01$ $\sigma_p = 0.80 \pm 0.01$

3.6 Correlation between dipole moment (μ) and Hammett's substituent constants (σ)

As shown in **Table 8** and **Figure 15**, the dipole moments (μ) calculated at DFT//B3LYP/6-31+g(d,p) for these series of benzofurazans **1a-f** are highly sensitive to the electronic nature of the substituent X, e.g., it decreases from μ = 17.08 D to μ = 7.23 D as the substituent X changes from a strong electron-withdrawing group (X = Cl) to a strong electron-donating group (X = morpholine). The linear relationship as expressed by Eq. (9) relates the dipole moment μ for **1a-f** directly to the Hammett's substituent constant σ.

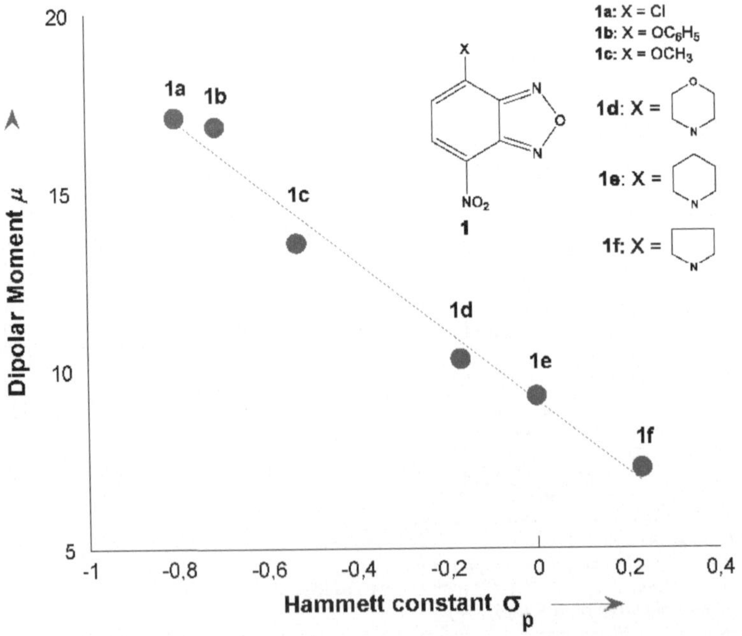

Figure 15.
*Correlation between calculated dipole moment (μ) for the benzofurazans **1a-f** and Hammett's substituent constant σ calculated at DFT//B3LYP/6-31+ g(d,p) in acetonitrile.*

$$\mu = 9.135\text{–}9.839\ \sigma\ (r^2 = 0.9908) \tag{9}$$

On the other hand, comparing the electrophilicity parameter E and global electrophilicity index w with the dipole moment (μ), we found that the E and values w correlate well with the μ as shown in **Figure 16.** In addition, the high dipole moment values being related to the strong effect exerted by the electron-donating group X clearly indicate that our benzofurazans **1b-j** can offer certain potential for nonlinear optical (NLO) applications.

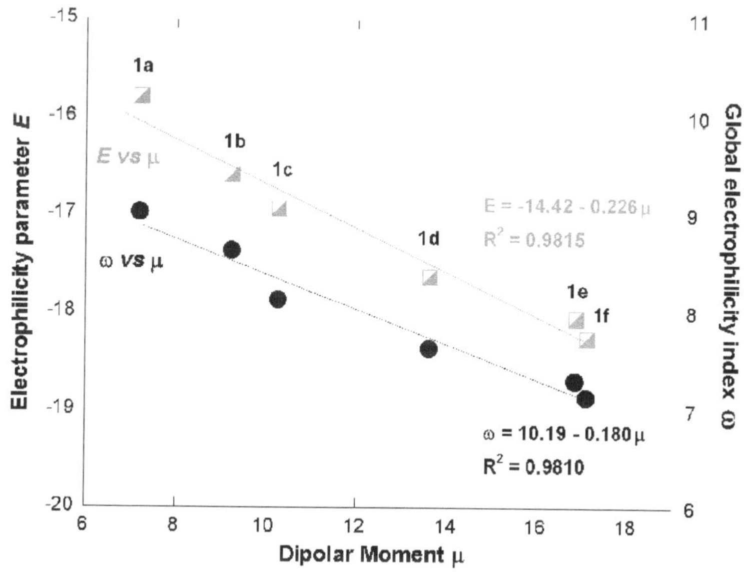

Figure 16.
*Correlation E vs. μ and w vs. μ for the benzofurazans **1a-f** calculated at DFT//B3LYP/6-31+g(d,p) in acetonitrile.*

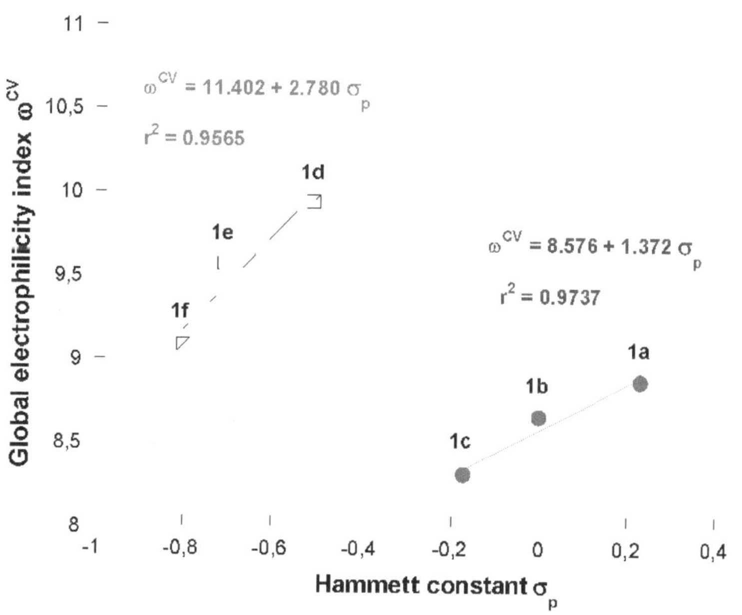

Figure 17.
*Correlations of the global electrophilicity index w^{VC} of the benzofurazans **1a-f** versus the Hammett's substituted σ_p values. The values of w^{CV} are given in **Table 1**. The σ_p values were taken from Ref. [63, 64].*

3.7 Correlation between global electrophilicity index w^{CV} and Hammett's substituent constants (σ_p)

To further extend the generality of the structure–property relationships to the global electrophilicity index w^{CV} calculated from the Cyclic Voltammetry (CV) listed in **Table 1**, we examined correlations w^{CV} vs. σ_p, w^{CV} vs. E and w^{CV} vs. w. As

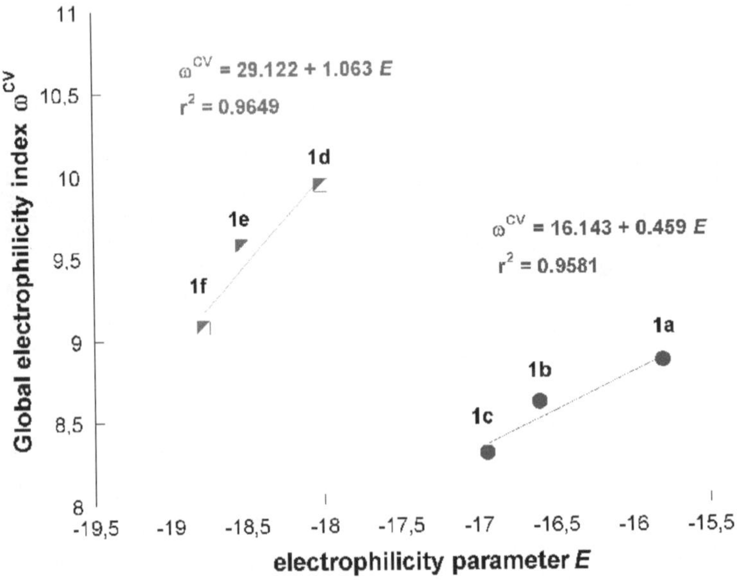

Figure 18.
*Correlations between global electrophilicity index w^{CV} and experimentally determined electrophilicities E for the benzofurazans **1a-c**. the values of w^{CV} are given in **Table 1** and values of E are given in **Table 6**.*

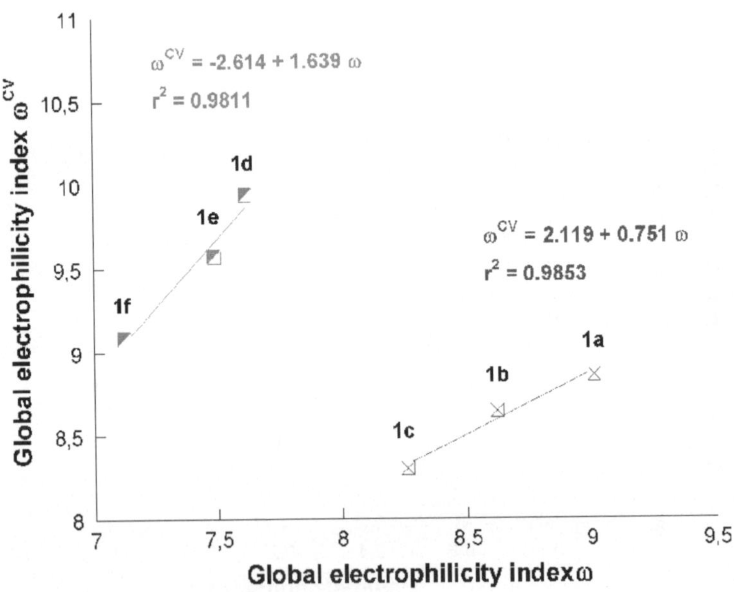

Figure 19.
*Correlations of the global electrophilicity index w^{CV} and global electrophilicity index ω calculated at DFT//B3LYP/6-31 + g(d,p) in acetonitrile for the benzofurazans **1a-f**. the values of w^{CV} are given in **Table 1** and values of w are given in **Table 6**.*

can be seen, the plotting of the w^{CV} values determined above (see **Table 1**) versus the known Hammett's constant σ_p values of substituent X gives rise to two separate linear correlations corresponding to a different behavior of Cl, OCH_3, OC_6H_5 groups and morpholinyl, piperidinyl and pyrrolidinyl groups (**Figure 17**).

Similar behaviors have also been obtained when the global electrophilicity index w^{CV} values were plotted against the electrophilicity parameter E or the global electrophilicity index w (**Figures 18** and **19**). The reason for this apparent disparity is not clear at this stage. Clearly, more investigate is necessary to explicate this behavior.

4. Conclusion

The reaction between NBD-Cl and morpholine, piperidine or pyrrolidine, as nucleophile group, proceeds normally by a S_NAr attack affording 4.7-di-substituted benzofurazans. They were thoroughly analyzed electrochemically and optically by means of Cyclic Voltammetry (CV) and UV–Vis and time resolved photoluminescence (TR-PL), respectively.

Complementary study based on density function theory (DFT) and its extent TD-DFT were also conducted to unravel the electronic structure, the simulated optical spectra and to further understand the structure property relationships of the investigated compounds. The used procedure based theoretical calculations allows deducing the electronic parameters related to the HOMO and LUMO energy levels and their energy differences. These levels are accessible by measuring the peak potentials of the oxidation and reduction reactions. It is found that the measured energy levels are in close agreement with the values computed from DFT method. Specific nucleophile-electrophile interactions in S_NAr reactions were expected to be governed by three main types of non-covalent bonds (O—H, N—O and N—H). Importantly, the compounds absorb visible light at longer wavelengths. In addition, the S_NAr attack leads to intra-molecular charge transfer of formed compounds and thus provides a remarkably slow non-radiative decay from the excited state. Investigation of photo-physical properties revealed importance of intra-molecular twisting on excited (S_1) states.

As verified by experimental and calculated results, the obtained molecules, by S_NAr reactions, were modulated by intra-molecular non-covalent interactions that force the co-planarity and the rigidity of the structures yielding an efficient intra-molecular charge transfer (ICT). Thus, the nucleophile substituents acting as electron-donating groups in S_NAr reaction afford to obtain materials with interesting nonlinear optical (NLO) response. In addition, the photo-optical data corroborated with TD-DFT approach were discussed in correlation with the structural architecture of each compound. We have shown the high efficiency fluorescent NBD-Morph compound that emits green to orange color at wavelength maximum of 559-603 nm.

The theoretical optimized parameters have been found to be in good agreement with the corresponding experimental data and results in the literature. Satisfactory linear correlation has been demonstrated between the parameters *E* and the global electrophilicity index w for the electrophilic reactivity at the C-X position of nitrobenzofurazans **1a-d**. Accordingly, structure–property relationships were found to be able to evaluate the unknown electrophilicity parameter *E* of 7-X-4-nitrobenzofurazans **1d-f**. Alternatively, the validity of Eqs. (7) and (8) has been satisfactorily verified by comparison of calculated and previously reported, in the literature, of the Hammett constant σ_p values of HNOH, $HNNH_2$, HNH and $HNCH_3$ groups.

References

[1] F. Qian, C. Zhang, Y. Zhang, W. He, X. Gao, P. Hu, Z. Guo, J. Am. Chem. Soc. 131 (2009) 1460-1468.

[2] T. Yamaguchi, M. Asanuma, S. Nakanishi, Y. Saito, M. Okazaki, K. Dodo, M. Sodeoka, Chem. Sci. 5 (2014) 1021-1029.

[3] P. B. Ghosh, M.W.Whitehouse. Biochem. J. 108 (1968) 155-156.

[4] P. B. Ghosh, M.W.Whitehouse. J. Med. Chem. 11(1968) 305-311.

[5] D. Castagnolo, M. Pagano, M. Bernardini, M. Botta, Tetrahed. Lett. 53 (2012) 5008-5011.

[6] C. Toriumi and K. Imai, Anal. Chem., 74 (2002) 2321-2327.

[7] T. Kajiro, Y. Nakajima, T. Fukushima and K. Imai, Anal. Chem., 74 (2002) 4519-4525.

[8] H.C. Ting, Y.T. Yang, C.H. Chen, J.H. Lee, J. H. Chang, C.I. Wu, T. L. Chiu, C. F. Lin, C. L. Chung, K. T. Wong, ChemSusChem 9 (2016) 1433-1441.

[9] J. Mori, T. Kaino, Phys. Lett. A 127 (1988) 259.

[10] L. Johnson, S. Lagerkvist, P. Lindroth, M. Ahnoff, K. Martinsson, Anal. Chem., 54 (1982) 939-942.

[11] D. Dal Monte, E. Sandri, L. Di Nunno, S. Florio, P. Todesco, Chimica e Industria, 53 (1971) 940-942.

[12] R.A. Manderville and E. Buncel, J. Chem. Soc. Perk. T. 2, (1993)1887-1894.

[13] S. Uchiyama, T. Santa, T. Fukushima, H. Homma, K. Imai, J. Chem. Soc. Perk. T. 2, (1998) 2165-2173.

[14] M. Bem, M. T. Caproiu, D. Stoicescu, T. Constantinescu, and A. T. Balaban, Cent. Eur. J. Chem., 2003, 1, 260.

[15] I. Jamaoui, T. Boubaker, R. Goumont, Int. J. Chem. Kinet. 45 (2013) 152-160.

[16] W.J. Wang, J.M. Xia, X. Hai, M.L. Chen, J.H. Wang, Environ. Sci. Nano, 4 (2017) 1037-1044.

[17] R. Poulain, D. Horvath, B. Bonnet, C. Eckhoff, B. Chapelain, M.-C. Bodinier, and B. Deprez, J. Med. Chem., 44 (2001) 3378.

[18] N. Lavignac, C. J. Allender, and K. R. Brain, Tetrahedron Lett., 45 (2004) 3625.

[19] M. Bem, M. Vasilescu, M. T. Caproiu, C. Draghici, A. Beteringhe, T. Constantinescu, M. D. Banciu, and A. T. Balaban,Cent. Eur. J. Chem., 2 (2004) 672-685.

[20] Jun-ichi Aihara, Reduced HOMO-LUMO Gap as an Index of Kinetic Stability for Polycyclic Aromatic Hydrocarbons, J. Phys. Chem. A 103, 37 (1999) 7487–7495.

[21] Jun-ichi Aihara, Phys. Chem. Chem. Phys., 2000, 2, 3121-3125.

[22] R. Zaier, S. Ayachi, Optik, 239 (2021) 166787.

[23] R. Zaier, S. Hajaji, M. Kozaki, S. Ayachi, Opt. Mat. 91C (2019) 108-114.

[24] A. J. Boulton, A. R. Katrizky, A. M. Hamid, J. Chem. Soc., 1967 (2005) 2007.

[25] J. C. Hallé, M. Mokhtari, P. Soulié, M. J. Pouet, Can. J. Chem 75 (1997) 1240.

[26] M. R. Crampton, L. M. Pearce, L. C. Rabbitt, J. Chem. Soc., Perkin trans 2 (2002) 257.

[27] H. Raissi, H. Ayachi, F. Mahdhaoui, S. Ayachi, T. Boubaker, J. Mol. Str. 1224 (2021) 128843.

[28] H. Ayachi, H. Raissi, F. Mahdhaoui, T. Boubaker, Int. J. Chem. Kinet. 52 (2020) 655-668.

[29] H. Raissi, I. Jamaoui, R. Goumont, T. Boubaker, Int. J. Chem. Kinet. 49 (2017) 835-846.

[30] F. Mahdhaoui, R. Zaier, N. Dhahri, S. Ayachi, T. Boubaker, Int. J. Chem. Kinet. 51 (2019) 249-257.

[31] I. Messaoudi, I. Aribi, Z. Zouhour S. Ayachi, M. Othman, A.H. Said, J. Mol. Str. 1231 (2021) 129810.

[32] A. Abbotto, L. Beverina, *N. manfredi*, G. A. Pagani, G. Archetti, H. G. Kuball, C. Wittenburg, J. heck, J. Holtman, Chem. Eur. J. 15 (2009) 6175-6185.

[33] J. A. Davies, A. Elangovan, P. A. Sullivan, B. C. Olbricht, D. H. Bale, T.R. Ewy, C.M. Isborn, B.E. Eichinger, B.H. Robinson, P.J. Reid, X. Li, L. R. Dalton, J. Am. Chem. Soc. 130 (2008) 10565-10575.

[34] Z. Zaaboub, F. Hassen, H. Maaref, Sol. Stat. Comm. 314–315 (2020) 113913.

[35] A. D. Becke, J. Chem. Phys. 98 (1993) 5648-5652.

[36] C. Lee, W. Yang, R. G. Parr, Phys. Rev. 37 (1988), 785-789.

[37] M. J. Frisch., et al. Gaussian 09, Revision A.01, Gaussian: Wallingford, CT (2009).

[38] Dennington, R.D., Keith, T.A. and Millam, J.M. (2008) GaussView 5.0.8, Gaussian.

[39] J. L. Bredas, R. Silbey, D. S. Boudreaux, and R. R. Chance, J. Am. Chem. Soc 105 (22) (1983) 6555-6559.

[40] Heberer, H.; Matschiner, H., J. Prakt. Chem., 328 (1986) 261-274.

[41] S. Ayachi, S. Ghomrasni, M. Bouachrine, M. Hamidi, K. Alimi, J. Mol. Str. 1036, (2013) 7-18.

[42] S. Ayachi, S. Ghomrasni, K. Alimi, J. App. Polym.Sci. 123 (2012) 2684-2696.

[43] R. G. Parr, L. V. Szentpaly, S. Liu, J Am Chem Soc. 1999, 121, 1922-1924.

[44] Parr, R. G, Yang, W. Density Functional Theory of Atoms and Molecules; Oxford University Press: New York 1989.

[45] H. Mayr, M. Patz, Angew. Chem. Int. Ed. Engl. 33 (1994) 938-957.(b) H. Mayr, B. Kempf, A. R. Ofial, Acc. Chem. Res. 36 (2003) 66-77.

[46] S. Ben Salah, F. Necibi, R. Goumont, T. Boubaker. ChemistrySelect. 5 (2020) 7648-7657.

[47] S. Souissi, W. Gabsi, A. Echaieb, J. C. Hierso, P. Fleurat-Lessard, T. Boubaker. RSC Advances. 10 (2020) 28635-28643.

[48] Z. Li, R. J. Mayer, A. R. Ofial, and H. Mayr. J Am Chem Soc. 142 (2020) 18, 8383-8402.

[49] R. G. Parr, R. G. Pearson, J. Am. Chem. Soc. 105 (1983) 7512-7516.

[50] R. G. Parr, L. V. Szentpaly, S. Liu. J. Am. Chem. Soc. 121 (1999) 1922-1924.

[51] D. S. Allgäuer, H. Jangra, H. Asahara, Z. Li, Q. Chen, H. Zipse, A. R. Ofial, H. Mayr. J. Am. Chem. Soc., 139 (2017) 13318–13329.

[52] Z. Li, H. Jangra, Q. Chen, P. Mayer, A. R. Ofial, H. Zipse, H. Mayr. J. Am. Chem. Soc. 140 (2018) 5500–5515.

[53] W. Gabsi, T. Boubaker, R. Goumont. Int. J. Chem. Kinet. 48 (2016) 266-273.

[54] P. Pérez, A. Labbé. J. Org. Chem. 67 (2002) 4747-4752.

[55] L. R. Domingo, P. Pérez, R. Contreras. L. R. Domingo, P. Pérez, R. Contreras. Tetrahed. 60 (2004) 6585-6591.

[56] D. S. Allgäuer, H. Jangra, H. Asahara, Z. Li, Q. Chen, H. Zipse, A. R. Ofial, H. Mayr. J. Am. Chem. Soc. 139 (2017) 13318-13329.

[57] Q. Chen, P. Mayer, H. Mayr. Angew Chem Int Ed 55 (**2016**) 12644-12667.

[58] R. J. Mayer, P. W. A. Allihn, N. Hampel, P. Mayer, S. A. Sieber, A. R. Ofia. Chem. Sci. 12 (2021) 4850-4865.

[59] S. Ben Salah, T. Boubaker, R. Goumont. Int J Chem Kinet 49 (2017) 576-583.

[60] E. Follet, P. Mayer, H. Mayr, Eur J Org Chem. 2016, 4050-4058.

[61] N. Dhahri, T. Boubaker, R. Goumont. J Phys Org Chem. , 27 (2014) 484.

[62] E. Chamorro, M. Duque-Noreña, P. Pérez. J Mol Struct. 896 (2009) 73-79.

[63] C. Hansch, *A. Leo*, R. W. Taft. Chem. Rev. 91 (1991) 185-195.

[64] Hammett L. P. Reaction Rates, Equilibria and Mechanism. Mc Graw-Hill, New York, 2nd edn, 1970.

[65] M. Rammah, F. Mahdhaoui, W. Gabssi, T. Boubaker. ChemistrySelect. 6 (2021) 4424-4431.

[66] O. Kaumanns, H. Mayr. J Org Chem. 73 (2008) 2738-2745.

[67] H. Asahara, H. Mayr. Chem Asian J. 7 (2012) 1401, 1407.

[68] K. Azaiez, N. Dhahri, T. Boubaker. Int J Chem Kinet. 49 (2017) 847-858.

[69] I. Zenz, H. Mayr. J Org Chem. 76 (2011) 9370-937.

[70] J. Ammer, C. Nolte, H. Mayr. J Am Chem Soc. 134 (2012) 13902-13911.

[71] K. Troshin, C. Schindele, H. Mayr. J Org Chem. 76 (2011) 9391-9408.

[72] D. Richter, N. Hampel, T. Singer, A. R. Ofial, H. Mayr. Eur J Org Chem. 2009, 3203-3211.

INDEX

9 789390 013982

Printed at Repro India Ltd.